Physical and Biological Properties of Waters in the Region of the Antarctic Peninsula and Adjacent Basins of the South Atlantic

Physical and Biological Properties of Waters in the Region of the Antarctic Peninsula and Adjacent Basins of the South Atlantic

Editor

Eugene Morozov

MDPI • Basel • Beijing • Wuhan • Barcelona • Belgrade • Manchester • Tokyo • Cluj • Tianjin

Editor
Eugene Morozov
Hydrological Processes
Shirshov Institute of
oceanology, Russian
Academy of Sciences
Moscow
Russia

Editorial Office
MDPI
St. Alban-Anlage 66
4052 Basel, Switzerland

This is a reprint of articles from the Special Issue published online in the open access journal *Water* (ISSN 2073-4441) (available at: www.mdpi.com/journal/water/special_issues/Physical_Biological_Waters_Antarctic).

For citation purposes, cite each article independently as indicated on the article page online and as indicated below:

LastName, A.A.; LastName, B.B.; LastName, C.C. Article Title. *Journal Name* **Year**, *Volume Number*, Page Range.

ISBN 978-3-0365-8269-6 (Hbk)
ISBN 978-3-0365-8268-9 (PDF)

Cover image courtesy of Eugene Morozov

Contents

About the Editor

Eugene Morozov

Eugene Morozov graduated in oceanography in 1970 from the Moscow Institute of Physics and Technology, and then received his Ph.D. in 1975 and a doctoral degree in Physics and Mathematics in 1990. He is a field oceanographer and his career developed at the Shirshov Institute of Oceanology (Russian Academy of Sciences) in Moscow where he heads the Hydrological Processes Laboratory. His research focuses on mesoscale dynamics, internal waves, and circulation. He has been President (2011–2015) of the International Association for the Physical Sciences of the Ocean. Eugene Morozov started his marine activity in 1967 by participating in a coastal expedition on the Black Sea. His first oceanic expedition was in 1969 exploring the western boundary currents in the Atlantic. During his life at sea, Eugene conducted research on many large-scale currents in the World Ocean: the Gulf Stream, the Antarctic Circumpolar Current, the Kuroshio, the California Current, the equatorial subsurface undercurrents in the Indian and Atlantic oceans, and the Falkland Current. He also studied mesoscale eddies and rings of the jet currents. Morozov has also been an Arctic oceanographer. He has explored the Barents and Kara seas and fjords in Spitsbergen. Since 1969, Eugene has made 53 cruise voyages in the ocean; among them, 7 were in the Southern Ocean. Eugene has spent 10 years of his life at sea. His publications include more than 300 journal papers and 7 books.

water

MDPI

Editorial

Research in the Atlantic Sector of the Southern Ocean and Propagation of Antarctic Bottom Water in the Atlantic

Eugene Morozov

Shirshov Institute of Oceanology, Russian Academy of Sciences, 117997 Moscow, Russia; egmorozov@mail.ru

Citation: Morozov, E. Research in the Atlantic Sector of the Southern Ocean and Propagation of Antarctic Bottom Water in the Atlantic. *Water* **2023**, *15*, 2348. https://doi.org/10.3390/w15132348

Received: 1 June 2023
Revised: 13 June 2023
Accepted: 20 June 2023
Published: 25 June 2023

This article belongs to the Special Issue "Physical and Biological Properties of Waters in the Region of the Antarctic Peninsula and Adjacent Basins of the South Atlantic,": a Special Issue that is dedicated to recent research in the Atlantic sector of the Southern Ocean and Antarctica Peninsula region. The papers in this issue are focused on the field and theoretical research of the physical properties and ecology of this key region in the Southern Ocean. The main objects of the study were the properties of seawater, currents, the ecosystem, and biological communities in the South Atlantic, the Powell Basin of the northwestern Weddell Sea, the Bransfield Strait, and the Antarctic Sound. The Antarctic marine ecosystem is unique was terms of its biodiversity and high productivity of basic trophic levels of the ecosystem. This ecosystem was efficiently used by organisms such as seabirds, seals, and whales that fed on Antarctic krill.

The Antarctic Circumpolar Current System (ACC) and oceanic circulation at higher latitudes near the Antarctic continent are the oceanographic basis for the functionality of the Antarctic marine ecosystem. The processes in this region of the World Ocean have been intensely studied in recent decades but still remain not completely understood because of an insufficient number of expeditions to this region. The processes in Antarctica require more and more investigation by scientists. In recent decades, the Southern Ocean has experienced significant changes associated with global climate trends.

These issues of the *Water* MDPI journal have been composed on the basis of analyzing the results of multi-disciplinary studies based on the physics and biology of the Atlantic sector of the Antarctic during two cruises in January-February 2020 and in January-February 2022 on the R/V *Akademik Mstislav Keldysh* (cruises 79 and 87) [1,2].

This issue presents the results of multidisciplinary studies in Antarctica. The issue includes articles on marine physics, marine chemistry, and marine biology, which jointly compose a comprehensive multidisciplinary approach to a modern view of the processes that govern the mean state of the ocean and its variability. Research has been performed in the region of the Antarctic Circumpolar Current, Weddell Gyre, Weddell Sea, and Bransfield Strait. The most important areas of study were the Bransfield Strait between the Antarctic Peninsula [3], the South Shetland Islands, and the Powell Basin east of the northern tip of the Antarctic Peninsula [2].

Research has also been conducted in the region of the South Orkney Islands and Orkney Passage east of the Powell Basin. The Orkney Passage is the main pathway for Antarctic Bottom Water flow from the Weddell Sea [4]. On the route of the ship to the Antarctic continent, the propagation of Antarctic Bottom Water has been studied in the Vema Channel [5]. The upwelling on the Patagonian shelf and Malvinas Current became a topic of multidisciplinary research during the expedition. We also studied the processes across an eddy dipole in the interaction zone between Subtropical and Subantarctic waters in the Southwest Atlantic [6].

Biological papers on this issue present an analysis of the recent changes in the composition and distribution of zooplankton [7,8] and Antarctic worms [9]. The biological analysis is based on the assessment of water quality and biota in this region.

Funding: This research was supported by the Russian Science Foundation grant 21-77-20004.

Conflicts of Interest: The author declares no conflict of interest.

References

1. Morozov, E.G.; Spiridonov, V.A.; Molodtsova, T.N.; Frey, D.I.; Demidova, T.A.; Flint, M.V. Investigations of the ecosystem in the Atlantic sector of Antarctica (Cruise 79 of the R/V Akademik Mstislav Keldysh). *Oceanology* **2020**, *60*, 721–723. [CrossRef]
2. Morozov, E.G.; Flint, M.V.; Spiridonov, V.A. *Antarctic Peninsula Region of the Southern Ocean*; Oceanography and Ecology, Springer Nature Oceanography; Springer International Publishing: Cham, Switzerland, 2021; 455p. [CrossRef]
3. Frey, D.I.; Krechik, V.A.; Morozov, E.G.; Drozd, I.D.; Gordey, A.S.; Latushkin, A.A.; Mekhova, O.S.; Mukhametianov, R.Z.; Murzina, S.A.; Stromata, S.A.; et al. Water exchange between deep basins of the Bransfield Strait. *Water* **2022**, *14*, 3193. [CrossRef]
4. Morozov, E.G.; Frey, D.I.; Zuev, O.A.; Velarde, M.G.; Krechik, V.A.; Mukhametianov, R.Z. Hydraulically controlled bottom flow in the Orkney Passage. *Water* **2022**, *14*, 3088. [CrossRef]
5. Morozov, E.G.; Zuev, O.A.; Frey, D.I.; Krechik, V.A. Antarctic Bottom Water jets flowing from the Vema Channel. *Water* **2022**, *14*, 3438. [CrossRef]
6. Salyuk, P.A.; Mosharov, S.A.; Frey, D.I.; Kasyan, V.V.; Ponomarev, V.I.; Kalinina, O.Y.; Morozov, E.G.; Latushkin, A.A.; Sapozhnikov, P.V.; Ostroumova, S.A.; et al. Physical and biological features of the waters in the outer Patagonian shelf and the Malvinas Current. *Water* **2022**, *14*, 3879. [CrossRef]
7. Bitiutskii, D.G.; Samyshev, E.Z.; Minkina, N.I.; Melnikov, V.V.; Chudinovskih, E.S.; Usachev, S.I.; Salyuk, P.A.; Serebrennikov, A.N.; Zuev, O.A.; Orlov, A.M. Distribution and demography of Antarctic krill and salps in the Atlantic sector of the Southern Ocean during austral summer 2021–2022. *Water* **2022**, *14*, 3812. [CrossRef]
8. Kasyan, V.V. Recent changes in composition and distribution patterns of summer mesozooplankton off the Western Antarctic Peninsula. *Water* **2023**, *15*, 1948. [CrossRef]
9. Chernyshev, A.V.; Polyakova, N.E. Distribution and phylogenetic position of the Antarctic ribbon worm *Heteronemertes longifissa* (Nemertea, Pilidiophora). *Water* **2023**, *15*, 809. [CrossRef]

water

MDPI

Article

Water Exchange between Deep Basins of the Bransfield Strait

Dmitry I. Frey [1,2,*], Viktor A. Krechik [1,3], Eugene G. Morozov [1,2], Ilya D. Drozd [1,4,5], Alexandra S. Gordey [1], Alexander A. Latushkin [2], Olga S. Mekhova [1,6], Rinat Z. Mukhametianov [1,7], Svetlana A. Murzina [1,8], Sofia A. Ostroumova [1,6], Vladimir I. Ponomarev [9], Pavel A. Salyuk [9], Daria A. Smirnova [1,4], Sergey A. Shutov [2] and Oleg A. Zuev [1]

1 Shirshov Institute of Oceanology, Russian Academy of Sciences, Nakhimovsky 36, 117997 Moscow, Russia
2 Marine Hydrophysical Institute, Russian Academy of Sciences, Kapitanskaya Str. 2, 299011 Sevastopol, Russia
3 Laboratory for Marine Environmental Management, Immanuel Kant Baltic Federal University, Ulitsa Aleksandra Nevskogo 14, 236041 Kaliningrad, Russia
4 Faculty of Geography, Moscow State University, Leninskie Gory 1, 119991 Moscow, Russia
5 Obukhov Institute of Atmospheric Physics, Russian Academy of Sciences, Pyzhevskiy Pereulok 3, 119017 Moscow, Russia
6 Laboratory for Marine Environmental Management, Russian State Hydrometeorological University, Malookhtinskiy Prospekt 98, 195196 St. Petersburg, Russia
7 Moscow Institute of Physics and Technology, Institutsky per. 9, 141700 Dolgoprudny, Russia
8 Institute of Biology of the Karelian Research Centre of the Russian Academy of Sciences, Ulitsa Pushkinskaya 11, 185910 Petrozavodsk, Russia
9 Ilichev Pacific Oceanological Institute, Far Eastern Branch of the Russian Academy of Science, Baltiyskaya Street 43, 690041 Vladivostok, Russia
* Correspondence: dima.frey@gmail.com; Tel.: +7-964-702-77-72

Abstract: The Bransfield Strait is a relatively deep and narrow channel between the South Shetland Islands and the Antarctic Peninsula contributing to the water transport between the Pacific and Atlantic sectors of the Southern Ocean. The strait can be divided into three deep separate basins, namely, the western, central, and eastern basins. The sources of deep waters in the three basins are different, leading to differences in thermohaline properties and water density between the basins. The difference in water density should in turn cause intense deep currents from one basin to another through narrow passages over the sills separating the basins. However, there are still no works dedicated to such possible overflows in the Bransfield Strait. In this study, we report our new CTD and LADCP measurements performed in 2022 over the watersheds between the basins. Quasisimultaneous observations of the main circulation patterns carried out at several sections allowed us to analyze the evolution of thermohaline and kinematic structures along the Bransfield Strait. Volume transports of waters in the strait were estimated on the basis of direct velocity observations. These new data also indicate the existence of intense and variable deep current between the central and eastern basins of the strait. The analysis of historical data shows that the mean flow is directed from the central to the eastern basin. In addition, LADCP data suggest the intensification of the flow in the narrow part of the sill between the basins, and the possible mixing of deep waters at this location.

Keywords: Bransfield Strait; deep overflow; CTD; LADCP; bottom circulation

Citation: Frey, D.I.; Krechik, V.A.; Morozov, E.G.; Drozd, I.D.; Gordey, A.S.; Latushkin, A.A.; Mekhova, O.S.; Mukhametianov, R.Z.; Murzina, S.A.; Ostroumova, S.A.; et al. Water Exchange between Deep Basins of the Bransfield Strait. *Water* **2022**, *14*, 3193. https://doi.org/10.3390/w14203193

Academic Editor: Changming Dong

Received: 7 August 2022
Accepted: 5 October 2022
Published: 11 October 2022

1. Introduction

The Bransfield Strait (BS) is an important passage for Antarctic waters in the region of the Antarctic Peninsula, which contributes to the zonal water transport between the Pacific and Atlantic sectors of the Southern Ocean. The strait extends over 460 km from the west–southwest to the east–northeast (true direction is 60° relative to the north), and is bounded by the South Shetland Islands from the northwest and the Antarctic Peninsula from the southeast. The BS region is characterized by strong climatic changes [1] that affect the oceanographic, meteorological, glaciological, and biological conditions [2]. The BS region is also important because of its high biological productivity [3]. In particular, the region is a key source of Antarctic krill (*Euphausia superba*) to the Southern Ocean [4–6], and the region is a significant spawning, breeding, and overwinter habitat of *E. superba* [7–9] and another abundant euphausiid in Antarctic waters, *Thyssanoessa macrura* [10]; the BS is one of the significant regions for commercial krill fishing [11–13].

The hydrography of the Bransfield Strait is highly dependent on the complicated bathymetry [1,14]. From a geomorphological point of view, the BS can be considered as a sequence of three basins (Figure 1), namely, the western (WB), central (CB), and eastern (EB) basins [15,16]. Modern bottom topography data GEBCO2021 suggest that the maximal depths of these basins are 1370, 1960, and 2750 m, respectively. The basins are separated by relatively shallow sills. The maximal depth of the sill between the WB and CB is 630 m; the depths of the sill between the CB and EB slightly exceed 1000 m. The WB is connected to the Bellingshausen Sea through the Gerlache Strait and other gaps between Smith, Low, and Hoseason islands, and to the Drake Passage through the Boyd Strait (Figure 1). The EB is open to the Weddell and Scotia seas through relatively shallow sills; their depths do not reach 800 m.

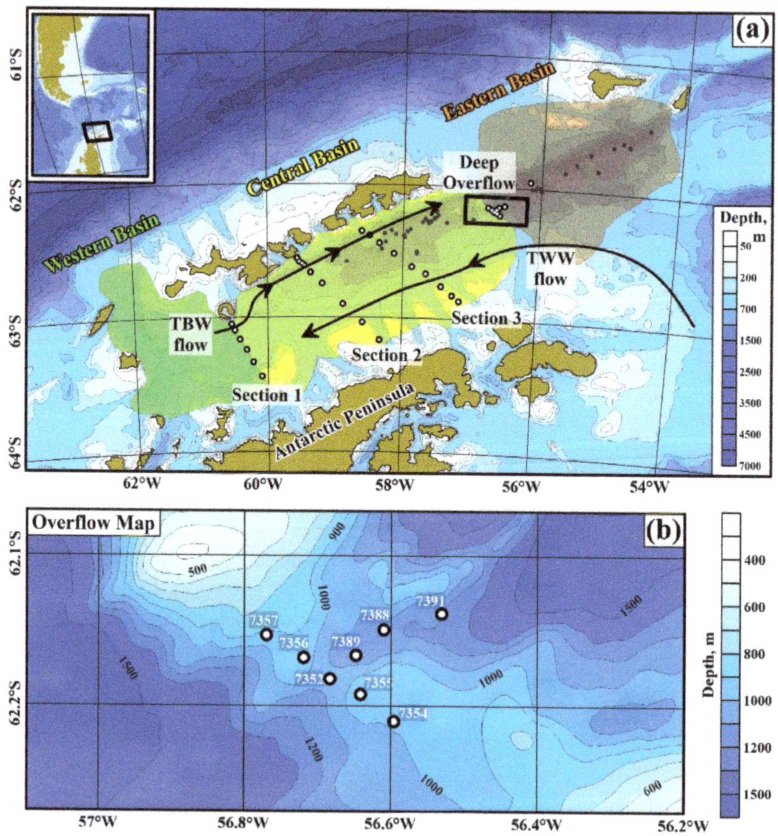

Figure 1. (**a**) Bathymetry of the Bransfield Strait and (**b**) the region of the overflow between the central and eastern basins. (**a**) The upper layer circulation schematic is shown by black arrows. Western, central, and eastern basins of the strait are shown with green, yellow, and orange, respectively. CTD/LADCP stations performed in 2022 are shown by white dots; historical CTD stations from World Ocean Database (WOD2018) used in this study are shown by grey dots. The bottom relief is shown according to the GEBCO2021 database; the shoreline is based on the GSHHS data [17]. Station 7390 (not shown) repeats station 7352.

The thermohaline structure of the upper BS layer is formed by two water masses, namely, transitional zonal water with Bellingshausen Sea influence (TBW) and transitional zonal water with Weddell Sea influence (TWW) [18,19]. According to [1], typical characteristics of these waters are $\theta > 1$ °C, $S < 34.1$ psu (TBW flow) and $\theta < 1$ °C, $S > 34.1$ psu (TWW flow). TBW waters propagate to the northeast along the South Shetland Islands in the form of a narrow high-velocity jet called the Bransfield Current [19–22]. Additionally, modified Circumpolar Deep Water (mCDW) with temperatures $\theta > 1$ °C and salinities >34.5 psu is stably observed within the Bransfield Current in a depth range of 200–450 m [23,24]. The maximal velocities of the Bransfield Current are observed at the sea surface and linearly decay towards the bottom [22,25,26]; its transport is approximately 1 Sv on the basis of direct velocity measurements [27]. The authors in [28] showed that the diurnal tide essentially affects the Bransfield Current; the same effect was observed in [29]. The TWW is located in the southern part of the strait and spreads southwestward along the Antarctic Peninsula [23]; the velocities of this flow are much lower than those in the Bransfield Current and usually do not exceed 20–30 cm/s [30]. Further inflows of TWW waters from the BS to the West Antarctic Peninsula slope are caused by wind forcing [31]. TWW and TBW waters are separated by two fronts: the Peninsula Front divides these waters at the sea surface and the Bransfield Front divides them in the deeper layers. The Bransfield Front is located much closer to the South Shetland islands than the Peninsula Front is.

The thermohaline structure of waters within the Bransfield Strait was repeatedly studied on the basis of in situ CTD data [15,24,32–36]. Deep layers of the BS basins are filled with relatively cold, saline, and dense waters from the continental shelf of the western Weddell Sea [32,33]. During the last few decades, the freshening and lightening of these waters have been observed [24,34,37]; their variability is caused by changes in source waters, and negatively correlated with Southern Annular Mode [38]. As the water density in the deep layers differs from one BS basin to another, one might expect that deep overflows can exist over the sills between the basins. However, there are no studies dedicated to direct measurements over the sill points between the WB and CB, and between the CB and EB. Regarding available data, the WOD2018 database contains 27 CTD profiles in the CB and 18 profiles in the EB, but there are no stations near the sill between these basins. The same lack of data is observed in the region of the sill between the WB and CB. LADCP data are even less available than CTD data. Regarding Shipboard ADCP measurements (for example, [27,30]), the maximal depth of such velocity profiles does not exceed 300–400 m, which is not sufficient for the studies of the bottom circulation between the deep basins of the BS. A recent work by [39] also showed that geostrophic velocity calculations based on hydrographic data do not reproduce actual ocean circulation patterns in the strait. This fact emphasizes the importance of direct velocity measurements in the BS. The objective of this paper is to study deep-water exchange between the basins of the BS based on the new data collected in January and February 2022. We used both CTD and LADCP profilers for the synchronous measurements of thermohaline and kinematic structures of currents.

The paper is structured as follows: we describe our in situ CTD and LADCP measurements in Section 2. In Section 3, we analyze the spatial thermohaline and kinematic structure of flows over the sill between the WB and CB (Section 3.1), and between the CB and EB (Section 3.2). The results are discussed in Section 4, followed by conclusions in Section 5.

2. Materials and Methods

This study is focused on the water exchange between the deep basins of the BS. For this purpose, we combined in situ measurements of temperature, salinity, and velocities performed during austral summer in January and February 2022. In this section, we describe our approach for station selection (Section 2.1), and the applied equipment and data processing techniques (Section 2.2).

2.1. Sections across the Strait

A total of 34 stations were performed within the BS from 21 January to 14 February 2022 (Table 1) in the 87th cruise of the research vessel *Akademik Mstislav Keldysh*. The measurements were performed almost at the bottom from the ship that maintained its position at the station with accuracy not worse than 200 m. The stations were organized in three sections, up to nine stations each across the strait (Figure 1); two relatively short but high-resolution sections were located at the sill point between the central and east basins. The GEBCO2021 bathymetry grid of 15″ resolution was used for the selection of the stations. The major part of the BS region has been covered by multibeam echosounder surveys. Their locations are shown with green in the right upper panel of Figure 2 based on the data from GEBCO2021 Type Identifier (TID) grid. Three BS basins are clearly seen along the thalweg of the strait (Figure 2). The first section was located over the ridge between the WB and CB; the second and third sections were located in the CB, allowing for us to trace how water properties change along the strait. Two additional sections were located across and along the sill between the CB and EB. These sections were performed with very high resolution (the distance between stations here was approximately 2 km), which allowed us to study deep overflow between the CB and EB. Because the GEBCO2021 grid contains multibeam echosounder data for the entire strait, the depths of our stations and own single-beam measurements using Kongsberg EA600 echosounder coincided very well with the GEBCO bathymetry.

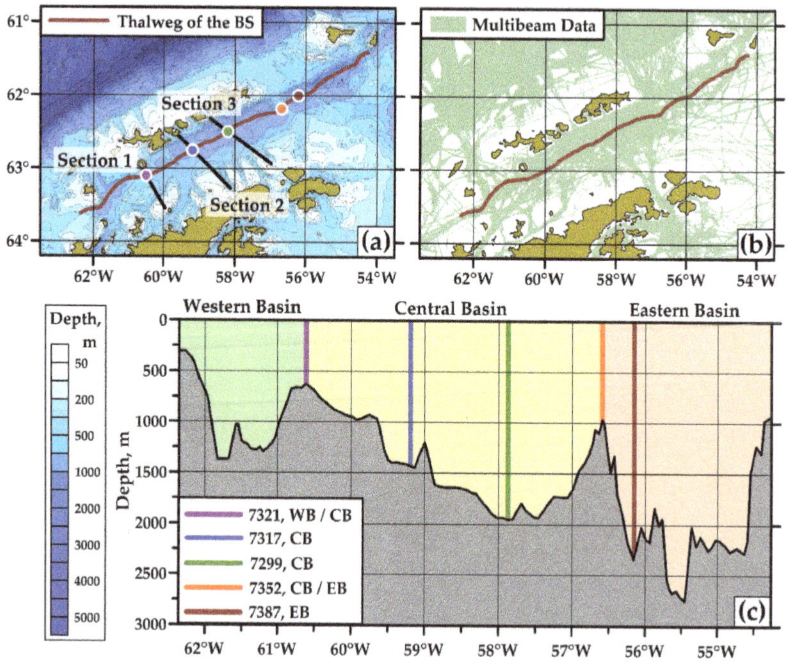

Figure 2. Depths along the thalweg of the Bransfield Strait and location of CTD/LADCP measurements. (**a**) Bottom topography according to the GEBCO2021 database, shoreline is based on GSHHS data [17], thalweg is indicated with dark red solid line, CTD/LADCP sections shown with black solid lines, colored dots show the locations of CTD stations along the thalweg presented in (**c**). (**b**) areas with ocean bottom measured with multibeam soundings indicated with green; the data were taken from TID file of GEBCO2021 database. (**c**) Ocean depth along the thalweg of the BS (shown with gray) and location of CTD stations along the strait; the deepest CTD/LADCP station was selected at each transect. Station numbers correspond to Table 1; colors are indicated in the bottom panel.

Table 1. Coordinates of stations carried out in the Bransfield Strait in January and February 2022.

Station Number	Date/ Time (UTC)	Coordinates	CTD Depth, m/Ocean Depth, m	Type of Measurements
		Section 1		
7318	26 January 2022 15:08	63°26.4′ S 60°05.7′ W	102/110	CTD/LADCP
7319	26 January 2022 17:09	63°20.0′ S 60°13.4′ W	613/616	CTD/LADCP
7320	26 January 2022 18:58	63°14.5′ S 60°19.4′ W	531/536	CTD/LADCP
7321	26 January 2022 21:41	63°09.9′ S 60°25.5′ W	660/666	CTD/LADCP
7322	26 January 2022 23:20	63°05.8′ S 60°29.9′ W	650/655	CTD/LADCP
7323	27 January 2022 00:48	63°03.1′ S 60°32.8′ W	368/373	CTD/LADCP
		Section 2		
7308	24 January 2022 09:12	63°10.0′ S 58°20.1′ W	186/191	CTD/LADCP
7309	24 January 2022 11:59	63°02.2′ S 58°35.7′ W	184/190	CTD/LADCP
7310	24 January 2022 14:00	62°54.0′ S 58°53.2′ W	742/746	CTD/LADCP
7311	24 January 2022 19:23	62°36.9′ S 59°27.2′ W	1271/1276	CTD/LADCP
7313	25 January 2022 16:58	62°33.5′ S 59°33.8′ W	392/397	CTD/LADCP
7314	25 January 2022 21:05	62°35.0′ S 59°31.8′ W	606/612	CTD/LADCP
7315	26 January 2022 00:04	62°35.9′ S 59°29.3′ W	988/994	CTD/LADCP
7316	26 January 2022 01:54	62°40.0′ S 59°22.0′ W	1403/1409	CTD/LADCP
7317	26 January 2022 06:23	62°44.9′ S 59°11.7′ W	1412/1422	CTD/LADCP
		Section 3		
7294	21 January 2022 11:29	62°52.0′ S 57°09.4′ W	148/152	CTD/LADCP
7295	21 January 2022 14:00	62°49.5′ S 57°15.8′ W	167/171	CTD
7296	21 January 2022 15:26	62°45.6′ S 57°25.8′ W	214/218	CTD
7297	21 January 2022 17:17	62°40.0′ S 57°39.6′ W	716/720	CTD
7298	21 January 2022 20:59	62°37.0′ S 57°52.0′ W	1227/1232	CTD/LADCP
7299	21 January 2022 23:25	62°31.0′ S 58°08.0′ W	1768/1773	CTD
7301	22 January 2022 21:00	62°26.4′ S 58°21.1′ W	1055/1063	CTD
7305	23 January 2022 15:55	62°21.0′ S 58°36.8′ W	748/751	CTD
7306	23 January 2022 20:29	62°23.0′ S 58°30.0′ W	1319/1323	CTD
		Overflow		
7352	31 January 2022 19:27	62°10.9′ S 56°41.0′ W	1016/1023	CTD/LADCP
7354	01 February 2022 14:41	62°12.6′ S 56°35.8′ W	968/972	CTD/LADCP
7355	01 February 2022 16:01	62°11.5′ S 56°38.5′ W	897/902	CTD/LADCP
7356	01 February 2022 17:18	62°10.1′ S 56°43.1′ W	925/931	CTD/LADCP
7357	01 February 2022 18:31	62°09.2′ S 56°46.2′ W	905/911	CTD/LADCP
7387	14 February 2022 05:26	61°57.0′ S 56°10.0′ W	2373/2379	CTD/LADCP
7388	14 February 2022 09:46	62°08.9′ S 56°36.5′ W	1160/1165	CTD/LADCP
7389	14 February 2022 11:16	62°10.0′ S 56°38.9′ W	1135/1139	CTD/LADCP
7390	14 February 2022 12:48	62°10.9′ S 56°40.7′ W	1000/1006	CTD/LADCP
7391	14 February 2022 14:37	62°08.4′ S 56°31.8′ W	1206/1212	CTD/LADCP

2.2. In Situ Measurements and Data Processing

The stations were performed using the lowered acoustic Doppler current (LADCP) and conductivity, temperature, depth (CTD) profilers mounted on a General Oceanics GO1018 rosette water sampler. The CTD measurements were performed along all three sections; the LADCP data are available only for Sections 1 and 2 due to technical reasons. Exact information about the type of measurements at each station is presented in Table 1. The water sampler was equipped with a Valeport VA500 altimeter allowing for measurements close to the ocean bottom (3–7 m above the seafloor; see Table 1 for more details). An Idronaut Ocean Seven 320plus CTD probe was used for the measurements together with an MKplus Deck Unit. CTD data were collected using standard package REDAS5 version 5.78. The declared accuracy of CTD measurements is 0.001 °C for temperature, and 0.001 mS/cm for conductivity sensors. The CTD data from the World Ocean Database (WOD2018) were used for addressing potential temperature and salinity variations in the bottom layer of the CB and EB. This database contains 27 CTD profiles in the CB, and 18 profiles in the EB carried out in different years and seasons (only profiles deeper than 1500 m were taken into account). Most stations were performed during the austral summer season. Thus, 18 stations of 27 in the CB, and 14 stations of 18 in the EB were occupied from November to February. The observation period covered the 1980s and 1990s; three stations in the EB were performed in 1975–1976. The LADCP data measured with a TRDI WorkHorse Monitor 300 kHz profiler were processed using programming package LDEO Software version IX.10 [40]. Data from the shipborne acoustic doppler current profiler (SADCP) TRDI Ocean Surveyor 75 kHz were used for more reliable data processing in the upper ocean layer. The accuracy of velocity measurements estimated by the processing program is usually 3–4 cm/s. In the bottom layers, due to the bottom track signals, the errors decreased to 1–2 cm/s. The results of LADCP processing were corrected by subtracting the tidal velocities. The barotropic tide was calculated on the basis of the TPXO 9.1 global inverse tide model [41].

3. Results

We analyzed the spatial kinematic and thermohaline structures of the currents within the BS at several transects across the strait; their locations are presented in Figure 1 and Table 1. The observations included CTD and LADCP stations from the surface to the bottom. The results of these measurements are discussed separately for the western (Section 1) and eastern (Section 2) margins of the CB.

3.1. Water Exchange between Western and Central Basins

The results of our measurements along three sections in the western part of the CB are shown separately on the basis of LADCP (Figure 3) and CTD (Figure 4) data. Previously, such measurements were performed in the CB, and allowed for investigations of the spatial structure of currents [26,27,30,35] and thermohaline properties of waters within the strait [16,23,37,42,43]. Some of these studies were focused on a single transect in different parts of the strait [25,26,35,39]. However, there are no studies focused on the sills between deep basins or where the structure of the TBW and TWW flows were analyzed on the basis of several quasisimultaneous crossings of the strait with synchronous CTD and LADCP measurements. Velocity data were projected to the direction along the strait (true direction is 60°). CTD data are available for all three sections; LADCP measurements were performed only at two western sections (Sections 1 and 2; see Figure 2 for their location). Two major circulation patterns were clearly observed at both sections: the fast and narrow Bransfield Current is located in the northwestern part of the strait, while the flow of waters from the Weddell Sea is observed along the Antarctic Peninsula. The main result is that TWW flow was observed at Section 1, the westernmost. It was previously suggested that TWW flow recirculates within the BS and does not propagate as far to the west through the entire strait [27,42]. The velocities of this flow do not decrease along the TWW path: the maximal velocities are 23 cm/s at Section 2 and 36 cm/s at Section 1. On the other hand, the Bransfield Current that transports TBW waters significantly changes between these sections. The maximal velocities changed from 24 cm/s at Section 1 to 52 cm/s at Section 2. The maximum was located at a 260 m depth at Section 1; at Section 2, the maximal velocities were observed at the sea surface. The total volume transports of TBW waters based on LADCP measurements were 1.19 Sv (Section 1) and 2.54 Sv (Section 2). The corresponding values of TWW transports were 0.77 Sv and 1.82 Sv. No significant currents were observed in the bottom layer of the sill between the WB and CB; the conditions in the bottom layer over the sill between CB and EB were significantly different and are discussed below in more detail.

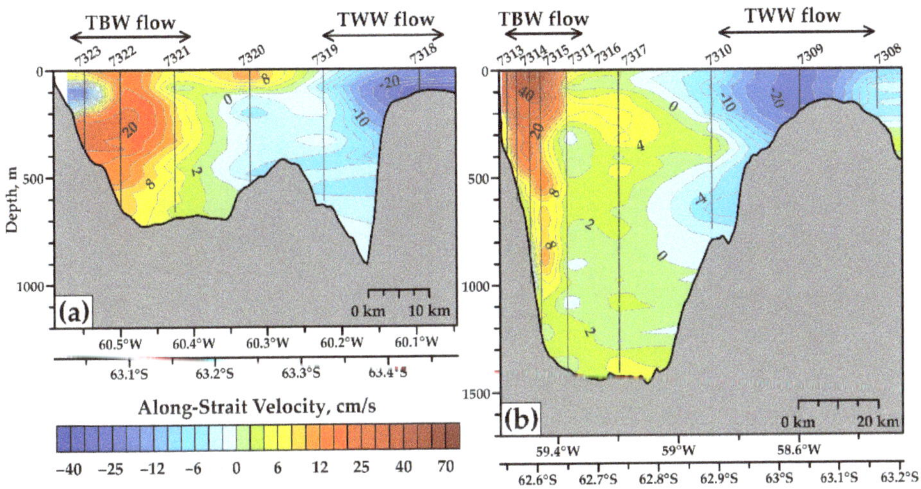

Figure 3. Along-strait LADCP velocity distributions at two sections across the BS: (**a**) Section 1 over the watershed between the WB and CB and (**b**) Section 2 in the CB. Locations of TBW and TWW flows are indicated at the top of each panel; stations are indicated by solid black lines. The bottom relief is shown according to the GEBCO2021 database.

The CTD data are available for all three sections across the CB of the strait (Figure 4). Potential temperature and salinity distributions showed quite similar structures; at all sections, TBW and TWW flows were clearly distinguished on the basis of different thermohaline properties. The warmest waters were observed within the Bransfield Current; the maximal water temperature changed along the strait from 1.27 °C (Section 1) to 1.56 °C (Section 2) and 1.52 °C (Section 3); the corresponding variations of the minimal salinity were 34.22, 34.14, and 34.24 PSU. At depths of 200–450 m, a core of mCDW waters was observed at all sections; their properties changed along the strait from 1.08 to 1.13 and 1.08 °C, and 34.73 to 34.73 and 34.71 psu. The minimal potential temperature in the upper 200 m was observed near the Antarctic Peninsula within the TWW flow. Along this flow, thermohaline properties changed from −0.73 °C and 34.60 psu (Section 3) to −0.62 °C and 34.57 psu (Section 2), and −0.66 °C and 34.58 psu (Section 1). The minimal potential temperature in the CB, −1.65 °C, was observed at the deepest point of the strait at station 7299; the minimal potential temperature over the watershed between the WB and CB was −0.95 °C (station 7321), confirming the well-known fact that the bottom waters of the CB originated directly from the waters of the Weddell Sea.

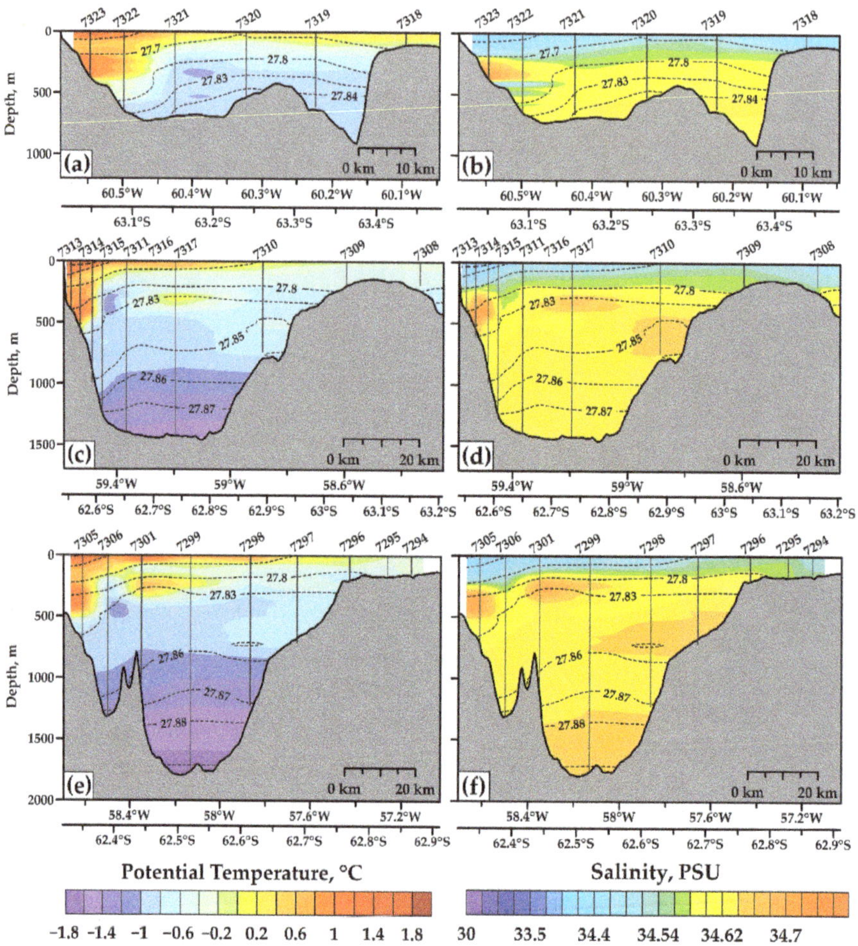

Figure 4. (**a**,**c**,**e**) Potential temperature and (**b**–**f**) salinity distributions at three sections across the Bransfield Strait. Data for (**a**,**b**) Section 1, (**c**,**d**) Section 2, and (**e**,**f**) Section 3. Isolines of potential density anomalies shown with dashed contours. Locations of CTD stations are indicated with solid lines. The bottom relief is shown according the GEBCO2021 database.

3.2. Deep-Water Overflow between Central and Eastern Basins

The overflow between the CB and EB was studied on the basis of a transect performed over the sill between the basins (Figure 5). The deepest point of the watershed between the basins is 1050 m based on the GEBCO2021 database. The transect included five stations and was oriented across the deep overflow. Along-flow velocity, potential temperature, salinity, and potential density relative to the sea surface are shown in Figure 5. Velocities over entire water column along the section were quite low (less than 10–15 cm/s). TBW and TWW flows are located at some distance away from the transect between two basins; the Bransfield Current was observed north of our stations, closer to the South Shetland Islands, while the inflow of waters from the Weddell Sea is located closer to the Antarctic Peninsula. On the basis of the data of our section, the intensification of currents was observed in the bottom layers. Two separate high-velocity jets were observed at station 7357 and 7352. The first jet with maximal velocity equal to 22 cm/s was located at the deepest point of the section. The coldest (minimal potential temperature was −1.31 °C) and densest (maximal potential density anomaly relative to the sea surface was 27.87 kg/m^3) waters were observed at this point. The second jet with the maximal velocity, equal to 21 cm/s, was observed at station 7357. Its location near the steep slope could be caused by the Coriolis force, which displaced the jet to the left in the Southern Hemisphere. A very thin bottom layer of cold waters was also observed at station 7354; this layer was formed by the overflow of cold waters from the central basin.

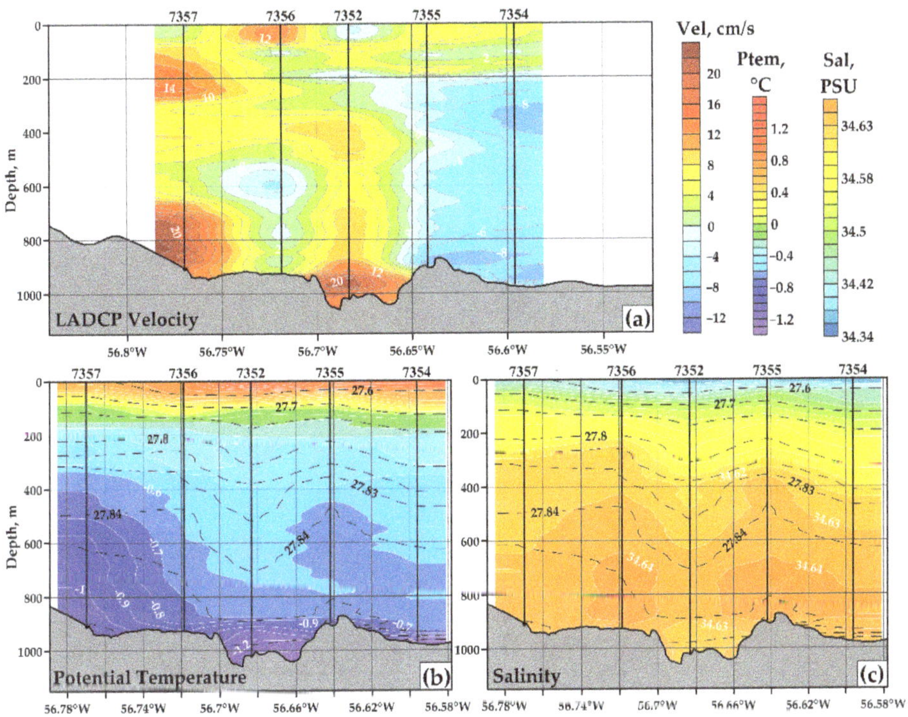

Figure 5. Structure of the deep overflow between the CB and EB of the strait: distribution of the (**a**) northeastern velocity (true direction is 45°) based on LADCP data, (**b**) potential temperature, and (**c**) salinity based on CTD data. Contours of potential density anomalies are shown with dashed black lines. Positions of stations are indicated with vertical solid lines together with the numbers of the stations.

The measurements in the region of the deep overflow between the CB and EB were performed twice with an interval of two weeks on 1 and 14 February 2022. The first survey included measurements at five stations across the bottom flow; the second survey consisted of four stations along the overflow. The station at the sill point between the basins was performed twice during both surveys. The direction of the deep overflow changed during these two weeks. The LADCP velocities measured in the bottom layer are shown in Figure 6. We present the maximal velocities in the 100 m bottom layer and averaged velocities in the bottom layers with thicknesses of 50 and 100 m. The magnitudes and directions of these velocity vectors were very close. For example, the maximal velocity of the overflow was 22 cm/s, while average velocities were 19 and 18 cm/s for the 50 and 100 m layers, respectively. As for the measurements at the same point performed two weeks later, the flow was directed in the opposite direction, from the EB to CB. The velocities of this opposite flow reached 18 cm/s (maximal value), 16 cm/s (averaged in the 50 m bottom layer), and 14 cm/s (averaged in the 100 m bottom layer). These changes in currents are shown in Figure 7 in more detail. Corresponding changes in the potential temperature and salinity between the measurements over the sill point between the basins were 0.47 °C (from −1.31 to −0.84 °C) and 0.02 psu (from 34.62 to 34.64 psu), respectively. A sharp thermocline was observed at depths of 880–920 m at station 7352; the vertical potential temperature gradient reached 0.01 °C/m (1 °C per 100 m depth). When the current changed direction in two weeks (station 7390), such a strong thermocline was not observed.

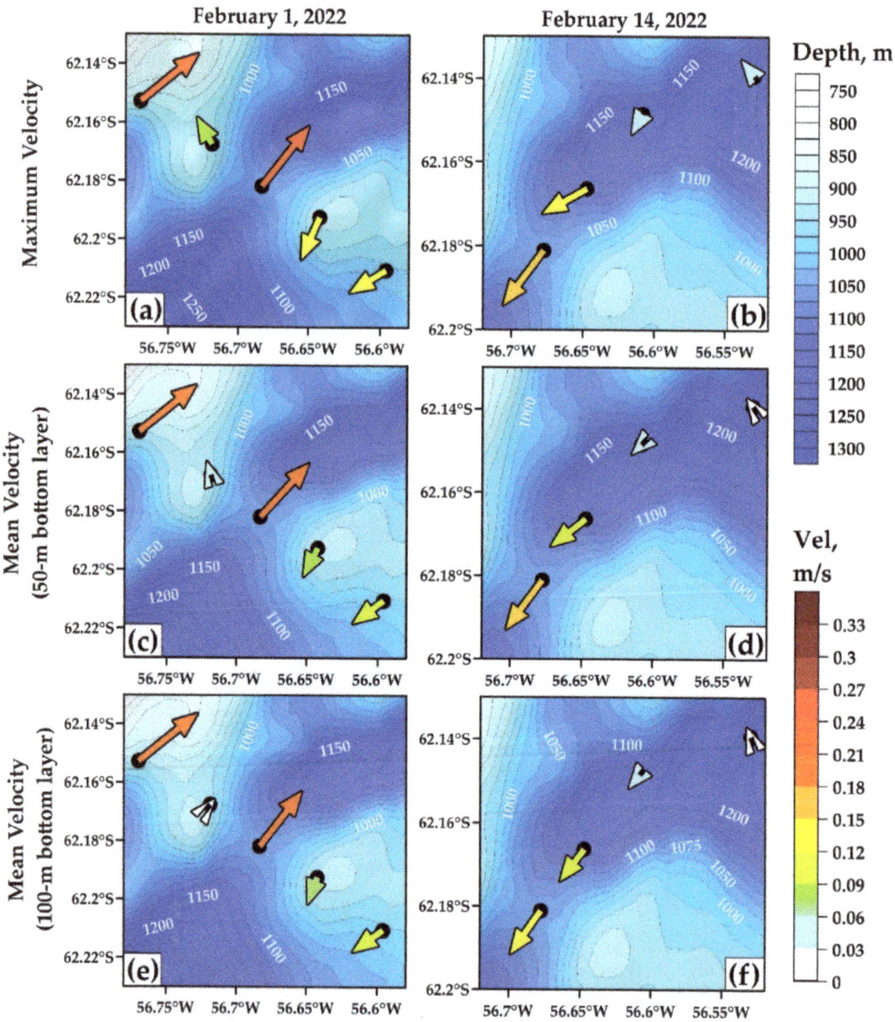

Figure 6. Measured LADCP velocities in the bottom layer over the sill point between the central and eastern basins of the BS. (**a**,**b**) Maximal velocities averaged (**c**,**d**) in the 50 m bottom layer and ((**e**,**f**)) the 100 m bottom layer. (**a**,**c**,**e**) Data measured on 1 February; (**b**–**f**) measurements performed two weeks later (on 14 February). The magnitude of currents is shown in both the colors and lengths of the arrows; all color and vector scales are the same in all panels. The relief is based on the GEBCO2021 database.

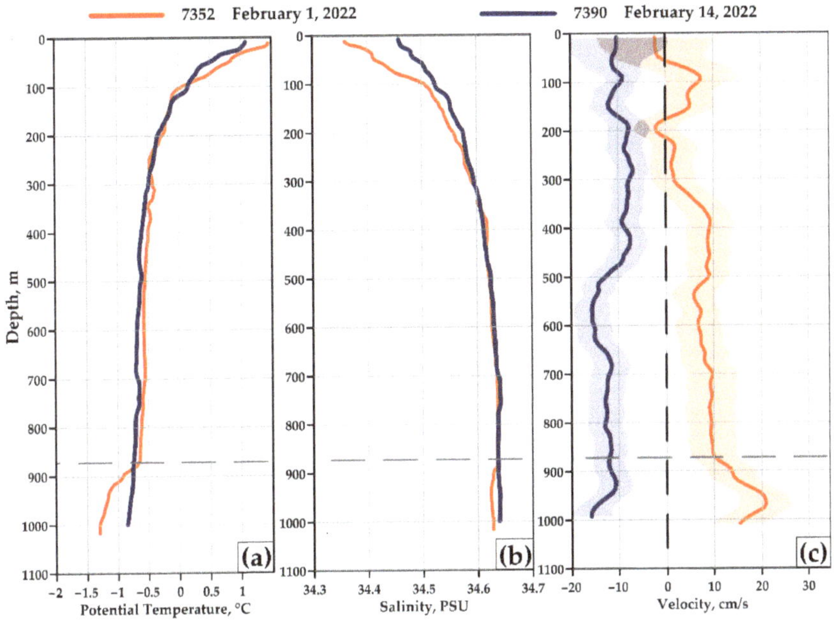

Figure 7. (**a**) Potential temperature, (**b**) salinity, and (**c**) along-flow velocity profiles measured at the same location over a sill point between the central and eastern basins on 1 February 2022 (station 7352, orange line) and 14 February 2022 (station 7390, blue line). Horizontal dashed line at a depth of 870 m indicates the upper boundary of the bottom layer with high velocities.

4. Discussion

Different properties of waters from the separate BS basins are caused by different pathways of propagation of Weddell Sea waters, which is the source of these waters. Differences in temperature and salinity define the density of deep waters in these basins, which in turn define the dynamics of deep layers within the strait. Temperature–salinity diagrams (Figure 8) were used for the analysis of thermohaline water structure at all studied sections. The densest waters were observed in the deepest part of the CB (Section 3, Figure 8c); the maximal recorded potential density anomaly at station 7299 was 27.89 kg/m^3. Such dense waters were not observed at Section 1 (Figure 8a) between the WB and CB, confirming the fact that the deepest layers of the CB are filled by TWW waters from the east. The minimal potential temperature in the bottom layer of the CB was −1.65 °C, which is quite close to the freezing-point temperature (−1.90 °C for waters with 34.65 PSU salinity). This fact indicates the possible importance of the local ice formation effects during austral winter. It is quite possible that cooling and ice formation over the shelves of the BS (which are sufficiently wide in the southern part of the strait near the Antarctic Peninsula) can cause the formation of such dense waters within the CB. More winter measurements are needed for studies of such processes in the strait.

For addressing the temporal variability of thermohaline properties of the deepest waters within the Bransfield Strait, we analyzed the modern World Ocean Database (WOD2018). Temperature–salinity diagrams allowed us to compare the maximal densities of deep waters in the CB (Figure 9a) and EB (Figure 9b).These data show that the deep waters in the CB are colder, saltier, and denser than waters in the EB. The minimal potential temperature in the CB usually reaches −1.75 °C, while waters in the EB are always warmer than −1.42 °C. Salinity in the deep layers of the CB significantly varies, but it is usually greater than that in the EB, which contributes to the observed density differences. The maximal density in the CB reaches 27.89 kg/m^3, but it does not exceed 27.82 kg/m^3 in the

EB. Thus, the mean bottom flow through the studied sill should be directed from the CB to the EB. Continuous observations are required to evaluate the magnitude of the deep overflow, and the question regarding the source of deep water in the central basin requires further studies.

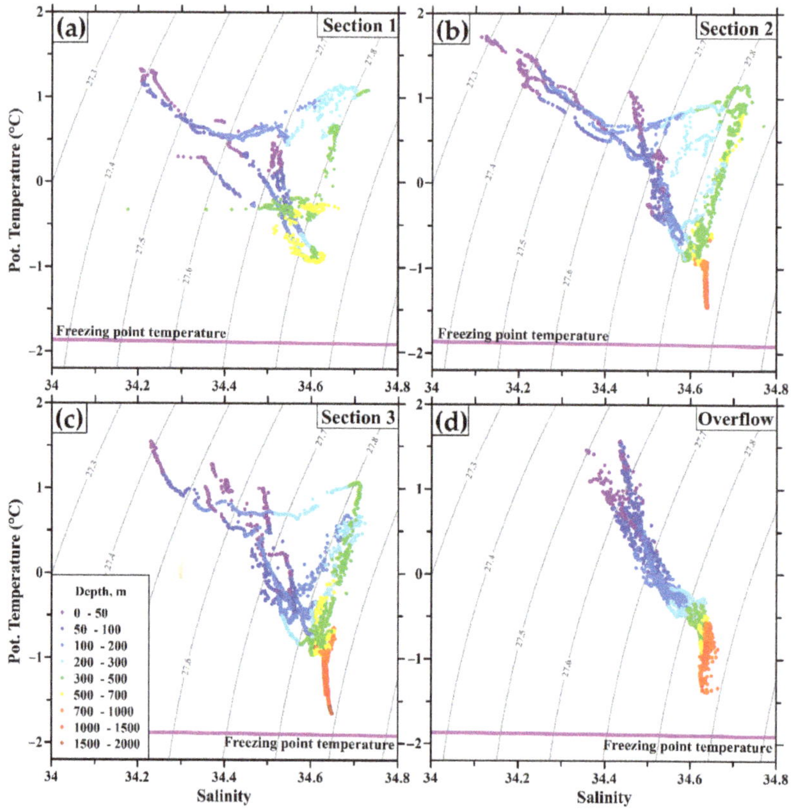

Figure 8. Temperature–salinity diagrams including the freezing-point temperature (magenta line) and sigma contours (gray lines) at the hydrographic stations at different sections in the BS. From west to east: (**a**) Section 1, (**b**) Section 2, (**c**) Section 3, (**d**) section across the deep overflow. The locations of the sections are shown in Figures 1 and 2. The colors of the dots indicate depths of measurements; see color scale in (**c**). The freezing-point temperature was calculated on the basis of EOS80 equations [44].

CTD data from several profiles are presented in Figure 10 for a more detailed analysis of water transformation along the strait. Stations 7321, 7317, 7299, 7352, and 7387 were established at the deepest points along the thalweg of the BS. The coldest (−1.65 °C) and densest (27.89 kg/m^3) waters were located in the CB (station 7299). Despite the fact that the EB is deeper (2750 against 1960 m in the CB), the waters in the EB are not as cold and dense as those in the CB. This indicates that the bottom water in the EB has not been directly transported from the TWW flow from the Weddell Sea, but appeared there due to the overflow from the CB. During propagation, these waters mix with overlaying warmer and more saline waters leading to the observed decrease in density. Regarding the overflow between the basins (station 7352), the bottom waters there are colder than any waters at these depths in the CB or EB. In fact, a potential temperature isotherm of −1.30 °C was observed at a depth of 1105 m in the CB (station 7299) and at a depth of 1000 at the sill point between the basins (station 7352), indicating that deeper layers of the CB are involved in the studied overflow.

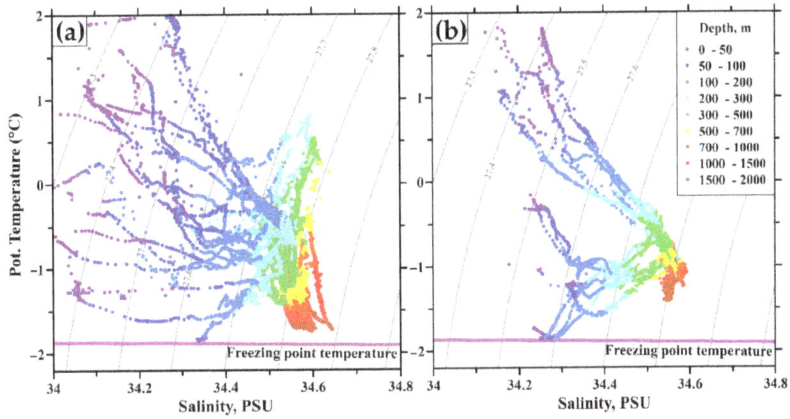

Figure 9. Temperature–salinity diagrams including the freezing-point temperature (magenta line) and sigma contours (gray lines) at the available hydrographic stations in the WOD2018 database for the (**a**) central and (**b**) eastern basins. Only full-depth profiles deeper than 1500 m were selected from the database. The colors of the dots indicate depths of measurements; see color scale in (**b**). The freezing-point temperature was calculated on the basis of EOS80 equations [44].

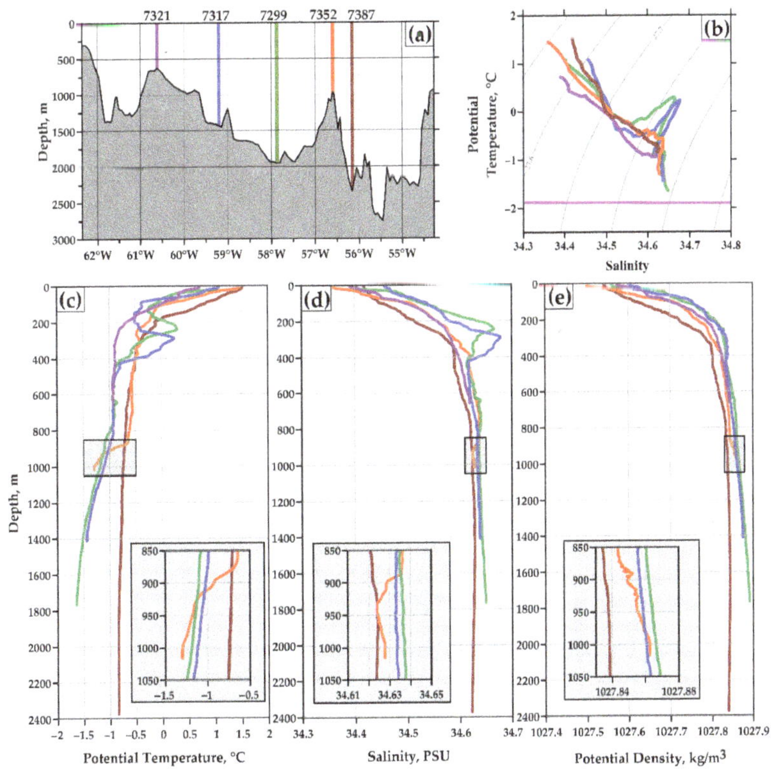

Figure 10. Potential temperature, salinity, and potential density along the thalweg of the BS: (**a**) location of stations relative to the basins of the strait, (**b**) temperature-salinity diagram, (**c**) profiles of potential temperature, (**d**) salinity, and (**e**) potential density. The colors of stations are the same in all panels. All profiles at depths of the overflow (850–1050 m) are shown in the insets in more detail.

Temporal variability in the flow between the CB and EB needs to be studied in more detail. Our two stations with a time interval of two weeks showed significant variability of currents between the basins. One of the possible reasons of such variability is tides; tidal motions are intensified over underwater ridges such as the watershed between the CB and EB. However, the calculations of tidal velocities at the exact time and positions of stations 7352 and 7390 based on the TPXO9.1 model show that variability cannot be completely explained by tidal motions. Thus, tidal velocities were 8.1 cm/s to the northeast (direction 48° relative to the north) at the time of station 7352, and 7.8 cm/s to the east (direction 88°) at the time of station 7390. The measured LADCP velocities are sufficiently higher. One of the other possible causes of such a variability are internal seiches. Some similar mechanisms were suggested for the variability in the overflow in the Denmark Strait [45]. In any case, further measurements of currents over this sill point are needed for better understanding the properties of the observed overflow, their temporal and spatial variability, and dynamic aspects that drive this intense current in this narrow gap between the basins.

5. Conclusions

Thermohaline structure and water dynamics in the BS during austral summer of 2022 were analyzed on a series of CTD/LADCP sections across the strait. The analysis was focused on the water exchange between the CB and adjacent basins; this exchange significantly differed in the western and eastern margins of the CB. The main result for the western margin is that there is a significant flow of TWW waters along the Antarctic Peninsula that continues up to the western basin. It was previously suggested that all these waters recirculate within the CB as a part of cyclonic circulation in the strait. Our direct velocity measurements show that the TWW flow transports 0.77 Sv to the WB. At the same time, there was no significant overflow of deep and dense waters between the WB and CB due to the relatively shallow sill separating the basins. The situation over the watershed between the CB and EB was completely different. The underwater ridge there was much deeper and reaches a depth of more than 1000 m at the sill point. At that point, a strong intensification of deep currents was observed; the maximal bottom velocities reached 22 cm/s. This bottom flow is very variable and sometimes changes its direction. More direct velocity measurements are needed at this point for the detailed analysis of the bottom overflow and its temporal variability.

Author Contributions: Conceptualization, D.I.F. and E.G.M.; methodology, V.A.K. and D.A.S.; software, A.S.G. and I.D.D.; validation, O.A.Z.; data curation, V.A.K. and S.A.S.; writing—original draft preparation, D.I.F.; writing—review and editing, E.G.M., A.A.L., S.A.M., V.I.P. and P.A.S.; visualization, O.S.M., R.Z.M. and S.A.O.; supervision, E.G.M.; funding acquisition, E.G.M. and D.I.F. All authors have read and agreed to the published version of the manuscript.

Funding: This work was supported by the State Task of Russia FMWE-2022-0001 (ship operations and expenses), 0211-2019-0007 (field studies), and FNNN-2022-0001 (bio-optical measurements). The LADCP data processing and analysis of circulation in the strait were supported with Russian Science Foundation grant 22-77-10004. CTD data processing by DIF was supported by a grant of the President of the Russian Federation for the state support of young Russian scientists—candidates of science, research project MK-1492.2021.1.5.

Institutional Review Board Statement: Not applicable.

Informed Consent Statement: Not applicable.

Data Availability Statement: All experimental data, including CTD and LADCP profiles, used in the publication are available in open access through the Mendeley Data service (http://dx.doi.org/10.17632/69v8599btr.1). GEBCO2021 data are available at the official service https://www.gebco.net/data_and_products/gridded_bathymetry_data/gebco_2021/ (accessed on 4 October 2022).

Conflicts of Interest: The authors declare no conflict of interest.

References

1. Huneke, W.; Huhn, O.; Schroeder, M. Water masses in the Bransfield Strait and adjacent seas, austral summer 2013. *Polar Biol.* **2016**, *39*, 789–798. [CrossRef]
2. Vaughan, D.; Marshall, G.; Connolley, W.; Parkinson, C.; Mulvaney, R.; Hodgson, D.; King, J.; Pudsey, C.; Turner, J. Recent rapid regional climate warming on the Antarctic Peninsula. *Clim Chang.* **2003**, *60*, 243–274. [CrossRef]
3. Veit-Köhler, G.; Durst, S.; Schuckenbrock, J.; Hauquier, F.; Durán Suja, L.; Dorschel, B.; Vanreusel, A.; Martínez Arbizu, P. Oceanographic and topographic conditions structure benthic meiofauna communities in the Weddell Sea, Bransfield Strait and Drake Passage (Antarctic). *Prog. Oceanogr.* **2018**, *162*, 240–256. [CrossRef]
4. Spiridonov, V. A Scenario of the Late-Pleistocene-Holocene Changes in the Distributional Range of Antarctic Krill (*Euphausia superba*). *Mar. Ecol.* **1996**, *17*, 519–541. [CrossRef]
5. Atkinson, A.; Siegel, V.; Pakhomov, E.; Rothery, P. Long-term decline in krill stock and increase in salps within the Southern Ocean. *Nature* **2004**, *432*, 100–103. [CrossRef] [PubMed]
6. Loeb, V.; Hofmann, E.; Klinck, J.; Holm-Hansen, O.; White, W. ENSO and variability of the Antarctic Peninsula pelagic marine ecosystem. *Antarct. Sci.* **2009**, *21*, 135–148. [CrossRef]
7. Reiss, C.; Cossio, A.; Santora, J.; Dietrich, K.; Murray, A.; Greg Mitchell, B.; Walsh, J.; Weiss, E.L.; Gimpel, C.; Jones, C.; et al. Overwinter habitat selection by Antarctic krill under varying sea-ice conditions: Implications for top predators and fishery management. *Mar. Ecol. Prog. Ser.* **2017**, *568*, 1–16. [CrossRef]
8. Perry, F.; Atkinson, A.; Sailley, S.; Tarling, G.; Hill, S.; Lucas, C.; Mayor, D. Habitat partitioning in Antarctic krill: Spawning hotspots and nursery areas. *PLoS ONE* **2019**, *14*, e0219325. [CrossRef]
9. Wang, R.; Song, P.; Li, Y.; Lin, L. An integrated, size-structured stock assessment of Antarctic krill, *Euphausia superba*. *Front. Mar. Sci.* **2021**, *8*, 710544. [CrossRef]
10. Nordhausen, W. Distribution and growth of larval and adult *Thysanoessa macrura* (*Euphausiacea*) in the Bransfield Strait Region, Antarctica. *Mar. Ecol. Prog. Ser.* **1992**, *83*, 185–196. [CrossRef]
11. Siegel, V. Distribution and population dynamics of *Euphausia superba*: Summary of recent findings. *Polar Biol.* **2005**, *29*, 1–22. [CrossRef]
12. Siegel, V.; Reiss, C.; Dietrich, K.; Haraldsson, M.; Gerhard, R. Distribution and abundance of Antarctic krill (*Euphausia superba*) along the Antarctic Peninsula. *Deep Sea Res. Part I Oceanogr. Res. Pap.* **2013**, *77*, 63–74. [CrossRef]
13. Meyer, B.; Atkinson, A.; Bernard, K.; Brierley, A.; Driscoll, R.; Hill, S.; Marschoff, E.; Maschette, D.; Perry, F.; Reiss, C.; et al. Successful ecosystem-based management of Antarctic krill should address uncertainties in krill recruitment, behavior and ecological adaptation. *Commun. Earth Environ.* **2020**, *1*, 28. [CrossRef]
14. Dorschel, B.; Gutt, J.; Huhn, O.; Bracher, A.; Huntemann, M.; Huneke, W.; Gebhardt, C.; Schröder, M.; Herr, H. Environmental information for a marine ecosystem research approach for the northern Antarctic Peninsula (RV *Polarstern* expedition PS81, ANT-XXIX/3). *Polar Biol.* **2016**, *39*, 765–787. [CrossRef]
15. Gordon, A.L.; Nowlin, W.D. The basin waters of the Bransfield Strait. *J. Phys. Oceanogr.* **1978**, *8*, 258–264. [CrossRef]
16. López, O.; Garcia, M.; Gomis, D.; Rojas, P.; Sospedra, J.; Sanchez-Arcilla, A. Hydrographic and hydrodynamic characteristics of the eastern basin of the Bransfield Strait (Antarctica). *Deep Sea Res. Part I Oceanogr. Res. Pap.* **1999**, *46*, 1755–1778. [CrossRef]
17. Wessel, P.; Smith, W. A global, self-consistent, hierarchical, high-resolution shoreline database. *J. Geophys. Res.* **1996**, *101*, 8741–8743. [CrossRef]
18. Tokarczyk, R. Classification of water masses in the Bransfield Strait and Southern part of the Drake Passage using a method of statistical multidimensional analysis. *Pol. Polar Res.* **1987**, *8*, 333–336.
19. García, M.; López, O.; Sospedra, J.; Espino, M.; Gràcia, V.; Morrison, G.; Pojas, P.; Figa, J.; Puigdefabregas, J.; Arcilla, A. Mesoscale variability in the Bransfield Strait region (Antarctica) during Austral summer. *Ann. Geophys.* **1994**, *12*, 856–867. [CrossRef]
20. Niiler, P.; Amos, A.; Hu, J.-H. Water masses and 200 m relative geostrophic circulation in the western Bransfield Strait region. *Deep Sea Res. Part A Oceanogr. Res. Pap.* **1991**, *38*, 943–959. [CrossRef]
21. Zhou, M.; Niiler, P.; Zhu, Y.; Dorly, R. The western boundary current in the Bransfield Strait, Antarctica. *Deep Sea Res. Part I Oceanogr. Res. Pap.* **2006**, *53*, 1244–1252. [CrossRef]
22. Poulin, F.; Stegner, A.; Hernández-Arencibia, M.; Marrero-Díaz, A.; Sangrà, P. Steep shelf stabilization of the coastal Bransfield Current: Linear stability analysis. *J. Phys. Oceanogr.* **2014**, *44*, 714–732. [CrossRef]
23. Sangrà, P.; Gordo, C.; Hernández-Arencibia, M.; Marrero-Díaz, A.; Rodríguez-Santana, A.; Stegner, A.; Martínez-Marrero, A.; Pelegrí, J.L.; Pichon, T. The Bransfield Current system. *Deep Sea Res. Part I Oceanogr. Res. Pap.* **2011**, *58*, 390–402. [CrossRef]
24. Ruiz Barlett, E.; Tosonotto, G.; Piola, A.; Sierra, M.; Mata, M. On the temporal variability of intermediate and deep waters in the Western Basin of the Bransfield Strait. *Deep Sea Res. Part II* **2017**, *149*, 31–46. [CrossRef]
25. Morozov, E.; Krechik, V.; Frey, D.; Polukhin, A.; Sapozhnikov, P. Water Masses, Currents, and Phytoplankton in the Bransfield Strait in January 2020. In *Antarctic Peninsula Region of the Southern Ocean*; Advances in Polar Ecology; Morozov, E., Flint, M., Spiridonov, V., Eds.; Springer: Cham, Switzerland, 2021; Volume 6, pp. 55–64.
26. Polukhin, A.; Morozov, E.; Tishchenko, P.; Frey, D.; Artemiev, V.; Borisenko, G.; Vidnichuk, A.; Marina, E.; Medvedev, E.; Popov, O.; et al. Water Structure in the Bransfield Strait (Antarctica) in January 2020: Hydrophysical, Optical and Hydrochemical Features. *Oceanology* **2021**, *61*, 632–644. [CrossRef]

27. Veny, M.; Aguiar-González, B.; Marrero-Díaz, Á.; Rodríguez-Santana, Á. Seasonal circulation and volume transport of the Bransfield Current. *Prog. Oceanogr.* **2022**, *204*, 102795. [CrossRef]
28. Zhou, X.; Zhu, G.; Hu, S. Influence of tides on mass transport in the Bransfield Strait and the adjacent areas, Antarctic. *Polar Sci.* **2020**, *23*, 100506. [CrossRef]
29. Khimchenko, E.; Frey, D.; Morozov, E.G. Tidal internal waves in the Bransfield Strait, Antarctica. *Russ. J. Earth Sci.* **2020**, *20*, ES2006. [CrossRef]
30. Savidge, D.; Amft, J. Circulation on the West Antarctic Peninsula derived from 6 years of shipboard ADCP transects. *Deep Sea Res. Part I Oceanogr. Res. Pap.* **2009**, *56*, 1633–1655. [CrossRef]
31. Wang, X.; Moffat, C.; Dinniman, M.; Klinck, J.; Sutherland, D.; Aguiar-González, B. Variability and Dynamics of Along-Shore Exchange on the West Antarctic Peninsula (WAP) Continental Shelf. *J. Geophys. Res.* **2022**, *127*, e2021JC017645. [CrossRef]
32. Wilson, C.; Klinkhammer, G.; Chin, C. Hydrography within the central and east basins of the Bransfield Strait, Antarctica. *J. Phys. Oceanogr.* **1999**, *29*, 465–479. [CrossRef]
33. Gordon, A.; Mensch, M.; Dong, Z.; Smethie, W.; de Bettencourt, J. Deep and bottom water of the Bransfield Strait eastern and central basins. *J. Geophys. Res. Atmos.* **2000**, *105*, 11337–11346. [CrossRef]
34. Dotto, T.; Kerr, R.; Mata, M.; Garcia, A. Multidecadal freshening and lightening in the deep waters of the Bransfield Strait, Antarctica. *J. Geophys. Res.* **2016**, *121*, 3741–3756. [CrossRef]
35. Krechik, V.; Frey, D.; Morozov, E. Peculiarities of water circulation in the central part of the Bransfield Strait in January 2020. *Dokl. Earth Sci.* **2021**, *496*, 92–95. [CrossRef]
36. Izhitskiy, A.; Romanova, N.; Vorobieva, O.; Frey, D. Impact of Ice Melting on Oceanographic and Hydrobiological Characteristics of Surface Waters in the Powell Basin, Weddell Sea, in January–February 2020. *Oceanology* **2022**, *62*, 439–446. [CrossRef]
37. Damini, B.; Kerr, R.; Dotto, T.; Mata, M. Long-term changes on the Bransfield Strait deep water masses: Variability, drivers and connections with the northwestern Weddell Sea. *Deep Sea Res. Part I Oceanogr. Res. Pap.* **2022**, *179*, 10367. [CrossRef]
38. Marshall, G.; Stott, P.; Turner, J.; Connolley, W.; King, J.; Lachlan-Cope, T. Causes of exceptional atmospheric circulation changes in the Southern Hemisphere. *Geophys. Res. Lett.* **2004**, *31*, L14205. [CrossRef]
39. Mukhametyanov, R.; Frey, D.; Morozov, E. Currents in the Bransfield Strait Based on Geostrophic Calculations and Data of Instrumental Measurements. *Izv. Atmos. Ocean. Phys.* **2022**, *58*, 500–506. [CrossRef]
40. Visbeck, M. Deep velocity profiling using Lowered Acoustic Doppler Current Profiler: Bottom track and inverse solution. *J. Atmos. Ocean. Technol.* **2002**, *19*, 794–807. [CrossRef]
41. Egbert, G.; Erofeeva, S. Efficient inverse modeling of barotropic ocean tides. *J. Atmos. Ocean. Technol.* **2002**, *19*, 183–204. [CrossRef]
42. Sangrà, P.; Stegner, A.; Hernández-Arencibia, M.; Marrero-Díaz, A.; Salinas, C.; Aguiar-González, B. The Bransfield Gravity Current. *Deep Sea Res. Part I Oceanogr. Res. Pap.* **2017**, *119*, 1–15. [CrossRef]
43. Morozov, E.; Zuev, O.; Zamshin, V.; Krechik, V.; Ostroumova, S.; Frey, D. Observations of icebergs in Antarctic cruises of the R/V "Akademik Mstislav Keldysh". *Russ. J. Earth. Sci.* **2022**, *22*, ES2001. [CrossRef]
44. Fofonoff, P.; Millard, R. *Algorithms for Computation of Fundamental Properties of Seawater*; UNESCO Technical Papers in Marine Science; Unesco: Paris, France, 1983; Volume 44, 53p.
45. Morozov, E.; Frey, D.; Gladyshev, S.; Gladyshev, V. Hydrodynamics of bottom water flow from the Arctic to the Atlantic through the Denmark Strait. *Izv. Atmos. Ocean. Phys.* **2020**, *56*, 479–487. [CrossRef]

water

Article

Antarctic Bottom Water Jets Flowing from the Vema Channel

Eugene G. Morozov *, Oleg A. Zuev, Dmitry I. Frey and Viktor A. Krechik

Shirshov Institute of Oceanology, Nakhimovsky Prospect 36, 117997 Moscow, Russia
* Correspondence: egmorozov@mail.ru

Abstract: Properties of the abyssal current of Antarctic Bottom Water (AABW) from the Vema Channel are studied based on temperature, salinity, and velocity profiler (CTD/LADCP) data. Previous studies over a period of almost 30 years revealed that very intense current of AABW exists in the Vema Channel. Later, it was found that this current consists of two branches. One branch spreads over the bottom of the channel; the other branch is elevated over the western wall of the channel. The deepest branch decays after it passes approximately 100 km while the upper one continues further to the North Atlantic and is the source of abyssal waters in the Canary and Cabo Verde basins of the North Atlantic. Data analysis suggested that the upper jet splits into two. One of these descends down a canyon at 24°30′ S, while the other (the third one) remains on the continental slope, and indications of its existence are also found at 24°00′ S. This research analyzes the existence and pathway of this third branch that can be traced up to latitude 24° S. Velocity measurements in 2022 allowed us to confirm the existence of this third branch.

Keywords: Antarctic Bottom Water; Vema Channel; CTD/LADCP measurements; three jets of bottom current

Citation: Morozov, E.G.; Zuev, O.A.; Frey, D.I.; Krechik, V.A. Antarctic Bottom Water Jets Flowing from the Vema Channel. *Water* **2022**, *14*, 3438. https://doi.org/10.3390/w14213438

Academic Editor: Changming Dong

Received: 23 August 2022
Accepted: 25 October 2022
Published: 28 October 2022

1. Introduction

Antarctic Bottom Waters originate from the Weddell Sea and spread to the north in the abyssal depths of the Atlantic Ocean. The Vema Channel is a pathway for the bottom waters through the Rio Grande Rise and Santos Plateau before they spread to the Brazil Basin and further to the north. Weddell Sea Deep Water (the densest abyssal water) flows only through the deep Vema Channel, which is approximately 4600–4800-m deep. The channel is a narrow passage situated between two terraces that are located on both sides. The narrowest channel width is about 15 km, and the channel length is about 700 km (Figure 1) [1].

Oceanographic studies of the Vema Channel have been carried out since the 1970s [2]. The largest number of CTD measurements was made over the section along 31°12′ S and at a point on this section at 39°18.3′ W. Since 1972, 29 visits have been made to the region; our group joined these studies in 2002 [3]. After 2005, CTD casts were supplemented with LADCP velocity profilers. The current of bottom water from the Vema Channel and circulation in the southern part of the Brazil Basin were studied for the first time in [4]. These studies revealed a strong current of AABW and displacement of the coldest core to the eastern wall of the channel [5]. A warming trend of Antarctic Bottom Water has been found in the Vema Channel [6–8]. The long-term temperature trend revealed on the basis of these data is shown in Figure 2. It can be observed that the temperature increase continues: our measurements taken in 2020 and 2022 reveal a continuing gradual increase in the bottom potential temperature at this point. The temperature increase is caused by the warming of Weddell Sea Deep Water in the Weddell Sea. The signal from the Weddell Sea to the Vema Channel has been propagating for more than 35 years [9]. Hence, a signal from the Weddell Sea, caused by warming approximately 40 years ago, has now been recorded. There are insufficient data to explain the abrupt change in temperature in the early 1990s as this process is related to the warming in the Weddell Sea in the 1950s–1980s.

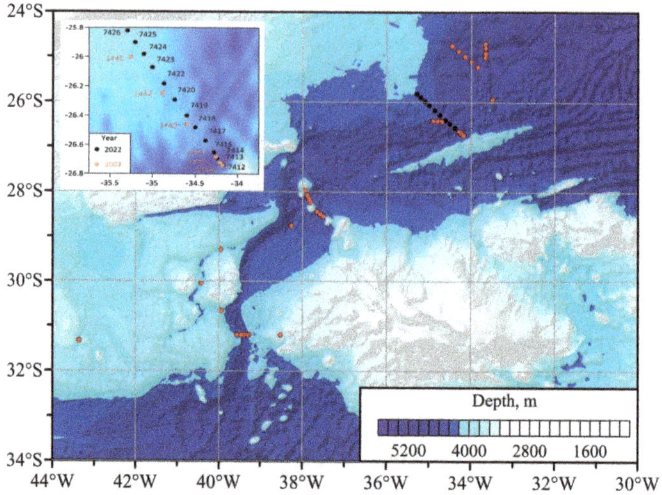

Figure 1. Bottom topography in the region of the Vema Channel (based on the GEBCO2019 data). Our stations in 2020 are indicated by red dots and those in 2022 are indicated by black dots. The inset shows stations in 2022 on a larger scale and stations in 2003 (dark yellow dots on the inset, not shown on the main map) approximately along the same line crossing the outflowing currents from the Vema Channel.

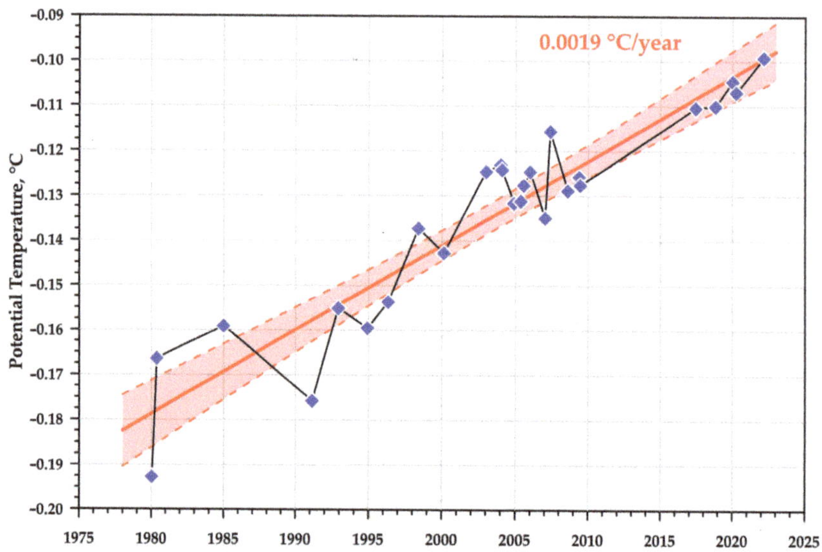

Figure 2. Long-term trend of potential temperature in the bottom layer of the Vema Channel. Blue dots indicate measurements during the visits for distinguishing the warming of AABW in the coldest part of the jet at 31°12° S, 39°18.3 W. The fit curve indicates that the warming rate is 0.0019 °C per year. Red color between two dashed lines shows the 95% confidence interval.

The existence of two branches of Antarctic Bottom Water current in the Vema Channel was later reported in [10–12]. The AABW outflow from the Vema Channel was also studied based on a regional version of the ocean circulation model [13,14]. However, studies of the fine structure of the outflow require direct observations, which are extremely rare in this region. The pathways of bottom water propagation into the Brazil Basin north of the

Vema Channel have been studied based on the measurements of temperature, salinity, and velocity profilers (CTD/LADCP). It was suggested in [12] based on historical potential temperature measurements at the bottom that there is a third jet of the bottom water outflow. The objective of this work is to confirm the existence of this shallower jet (4200–4600 m) based on velocity measurements and map its pathway to the north up to 24° S.

2. Data

A long-term series of observations have been performed since the 1970s [3]. These measurements allow us to study long-term trends in the properties of the abyssal flow. Two regions of the channel were studied during the WOCE experiment in the 1990s. It is the standard section along 31°12′ S from 39°18′ W to 39°27′ W and the region in the northern part of the channel [15]. In addition, WOCE section A17 occupied by French scientists in 1994 and our stations in 2003 can be related to historical stations in the region. Complete information about the oceanographic stations with tables of coordinates and time is reported in [3]. Since this research concerns the northern part of the region, we emphasize that we performed investigations in the northern region of the Vema Channel in 2003, 2009, 2010, 2012, 2013, and, recently, in 2018, 2019, 2020, and 2022.

In cruise 87 of the R/V *Akademik Mstislav Keldysh* (AMK87) in 2022, we occupied one station over the section along 31°12′ S, which has been repeatedly occupied in the coldest jet at (31°12′ S; 39°18.3′ W). We also occupied a section of 13 stations north of the Vema Channel, which was planned to intersect all outflow jets and determine the position of individual stream jets of Antarctic Bottom Water outflow from the Vema Channel at latitudes 25.8–26.8° S. This section also made it possible to determine the flow velocities in the jets (Figures 1 and 3). The coordinates of stations in the region are given in Table 1.

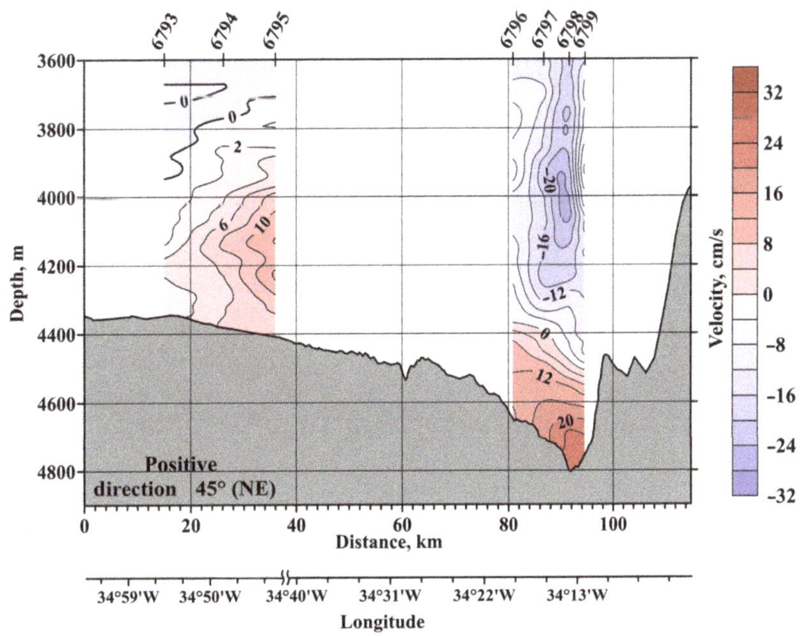

Figure 3. Distributions of velocity normal to the sections at 26°25′ S, 34°50′ W (western) and 26°40′ S, 34°15′ W (eastern) (see red dots in Figure 4) in April 2020. Numbers of stations are shown along the top axis [12].

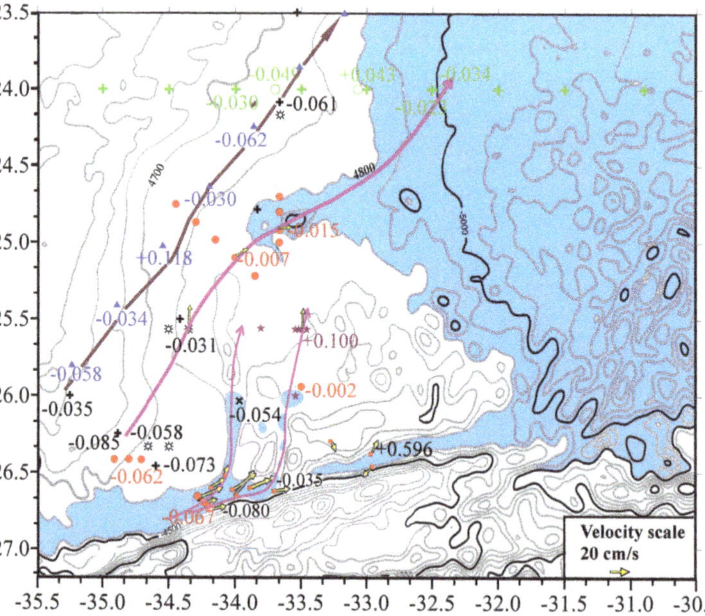

Figure 4. Bottom topography and stations in the study site. Bathymetry is based on GEBCO 2019, supplemented with our echo sounder measurements from 2010 to 2020. Locations of CTD/LADCP stations are indicated with different symbols. The bottom potential temperatures are indicated near the stations. Blue triangles show stations of the WOCE A17 transect in 1994. Black crosses (+) show stations in 2003. Sign (x) shows the stations in 2009. Green color shows stations and data of section A09 in 2009 and 2018 (24° S). Small red squares show stations in 2012 and 2013. Purple stars show stations in 2018. Circles with beams (☼) show stations in 2019. Red circles show stations in 2020. Yellow arrows show directions of currents and their speed (vectors). Deep water areas are shown in blue, especially the canyon at 24°40′ S. Two thin magenta lines show the previously analyzed AABW flows from the deep bottom of the channel. The thick magenta line shows the flow of the branch that initially was found above the western wall. The brown line shows the possible pathway of the shallowest jet (~4600–4700 m), which is plotted using only the CTD data from 1994 and 2003 [updated from [12]]. The existence of this flow was confirmed in 2022.

The Sea-Bird 19plus (Sea-Bird electronics, Bellevue WA, USA), Idronaut 320 plus (Idronaut S.R.L., Brugherio MS, Italy), and RDI Monitor 300 kHz (LADCP) (Teledyne RDI, San Diego, CA, USA) instruments were used for the measurements. The cold current jet was tracked based on the coldest bottom potential temperatures and velocity measurements using the LADCP profiler (Teledyne RDI, San Diego, CA, USA)at the stations where these data were available. It was also assumed that the current of bottom water should flow along the isobaths or go deeper.

The bottom topography used in this paper is based on satellite altimetry (GEBCO 2019) to plot our topographic maps. Some data of this digital bathymetry were corrected based on the measurements along the routes of our ships using the data of our echo sounders(Kongsberg Maritime, Kongsberg Viken, Norway).

Table 1. CTD/LADCP stations at the exit from the Vema Channel in 1994–2022.

Stations	Date	Coordinates
		R/V *Maurice Ewing* (only CTD)
81	02.02.1994	27°21.1′ S, 36°36.7′ W
82	02.02.1994	26°58.6′ S, 36°16.1′ W
83	02.02.1994	26°34.6′ S, 35°55.7′ W
84	02.02.1994	26°11.2′ S, 35°34.2′ W
85	03.02.1994	25°47.8′ S, 35°14.0′ W
86	03.02.1994	25°24.4′ S, 34°53.4′ W
87	03.02.1994	25°01.1′ S, 34°32.7′ W
88	03.02.1994	24°37.6′ S, 34°11.9′ W
89	04.02.1994	24°14.2′ S, 33°51.8′ W
90	04.02.1994	23°50.8′ S, 33°30.9′ W
91	04.02.1994	23°27.5′ S, 33°10.3′ W
92	04.02.1994	23°04.0′ S, 32°49.4′ W
		R/V *Akademik Sergey Vavilov* (only CTD)
1441	02.11.2003	26°00.0′ S, 35°15.0′ W
1442	02.11.2003	26°14.9′ S, 34°53.0′ W
1443	02.11.2003	26°27.5′ S, 34°35.8′ W
1444	03.11.2003	26°40.9′ S, 34°15.4′ W
1445	03.11.2003	26°43.2′ S, 34°12.1′ W
1446	03.11.2003	26°44.4′ S, 34°10.8′ W
1454	09.11.2003	27°05.5′ S, 35°54.7′ W
		R/V *Akademik Ioffe*
2079	18.04.2009	26°42.9′ S, 34°12.2′ W
2080	18.04.2009	26°01.9′ S, 33°58.0′ W
2437	06.11.2010	26°39.9′ S, 34°16.8′ W
2438	06.11.2010	26°42.2′ S, 34°13.8′ W
2439	06.11.2010	26°43.4′ S, 34°12.1′ W
		R/V *Akademik Sergey Vavilov*
2494	05.11.2012	26°36.8′ S, 33°59.3′ W
2495	05.11.2012	26°35.9′ S, 34°10.2′ W
2496	06.11.2012	26°31.1′ S, 34°03.3′ W
2497	06.11.2012	26°35.8′ S, 33°51.7′ W
2521	18.10.2013	26°20.5′ S, 32°00.1′ W
2522	18.10.2013	26°27.7′ S, 32°53.1′ W
2523	18.10.2013	26°23.0′ S, 32°53.1′ W
2525	19.10.2013	26°18.6′ S, 33°11.5′ W
2526	20.10.2013	26°35.7′ S, 33°51.6′ W
2527	20.10.2013	26°37.3′ S, 33°39.6′ W
2710	22.10.2018	25°34.1′ S, 33°29.5′ W
2711	22.10.2018	25°34.0′ S, 33°31.1′ W
2712	22.10.2018	25°34.0′ S, 33°27.4′ W

Table 1. *Cont.*

Stations	Date	Coordinates
2713	22.10.2018	25°34.0′ S, 33°32.6′ W
2714	23.10.2018	25°33.6′ S, 33°48.3′ W
2716	23.10.2018	26°00.0′ S, 33°32.6′ W
		R/V *Akademik Mstislav Keldysh*
6563	30.12.2019	24°10.1′ S, 33°37.9′ W
6564	30.12.2019	25°34.0′ S, 34°21.2′ W
6565	31.12.2019	25°33.9′ S, 34°30.0′ W
6566	31.12.2019	26°20.0′ S, 34°29.9′ W
6567	31.12.2019	26°20.0′ S, 34°40.0′ W
6793	03.04.2020	26°25.0′ S, 34°54.6′ W
6794	03.04.2020	26°25.0′ S, 34°48.0′ W
6795	03.04.2020	26°25.0′ S, 34°41.9′ W
6796	06.04.2020	26°39.2′ S, 34°16.8′ W
6797	06.04.2020	26°41.4′ S, 34°14.2′ W
6798	06.04.2020	26°43.1′ S, 34°12.1′ W
6799	06.04.2020	26°44.4′ S, 34°10.8′ W
6800	07.04.2020	25°56.4′ S, 33°30.0′ W
6801	07.04.2020	25°13.0′ S, 33°51.0′ W
6802	07.04.2020	24°00.0′ S, 33°40.0′ W
6803	07.04.2020	24°55.1′ S, 33°40.0′ W
6804	07.04.2020	24°48.0′ S, 33°40.0′ W
6805	08.04.2020	24°42.0′ S, 33°40.0′ W
6806	08.04.2020	25°06.0′ S, 34°00.0′ W
6807	08.04.2020	24°59.0′ S, 34°09.0′ W
6808	08.04.2020	24°52.0′ S, 34°17.9′ W
6809	08.04.2020	24°45.0′ S, 34°27.0′ W
7411	02.03.2022	26° 54.1 S, 34° 18.8′ W
7412	03.03.2022	26° 44.3 S, 34° 10.1′ W
7413	03.03.2022	26° 43.2 S, 34° 12.0′ W
7414	03.03.2022	26° 41.4 S, 34° 14.3′ W
7415	03.03.2022	26° 39.0 S, 34° 16.7′ W
7417	04.03.2022	26° 34.1 S, 34° 22.7′ W
7418	04.03.2022	26° 28.7 S, 34° 30.0′ W
7419	04.03.2022	26° 24.0 S, 34° 36.0′ W
7420	04.03.2022	26° 17.4 S, 34° 44.5′ W
7422	05.03.2022	26° 10.8 S, 34° 52.0′ W
7423	05.03.2022	26° 04.2 S, 35° 00.0′ W
7424	05.03.2022	25° 58.8 S, 35° 06.0′ W
7425	05.03.2022	25° 54.0 S, 35° 12.0′ W
7426	05.03.2022	25° 49.2 S, 35° 17.0′ W

Our analysis is based on our CTD/LADCP measurements in 2003, 2004, 2009, 2010, 2012, 2013, 2018, 2019, 2020, and 2022. The casts reached depths approximately 5 m above the bottom. We also used the data from the WOCE A17 section (WHP, 2002) and the A09 section in 2009 and 2018 along 24° S (http://cchdo.ucsd.edu (accessed on 27 October 2022)). The data from several stations from the WOD18 database were also used. Velocity measurements were corrected by removing the currents of the barotropic tide using the TPXO 9 model [16].

3. Results

The CTD/LADCP measurements from the Vema Channel over sections in 2020 and earlier show the presence of the two cold AABW flows of Weddell Sea Deep Water (with potential temperatures below 0.2 °C). Both cold cores of the currents were observed in the sections at latitudes 26°30′–27°00′ S, measured in 2020 (Figure 3) [3,12]. Let us follow each jet and consider them separately. Each jet was identified based on the bottom topography (a channel deeper than the surroundings by 50–100 m), lower potential temperature, and velocities of the current.

The best-known AABW jet flows along the deep bed of the Vema Channel [5,17]. The deep core of the current is located between 4400 and 4700 m. Let us consider this deep outflowing current of AABW using the data gathered at our CTD/LADCP stations in 2003–2020. The stations were located in the region 26°30′–26°50′ S, 33°30′–34°20′ W (Figure 4). In 2003, the minimum measured potential temperature in this region of the northern part of the Vema Channel (Vema extension at 26°43.2′ S, 34°12.1′ W) was $\theta = -0.094$ °C. In 2020, the potential temperature at the same point increased to $\theta = -0.067$ °C. We found previously that the continuation of the coldest AABW flow splits into two jets [10,11]. These near-bottom jets rapidly warm and decay over a distance of about 100 km at latitude 25°30′ S (Figure 4).

Another AABW branch was detected over the northwestern slope of the Vema Channel. This flow is elevated by 600 m over the deepest branch of the AABW current [12]. The highest velocities in the core of this branch were recorded at depths of 4100–4200 m (150 m above the bottom). Such an isolated core of Weddell Sea Deep Water over the western terrace was repeatedly observed (in a section along 31°12′ S) from 1984 to 2020. The potential temperature of this branch has been increasing over time. The measurements of velocities within 30–36 cm/s using LADCP over the repeated section confirm the stable existence of this branch [3,18]. The continuation of this jet was found at the outflow of the Vema Channel in 2020 at 26°25′ S, 34°42′ W above a depth of 4420 m (Figure 3, western section). Then, this jet continues to the north, approaches the upper part of a zonal canyon at 24°30′ S (Figure 4), and flows down the canyon to depths deeper than 4800 m [12].

In 2009, scientists from Great Britain occupied a WOCE CTD section (A09) along 24° S. They repeated experiments on this section in 2018 (https://cchdo.ucsd.edu, accessed on 27 October 2022). In 2009, two cores of cold water were detected in this section: the bottom potential temperatures at longitude 31°50′ W (at depths of approximately 4650–4750 m) ranged from −0.04 °C to −0.05 °C, while at 33°50′ W (in the depth range of 5000–5200 m), potential temperatures ranged from −0.03 °C to −0.04 °C. These cores were slightly displaced in 2018. Potential temperatures in the cold cores in 2018 were close to −0.03 °C. Sections of potential temperature along 24° S in 2009 and 2018 are shown in Figure 5. No measurements of currents over this section are available.

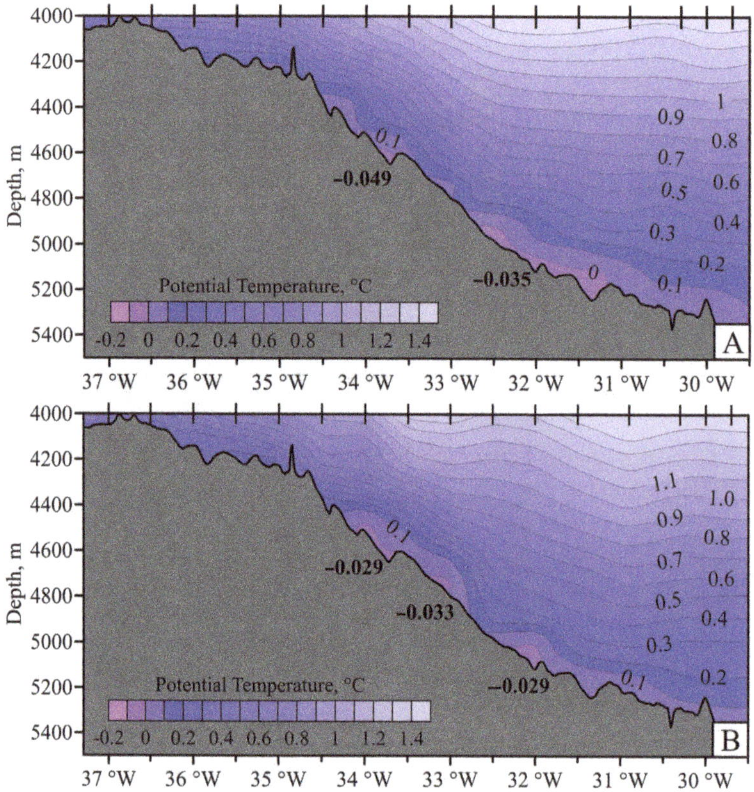

Figure 5. Sections of potential temperatures deeper than 4000 m in 2009 (**A**) and 2018 (**B**) (WOCE line A09 along 24° S). The coldest bottom potential temperatures are indicated over the background of the bottom.

In Figure 5, two cores of cold water are observed at depths of 4600–4800 m and 5100 m. Previously, it was found that two jets of bottom current became warmer and decayed at 25°30′ S at depths of ~4700 m [12]. Thus, they could not transport very cold water to latitude 24° S. The jet that descended the canyon could have been recorded at 24° S at a depth of 5100 m. However, a cold core at 33°–34° W at depths of 4700 m could appear only if this water has been transported by another jet. Finding this jet is the goal of this research. Below, the existence of this jet is considered in more detail. The potential temperature of cold water detected at 24°00′ S, 33°42′ W in 2009 at depths of ~4600 m was −0.049 °C. By 2018, it had become warmer (−0.029 °C).

There is no other source of such cold water than the upper branch of the current from the Vema Channel. The section at 24° S that was occupied in 2009 and 2018 is located approximately 300 km north of the Vema Channel. The continental slope of South America here is not at all steep: the slope is estimated as 1/(100–120). A moving flow along the isobaths can keep its path above such a slope due to the balance of the gravity force directed down the slope and the Coriolis force that is directed to the left of the flow upslope. Small channels on the continental slope with a depth of about 50–100 m over a section at 24° S were found exactly coinciding with the cold potential temperatures above the bottom. Thus, these flows could have eroded them in the sediments. In our expeditions after 2018, we tried to find the pathway of the continuation of the upper jet north of the Vema Channel. The criterion was the existence of low bottom potential temperatures and a general northward direction of velocity, with possible deviations along the isobaths to the east.

The upper branch of the AABW current was traced on the basis of low potential temperatures and the northern-northeastern direction of velocity vectors at the bottom. These vectors are shown by the yellow arrows in Figure 4. In 2020, we found that this bottom branch descends into an underwater abyssal canyon (Figure 4), which is at approximately 24°25′–24°40′ S [12]. At longitude 33°10′ W, this canyon becomes wider. One can see from the historical bottom temperature distribution that this canyon diverts part of this cold AABW jet downslope. The branch that descends the canyon was detected in 2020 based on low potential temperatures and vectors of velocities at the bottom at five locations between 26°25′ S and 24°55′ S. This branch forms the cold core at a depth of 5100 m at 24° S.

At latitude 24° S, one can see another cold core at depths of 4600 at longitudes 33°–34° W. This core could have been formed by a jet located shallower than the one that descended the canyon. Therefore, there is a third jet at depths of ~4600 m that has been detected from low potential temperatures over the section. This branch has been traced from low temperatures based on CTD casts in 1994 and 2003; however, no LADCP measurements were made in 1994 or 2003. In 1994–2003, the bottom potential temperatures were much lower than now. This jet is drawn in Figure 4 with a brown line connecting the points, with cold potential temperatures at the bottom. When plotting this line, we tried to connect measurement points with low temperatures. In addition, the pathway was drawn so that over steep slopes the flow was descending more rapidly than over flatter slopes. The potential temperature of the bottom flow has been strongly increasing since 1994. The very cold temperatures that were measured in 1994 and 2003 have not since been recorded near the bottom. Thus, we detected a third branch of AABW flow directed to the north, which is shown in Figure 5 as a cold core at 24° S at depths of 4600 m. The potential temperature of this upper branch at a depth of 4600 m is below zero at 24° S. Based on the measurements in the databases north of 24° S, this branch spreads further to the north. The continuation of this jet north of 24° S is presumably the flow of AABW described in [19]. Thus, on the basis of our targeted measurements and historical data, we have constructed a scheme of pathways for the continuation of AABW flow (Figure 4).

To detect this jet at the outflow from the Vema Channel, we constructed a section in 2022 to cross all possible flows from the Vema Channel. Before measurements were taken in 2022, a CTD section (without LADCP) was constructed in 2003 of only five stations (1441–1446) (Figures 6 and 7) approximately along the same line (inset in Figure 1). Section 2003 does not allow us to resolve three jets of currents because of a small number of stations and lack of LADCP measurements. Because of the general warming of abyssal waters in the Atlantic, potential temperatures increased significantly [8].

The section of 13 stations occupied in 2022 revealed three branches of AABW current from the Vema Channel, which were presumably detected from previous measurements. These jets are observed both on the temperature section and on the sections of the current components along the meridian and normal to the section. Jets of currents on a gentle slope are observed at longitudes of 35.1° W and 34.7° W.

Three different components of currents across the section north of the Vema Channel are shown in Figure 8. The currents are generally directed to the east-northeast. High current velocities are observed at the bottom in the channel bed closer to the western wall: up to 21 cm/s with an eastern-northeastern direction at station 7415. At the neighboring station 7417, the current accelerates to 28 cm/s and acquires a pronounced northern direction near the bottom. A fairly wide flow with velocities exceeding 10 cm/s and a predominant direction to the east-northeast was noted on a gentle slope. The maximum velocities (20 cm/s) near the bottom were simultaneously recorded at station 7424 (35°06′ W) in the same place where the third jet with the minimum potential temperature was observed. Even higher current velocities, up to 29 cm/s, were noted in the depth range of 3800–3900 m at stations 7420, 7424, and 7426. Our previous numerous LADCP casts that returned profiles of bottom currents in the Vema Channel suggest that the maximum velocities were found not at the very bottom but 100–150 m above it, while the water with the coldest potential temperatures was detected exactly at the bottom.

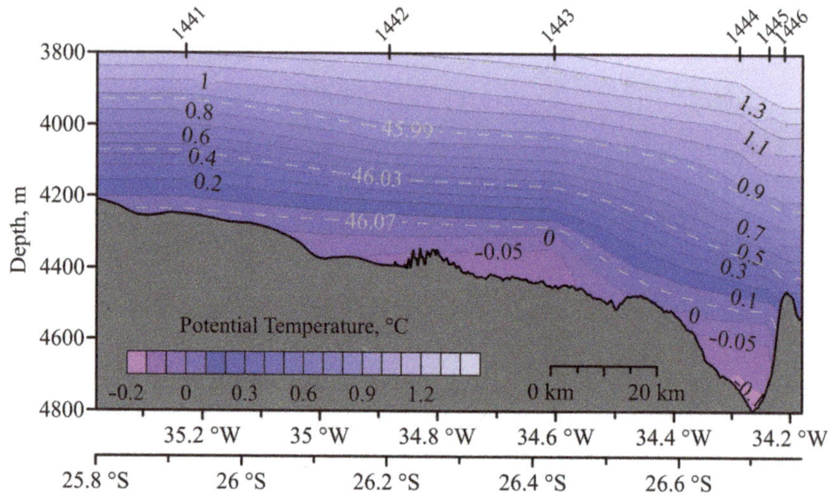

Figure 6. Potential temperature section north of the Vema Channel in 2003. Contour lines of density (~46 units) are shown.

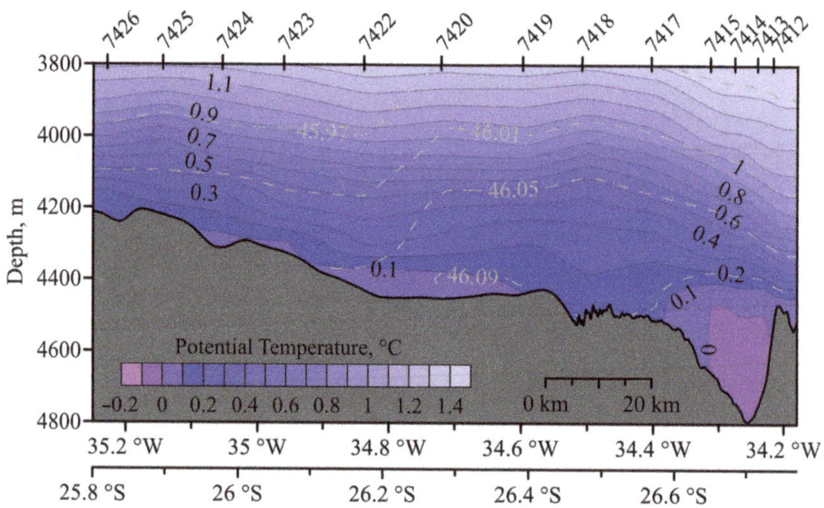

Figure 7. Potential temperature section of AABW outflow from the Vema Channel in 2022. The location of the section was slightly different from its position in 2003 (Figure 1). Contour lines of density (~46 units) are shown.

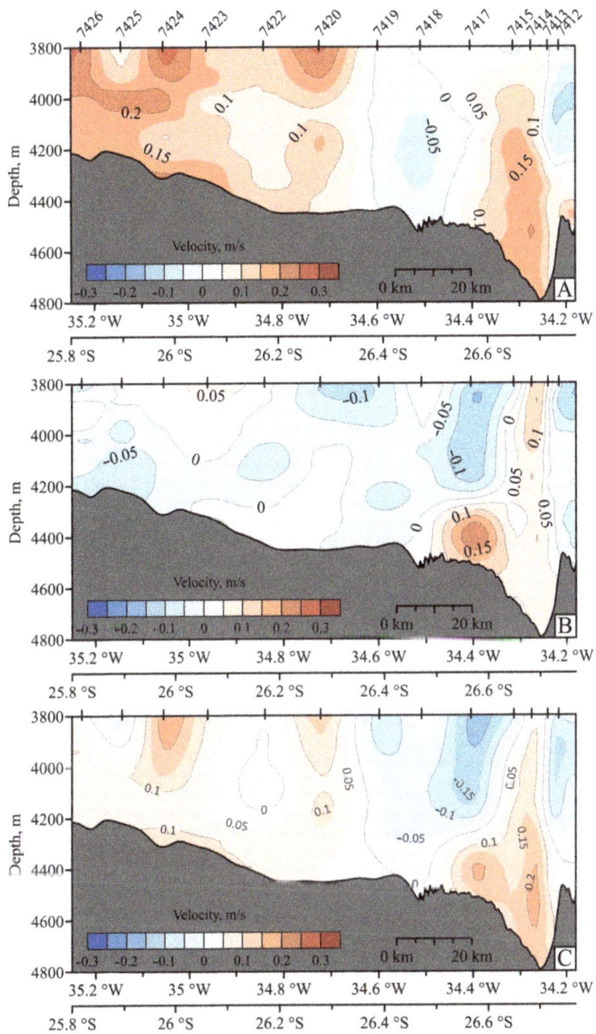

Figure 8. Sections of the velocity components at the outflow of the AABW from the Vema Channel in 2022. Top panel (**A**): positive direction to the east; middle panel (**B**): positive direction to the north; bottom panel (**C**): positive direction to the northeast, normal to the section.

Thus, our analysis reveals the existence of three jets of bottom flow that continue the spreading of AABW in the Vema Channel. Previously, the existence of a third jet before measurements were taken in 2022 could only be judged from CTD data. Presumably, the upper jet, which initially flows above the western wall of the Vema Channel, splits into two branches of continuation, and both of them spread further north than the lower jet from the deep bottom of the Vema Channel. Therefore, the upper jet is the current that transports AABW further north and is the source of abyssal bottom water in the deep basins in the North Atlantic. The shallower part of this upper flow (the third jet in our context) transports cold water to 24° S and is observed as the cold core at depths of 4600 m. The deeper part of the upper jet descends the canyon and fills the deepest abyssal regions of the Atlantic.

Author Contributions: E.G.M.: conceptualization and original draft preparation; D.I.F., V.A.K. and O.A.Z.: field data, original draft preparation, writing, and figures. All authors have read and agreed to the published version of the manuscript.

Funding: The work was supported by the Russian Science Foundation grant 21-77-20004.

Institutional Review Board Statement: Not applicable.

Informed Consent Statement: No studies involving humans and animals were performed.

Data Availability Statement: Data are available upon request.

Conflicts of Interest: The authors declare no conflict of interest.

References

1. Zenk, W.; Speer, K.G.; Hogg, N.G. Bathymetry at the Vema Sill. *Deep Sea Res.* **1993**, *40*, 1925–1933. [CrossRef]
2. Johnson, D.A.; McDowel, S.E.; Sullivan, L.G.; Biscaye, P.E. Abyssal hydrography, nephelometry, currents, and benthic boundary layer structure in the Vema Channel. *J. Geophys. Res.* **1976**, *81*, 5771–5786. [CrossRef]
3. Morozov, E.G.; Tarakanov, R.Y.; Frey, D.I. *Bottom Gravity Currents and Overflows in Deep Channels of the Atlantic Observations, Analysis, and Modeling*; Springer: Dordrecht, The Netherlands, 2021; p. 483. [CrossRef]
4. Speer, K.G.; Zenk, W. The flow of Antarctic Bottom water into the Brazil Basin. *J. Phys. Oceanogr.* **1993**, *23*, 2667–2682. [CrossRef]
5. Jungclaus, J.; Vanicek, M. Frictionally modified flow in a deep ocean channel: Application to the Vema Channel. *J. Geophys. Res. Ocean.* **1999**, *104*, 21123–21136. [CrossRef]
6. Zenk, W.; Hogg, N.G. Warming trend in Antarctic Bottom Water flowing into the Brazil Basin. *Deep Sea Res.* **1996**, *43*, 1461–1473. [CrossRef]
7. Zenk, W.; Morozov, E.G. Decadal warming of the coldest Antarctic Bottom Water flow through the Vema Channel. *Geophys. Res. Lett.* **2007**, *34*, L14607. [CrossRef]
8. Campos, E.J.D.; van Caspel, M.C.; Zenk, W.; Morozov, E.G.; Frey, D.I.; Piola, A.R.; Meinen, C.S.; Sato, O.T.; Perez, R.C.; Dong, S. Warming trend in the abyssal flow through the Vema Channel in the South Atlantic. *Geophys. Res. Lett.* **2021**, *48*, e2021GL094709. [CrossRef]
9. Smythe-Wright, D.; Boswell, S. Abyssal circulation in the Argentine Basin. *J. Geophys. Res. Ocean.* **1998**, *103*, 15845–15851. [CrossRef]
10. Tarakanov, R.Y.; Morozov, E.G. Flow of Antarctic Bottom Water at the output of the Vema Channel. *Oceanology* **2015**, *55*, 153–161. [CrossRef]
11. Tarakanov, R.Y.; Morozov, E.G.; Frey, D.I. Hydraulic continuation of the abyssal flow from the Vema Channel in the southwestern part of the Brazil Basin. *J. Geophys. Res.* **2020**, *125*, e2020JC016232. [CrossRef]
12. Morozov, E.G.; Frey, D.I.; Tarakanov, R.Y. Flow of Antarctic Bottom Water from the Vema Channel. *Geosci. Lett.* **2020**, *7*, 16. [CrossRef]
13. Frey, D.I.; Morozov, E.G.; Fomin, V.V.; Diansky, N.A.; Tarakanov, R.Y. Regional modeling of Antarctic Bottom Water flows in the key passages of the Atlantic. *J. Geophys. Res. Ocean.* **2019**, *124*, 8414–8428. [CrossRef]
14. Frey, D.; Borisov, D.; Fomin, V.; Morozov, E.; Levchenko, O. Modeling of bottom currents for estimating their erosional-depositional potential in the Southwest Atlantic. *J. Mar. Syst.* **2022**, *230*, 103736. [CrossRef]
15. Hogg, N.; Siedler, G.; Zenk, W. Circulation and variability at the Southern Boundary of the Brazil Basin. *J. Phys. Oceanogr.* **1999**, *29*, 145–157. [CrossRef]
16. Egbert, G.D.; Erofeeva, S. Efficient inverse modeling of barotropic ocean tides. *J. Atmos. Ocean. Tech.* **2002**, *19*, 183–204. [CrossRef]
17. Zenk, W. Abyssal and contour currents. In *Developments in Sedimentology, Contourites*; Rebesco, M., Camerlenghi, A., Eds.; Elsevier: Amsterdam, The Netherlands, 2008; Volume 60, pp. 37–57.
18. Morozov, E.G.; Demidov, A.N.; Tarakanov, R.Y.; Zenk, W. *Abyssal Channels in the Atlantic Ocean: Water Structure and Flows*; Springer: Dordrecht, The Netherlands, 2010; p. 266. [CrossRef]
19. Sandoval, F.J.; Weatherly, G.L. Evolution of the deep western boundary current of Antarctic Bottom Water in the Brazil Basin. *J. Phys. Oceanogr.* **2001**, *31*, 1440–1460. [CrossRef]

water

MDPI

Article

Physical and Biological Features of the Waters in the Outer Patagonian Shelf and the Malvinas Current

Pavel A. Salyuk [1,*], Sergey A. Mosharov [2], Dmitry I. Frey [2,3], Valentina V. Kasyan [4], Vladimir I. Ponomarev [1], Olga Yu. Kalinina [2], Eugene G. Morozov [2], Alexander A. Latushkin [3], Philipp V. Sapozhnikov [2], Sofia A. Ostroumova [2,5], Nadezhda A. Lipinskaya [1], Maxim V. Budyansky [1], Pavel V. Chukmasov [6], Viktor A. Krechik [2,7], Michael Yu. Uleysky [1], Pavel A. Fayman [1], Alexander Yu. Mayor [8], Irina V. Mosharova [2], Anton D. Chernetsky [2], Svetlana P. Shkorba [1] and Nikita A. Shved [4]

1 V.I. Il'ichev Pacific Oceanological Institute, Far Eastern Branch, Russian Academy of Sciences, 690041 Vladivostok, Russia
2 Shirshov Institute of Oceanology, Russian Academy of Sciences, 117997 Moscow, Russia
3 Marine Hydrophysical Institute, Russian Academy of Sciences, 299011 Sevastopol, Russia
4 A.V. Zhirmunsky National Scientific Center of Marine Biology, Far Eastern Branch, Russian Academy of Sciences, 690041 Vladivostok, Russia
5 Oceanology and Hydrology Institute, Russian State Hydrometeorological University, 195196 St. Petersburg, Russia
6 A.N. Severtsov Institute of Ecology and Evolution, Russian Academy of Sciences, 119071 Moscow, Russia
7 Laboratory for Marine Environmental Management, Immanuel Kant Baltic Federal University, 236041 Kaliningrad, Russia
8 Institute of Automation and Control Processes, Far Eastern Branch, Russian Academy of Sciences, 690041 Vladivostok, Russia
* Correspondence: pavel.salyuk@gmail.com

Citation: Salyuk, P.A.; Mosharov, S.A.; Frey, D.I.; Kasyan, V.V.; Ponomarev, V.I.; Kalinina, O.Y.; Morozov, E.G.; Latushkin, A.A.; Sapozhnikov, P.V.; Ostroumova, S.A.; et al. Physical and Biological Features of the Waters in the Outer Patagonian Shelf and the Malvinas Current. *Water* **2022**, *14*, 3879. https://doi.org/10.3390/w14233879

Academic Editor: Guangyi Wang

Received: 9 September 2022
Accepted: 25 November 2022
Published: 28 November 2022

Publisher's Note: MDPI stays neutral with regard to jurisdictional claims in published maps and institutional affiliations.

Abstract: The aim of this study is to trace how the fine-thermohaline and kinematic structure, formed over a section along 45.8° S in the interaction zone of the outer Patagonian Shelf (PS) and Malvinas (Falkland) Current (MC) System waters, affect the spatial distribution of bio-optical characteristics, phyto/zooplankton, birds, and marine mammals. For the first time, simultaneous multidisciplinary observations at high spatial resolution (~2.5 km) were performed in this region during the cruise of the R/V "Akademic Mstislav Keldysh" in February 2022. A fine structure of alternating upwelling and downwelling zones over the PS and slope was identified, which resulted from the interaction between the MC inshore branch (MCi), bottom topography, and wind. This interaction significantly affects all the physical, and optical characteristics analyzed in the work, as well as the biota of the region. It was found that the euphotic zone is larger in the downwelling zones than in the upwelling zones, and all spatially local maxima of phytoplankton photosynthetic efficiency are observed in the zones between upwelling and downwelling. Phytoplankton along the section were represented by 43 species. A total of 30 zooplankton species/taxa were identified. Three species of marine mammals and 11 species of birds were recorded in the study site. Most of the phytoplankton species list were formed by dinoflagellates, and picoplankton *Prasinoderma colonial* quantitatively dominated everywhere. Two floristic and three assemblage groups were distinguished among the analyzed phytoplankton communities. High phytoplankton biodiversity was observed above the PS and low above the PS edge and in the MCi core. Copepods mostly dominated in zooplankton. Subantarctic species/taxa of zooplankton concentrated in the nearshore waters of the PS, while Antarctic species/taxa were most abundant in the zone between the MCi and the MC offshore branch (MCo). The relative abundance of birds in the PS was several times higher than in the MCo. The minimum abundance of birds was in the MCi in the zone of the strongest upwelling identified above the PS edge.

Keywords: Malvinas Current; submesoscale; water mass; Lagrangian analysis; upper mixed layer; euphotic layer; bio-optical; fluorescence; primary production; SADCP; CTD

1. Introduction

The Malvinas Current (MC) in the Southwest Atlantic, which transports Subantarctic waters to the north, makes a significant contribution to the formation of zones of high productivity in the waters along the Patagonian Shelf [1]. The cool, nutrient-rich MC branches off from the Antarctic Circumpolar Current [2] and splits into two main currents: inshore and offshore [3–5]. The inshore branch of the current is formed along the edge of the Patagonian Shelf. The offshore branch flows in the deep ocean. The MC exists as two jets, which merge near 44° S as the continental slope steepens [6]. South of this location, significant variability of the current was observed based on direct velocity measurements on a mooring array [7].

The waters of the MC strongly interact with the warmer fresh waters of the Patagonian Shelf [8–10]. They play an important role in the spreading and migration of various marine organisms in the Southwest Atlantic [11]. The interaction of different water masses leads to a high species biodiversity in this region [12]. This may include species from mid-latitudes, Subantarctic, and Antarctic regions.

Alternating zones of upwelling and downwelling are formed as a result of the influence of the Patagonian slope on the dynamic characteristics of the baroclinic MC [13,14]. In addition, the effect of wind, which modifies upwelling processes and can enhance them, is an important factor affecting slope upwelling [15]. Upwelling leads to the transport of the waters from the underlying layers, which are rich in nutrients and microelements, to the euphotic layer. This is accompanied by an increase in the abundance of phytoplankton [16], which is the primary link in the food chain. This process is manifested in the satellite sea color data, as the formation of a pronounced band of increased concentration of chlorophyll-*a* (chl-*a*) observed approximately from (50° S, 62° W) to (38° S, 56° W) [17,18], which corresponds to a spatial scale of about 1300 km. Chl-*a* concentration maps based on satellite data make it possible to study the formation of a zone of high chl-*a* concentration west of the MC inshore branch [16,19].

Zooplankton is the most important intermediate link in the food chain between primary producers and higher trophic levels of the pelagic marine ecosystem [20]. The biomass of zooplankton determines the amount of resources available to fish, birds, and marine mammals [21]. Regular studies of phytoplankton and zooplankton communities are important because most planktonic organisms are characterized by short life cycles and quickly respond to environmental changes [22], including changes in anthropogenic impact [23] and climate [24].

The Patagonian Shelf is a region of intense fishing [25] and many oil and gas fields [26]. This may have an impact on biodiversity and status of all trophic levels in the marine ecosystem. At the same time, decadal variations in the circulation have been observed in the region, associated with a more distant propagation of the subtropical waters of the Brazil Current to the south [27,28]. A long-term increase in the temperature gradient on the slope of the Patagonian Shelf has been recorded, which can affect the environmental conditions of the development of phytoplankton cells [29]. Further climate change is predicted that would lead to reduced fish catches in the Southwest Atlantic [25]. An integrated view of Atlantic coastal Patagonian ecosystems, including the physical environment, biodiversity, and the main ecological processes, together with the ecosystem changes derived by them and anthropogenic impacts are presented in the book [30], where the past, present, and future of ecosystem functions and services in the Patagonian coastal ecosystems are described.

Thus, a significant influence of oceanographic processes on the functioning of marine ecosystems of various trophic levels occurs in the study site. Therefore, the ongoing biological processes must be studied together with the multiscale variability of physical processes in the ocean. Such multidisciplinary investigations were performed in February 2022 during cruise 87 of the R/V "Akademik Mstislav Keldysh", in which a section along 45.8° S has been made. The presence of scientists of various specialties allowed us to perform integrated research in the area linking the oceanographic variables of the MC over the Patagonian Shelf and slope with the optical and biological studies.

The novelty of the study is based on the analysis of the first simultaneous multidisciplinary observations of physical oceanography conditions, optical variables, and marine biology aspects in a high-productive region of the Patagonian Shelf. The data were collected at very high spatial resolution (~2.5 km between stations on average), which is much higher than all previously published surveys [3–5,14]. These new data allowed us to determine narrow upwelling and downwelling zones not fully revealed by satellite data and available numerical models.

The study aims to trace how the fine-thermohaline and kinematic structure formed in the interaction zone of the outer Patagonian Shelf and how inshore Malvinas Current branch waters affect the spatial distribution of bio-optical characteristics, phytoplankton, zooplankton, birds, and marine mammals.

2. Materials and Methods

Oceanographic, optical, and biological data for the analysis were collected on cruise 87 of the R/V "Akademik Mstislav Keldysh" on 21–22 February 2022 (end of summer in the Southern Hemisphere) over the section from the waters of the Patagonian Shelf to the intersection of the inshore and offshore jets of the MC at 45.8° S. A total of 16 stations were made from 60.4° W to 59.6° W and continuous hydrographic and bio-optical records of water properties in the upper layer were made using a flow-through system along the route of the ship from 60.4° W to 58.4° W. In addition, satellite data of chl-*a* concentration and sea surface temperature were used for the analysis. Summary information on the stations and continuous measurements is presented in Table 1.

Table 1. Summary of completed oceanographic and biological studies.

Station	Bottom. Depth, m	Region *	Day, Time, UTC	Lat, S	Lon, W	Δ ** Dist., km	(1) Continuous	(2) Profiling	(3) Phytopl. Act.	(4) Phytopl. Spec.	(5) Zoopl. Spec.	(6) Bird., Mam.
7393	137	PS	21.02 15:34	45°48′	60°24′		+	+		+	+	+
7394	140	PS	21.02 16:07	45°48′	60°22′	2.9	+	+				+
7395	136	PS	21.02 16:35	45°48′	60°19′	2.9	+	+		+	+	+
7396	134	PS	21.02 17:04	45°48′	60°17′	2.7	+	+	+			+
7397	134	PS	21.02 17:36	45°48′	60°15′	2.7	+	+		+	+	+
7398	139	PS, MCi	21.02 18:19	45°48′	60°13′	2.8	+	+			+	+
7399	139	PS, MCi	21.02 19:03	45°48′	60°11′	2.9	+	+	+			+
7400	155	PSe, MCi	21.02 19:34	45°48′	60°09′	2.3	+	+	+	+	+	+
7401	293	PSe, MCi	21.02 20:27	45°48′	60°07′	3.2	+	+	+	+	+	+
7402	365	MCi	21.02 21:07	45°48′	60°05′	2.7	+	+	+	+	+	+
7403	462	MCi	21.02 22:03	45°48′	60°03′	2.9	+	+	+			+
7404	631	MCi	21.02 22:36	45°48′	60°00′	2.9	+	+	+	+	+	+
7405	742	MCi	21.02 23:43	45°48′	59°57′	3.8	+	+	+			
7406	790	MCi	22.02 00:30	45°48′	59°53′	5.4	+	+	+	+	+	
7407	853	ZB	22.02 02:22	45°48′	59°47′	7.4	+	+	+			
7408	942	ZB	22.02 03:20	45°48′	59°40′	9.2	+	+	+		+	
–	>1000	ZB, MCo	>22.02 04:00	45°48′	<59°50′	>10	+					+

Note(s): * PS is the Patagonian Shelf, PSe is the PS edge, MC is the Malvinas Current, (MCi) is the MC inshore branch, (MCo) is the MC offshore branch, ZB is the zone between inshore and offshore branches of the MC. ** Distance to the previous station. (1) Continuous ADCP and flow-through measurements on the ship's route. (2) In-situ CTD and bio-optical profiling. (3) Phytoplankton activity and primary production analysis. (4) Phytoplankton species analysis. (5) Zooplankton species analysis. (6) Observations of birds and mammals.

2.1. *Field Oceanographic and Optical Measurements*

2.1.1. Vertical Profiling at Oceanographic Stations

Vertical profiles of oceanographic variables such as temperature (T in °C), salinity (S in practical salinity units (PSU)), and potential density (σ in kg/m^3) of seawater, have been measured at stations. Optical parameters were also determined: chl-*a* fluorescence intensity (F_{chl} in relative units), phycoerythrin fluorescence intensity (F_{pe} in relative units), fluorescence intensity of colored dissolved organic matter (CDOM) (F_{CDOM} in relative units), seawater turbidity (*Turb* in formazin turbidity units (FTU)), the beam attenuation coefficient at 660 nm (BAC_{660} in FTU), and the photosynthetically active radiation value (*PAR* in μmol photons/m^2/s). The *PAR* measurements were carried out only during daylight hours. The depth of the euphotic zone (Z_{eu}) was determined from the vertical profiles of *PAR* as 1% of the *PAR* incident on the sea surface.

The measurements were performed by an instrument set consisting of three probes: AML Oceanographic BaseX CTD probe for measuring oceanographic variables (T, S and calculated σ, and calculated potential temperature θ for θ-S diagrams plotting), measurements were carried out at a frequency of 4 Hz; Turner C6P for measuring F_{chl}, F_{CDOM}, and *Turb* at a frequency of 1 Hz; Kondor instrument to determine the BAC_{660} at 4 Hz and the *PAR* at 8 Hz. All instruments were calibrated before the expedition. The instrumental complex descended at a speed of 0.5 m/s to a depth of 250 m or to the bottom if the depth was shallower. Only the data measured during downcast of the instruments were used for the analysis.

2.1.2. Measurements of the Current Velocity Vector during the Ship Motion

Direct velocity measurements were carried out by Shipborne Acoustic Doppler Current Profiler (SADCP) system Teledyne RD Instruments Ocean Surveyor (TRDI OS) with a frequency of 76.8 kHz. During the survey the profiler was set in the narrowband mode, which increases the profiling range up to 700 m depth. We set 60 vertical bins 16 m each with an 8 m blank distance immediately below the transducer. The draught of the ship is 6 m, which gives 22 m depth for the center of the first bin (the depth of the uppermost layer of velocity measurements). Time averaging of the raw data was made over 120 s intervals. Since the ship speed varied between 8 and 10 knots, this time average represents an along-track averaging of roughly 500 m. Measurement errors in the amplitude of the horizontal velocities were small, approximately 1–2 cm/s [31]. The TPXO9 model [32] was used to subtract the barotropic tidal velocities at the time of measurements.

2.1.3. Distinguishing Upwelling and Downwelling Zones

The upwelling and downwelling zones were determined based on the analysis of six oceanographic and optical variables over the section: temperature, salinity, density, chl-*a* fluorescence, beam attenuation coefficient at 660 nm, and turbidity. The zones were identified based on the topography of the lower boundary of the upper mixed layer (UML), temperature, salinity, and density contours in the underlying layers. The UML depth (Z_{UML}) was estimated by four different methods: (1) is the depth where $\Delta T = 0.5$ °C relative to the sea surface [33]; (2) is the depth where $\Delta T = 0.8$ °C relative to 10 m depth [34]; (3) is the depth where potential density gradient is equal to 0.05 kg/m^3/m [35]; (4) is the depth where $\Delta\sigma_\theta$ is equal to the theoretical difference between σ_θ (T_{10}–0.2 °C, S_{10}) and σ_θ (T_{10}, S_{10}) for the atmospheric pressure, T_{10} is the measured temperature at 10 m, S_{10} is measured salinity at 10 m [36]. Using four different methods made it possible to increase the accuracy of Z_{UML} calculations.

Upwellings correspond to the ascent of the density interface layer and a decrease in the UML thickness, as well as the elevation of the isolines of other measured parameters. In downwelling zones, the opposite behavior of isolines has been observed. Only the most obvious cases of upwelling or downwelling in the UML were selected in the zones of local Z_{UML} maxima or minima along the section. Joint analysis of oceanographic and optical variables makes it possible to more accurately identify upwelling and downwelling zones, and to make the cross-validation of the results. Some zones of vertical motions are better manifested in oceanographic characteristics, while others are more clearly identified from the optical parameters.

2.2. Shipboard Biological Measurements

2.2.1. Chlorophyll-a and Primary Production

Water samples were collected at 11 stations in the study site from a depth of 5 m using the flow-through systems installed on the research vessel (Table 1). The samples were divided into subsamples, which were used for measurement of different productivity parameters (chlorophyll concentrations, experimental carbon fixation estimations (primary production), active fluorescence of chlorophyll).

Primary production (*PP*) was measured onboard using the ^{14}C uptake method [37]. Water samples were collected in 50-mL flasks and incubated for 3 h in an original laboratory phytoincubator with individually adjustable LED illumination and temperature maintenance using a HAILEA-100 laboratory cooler. After incubation, flasks were filtered onto a 0.45-µm "Vladipore" membrane (Vladimir, Russia). The radioactivity of the samples was determined using a Triathler (Hidex, Turku, Finland) liquid scintillation counter.

Chl-*a* concentration was measured fluorometrically with correction for phaeopigments [38]. Water samples (500 mL) were filtered onto Whatman GF/F (glass-fiber filters) under a low vacuum (0.2 atm). For extraction, Chl-*a* filters were placed in acetone (90%). After 24 h in the cold (4 °C) and dark conditions, the extracts were measured with a MEGA-25 fluorometer (Moscow, Russia) before and after acidification with 1 N HCl [39]. The fluorometer has been pre-calibrated using a chl-*a* standard (Burlington, VT, USA, Sigma-Aldrich).

Biomass-specific *PP* (P^B in [mgC/mg chl-*a*/day]) was calculated by dividing *PP* rates by the chl-*a* concentration in the respective sample [40].

Active chl-*a* fluorescence was measured using an ultrasensitive WATER-PAM fluorometer (Walz, Effeltrich, Germany). The samples were exposed in the fluorometer to light intensity (*E*) as to ^{14}C uptake and steady fluorescence (F_t) and maximum fluorescence ($F_m{}'$) measurements. The effective quantum efficiency of photosystem II (PSII) was determined as follows [41]:

$$\Delta F/Fm' = (F_m{}' - F_t)/F_m{}'$$

The relative electron transport rate (*rETR*, relative unit) was calculated from the equation:

$$rETR = \Delta F/F_m{}' \times E \times 0.5,$$

where coefficient 0.5 takes into account the distribution of photons between photosystems I and II [42]. *rETR* is a measure of the rate of linear electron transport through PSII, which is correlated with the overall photosynthetic performance of the phytoplankton [43]. *E* is the level of natural illumination of the sample in the chamber of the PAM fluorimeter (µmol photons/m^2/s), and coefficient 0.5 takes into account the distribution of photons between photosystems I and II [42]. To determine *rETR*, water samples were exposed at 8 levels of light intensity in the range from 0 to 1500 µmol photons/m^2 per second, including the light intensity corresponding to the natural light at the sampling point, and applied in the incubator for radiocarbon measurements. The *rETR* value reflects the rate of conversion of solar energy into chemically bound energy in cells, which provides the processes of organic matter biosynthesis by phytoplankton.

2.2.2. Phytoplankton Species

Phytoplankton samples were taken at stations using the Apshtein medium plankton net (mesh size 35 µm, the diameter of the upper ring 400 mm) from depths of 0–50 m. This method was used to obtain an integral sample covering a continuous range of depths. In contrast to the method of collection with the help of Niskin bottles, total sampling over a layer of the water column makes it possible to cover the layers where phenomena that are significant for the structural features of phytoplankton, but poorly detected by hydrophysical instruments, take place. We are talking, for example, about layers with a high concentration of heterotrophic flagellates. In order to avoid clogging the cells of the Capron tree, the plankton net was lifted at a speed of no more than 0.5 m/s. At those stations, network clogging did not occur: none of the microalgae species identified in the samples produced a large amount of exopolymers. The method of sampling the water column layer (individual depths) using a fine-mesh net makes it possible to quantify the structure of the community from the standpoint of the relative abundance of species [44,45]. In this case, this form of a result was relevant for us, since we did not set ourselves the task of identifying the commercial characteristics of communities. This study did not take into account the absolute values of the abundance and biomass of phytoplankton at individual depths.

Two samples were taken from each location. One of them was viewed live, immediately after selection noting the status of the abundance of various species of microphytes (planktonic microalgae). The primary count of microalgae differentially by species and photo-documentation of microalgae was performed based on the material of this sample using a ToupCam 5.1 MP digital microscope camera at magnifications 200×, 400×, and 1000×. The second sample was fixed with an ethanol solution at a final concentration of 70% for transportation to the onshore laboratory of the Shirshov Institute.

Here, 0.25 mL was taken from a concentrated sample with a volume of 25 mL and placed on a glass slide (75 × 25 mm, Citoglas, ApexLab, Moscow, Russia), covered with a large coverslip (24 × 50 mm, Premier Cover Glass, Leica Biosystems, Wetzlar, Germany). Microalgae were counted differentially by species, in strips along the glass slide, based on a total number of 400 specimens [46].

Identification of organisms to a particular species was established using modern atlases and guides, as well as articles on the phytoplankton of the Atlantic and a review of some genera inhabiting oceanic waters and containing species diagnoses [47–58].

2.2.3. Zooplankton Species

Zooplankton samples were obtained at 10 stations by a Multinet [59] opening/closing net system (0.25 m^2 aperture) equipped with five nets of 150 µm mesh-size. Additionally, larger and rarer taxa were collected by a pelagic Double Square net (DSN) (505 µm mesh, 1.0 m^2 inlet area) [60] equipped with a pterygoid deepener weighing 24 kg (Hydrobios, Altenholz, Germany) by performing oblique tows from 200 m to the surface or from near-bottom to the surface at depths <200 m at an average speed of 1.5 knots. The DSN net was equipped with a water flow counter (Hydrobios, Altenholz, Germany). Species were identified to the lowest possible taxonomic level using stereoscopes SZX7 and SZ51 (Olympus, Tokyo, Japan) using special identification keys [61–63]. Copepods, as a dominant group of zooplankton were subjected to a more detailed taxonomical processing. Other zooplanktonic crustaceans were grouped into a class or family. The abundance of zooplankton taxa was determined from each nets sample; it is expressed as individuals per cubic meter (ind./m^3) and calculated using the Zooplankton Methodology Collection and Identification [64].

2.2.4. Statistical Analysis of Phyto- and Zooplankton Communities

The algorithms of the Plymouth routines in multivariate ecological research (PRIMER-6 software package [65]) were used for the ecological analysis of the collected data. The Shannon-Weaver index (H') was used for calculating the diversity of phytoplankton at stations. The Pielou index (J') and the probability of interspecific encounters (*PIE*) were used to determine the evenness of species structure. The latter index shows the probability that two microalgae randomly found in the same field of view of the microscope at a working magnification of 400× belong to different species. The Sørensen index was used for calculating the similarity of communities at stations, according to the qualitative criteria, while the Bray-Curtis index was used according to the quantitative criteria. The construction of models with the identification of community groups was carried out using the MDS and clustering methods. The ANOSIM procedures (PRIMER-6) were used to check the reliability of the identification of floristic and assemblage groups. The SIMPER procedure (PRIMER-6) was used for identifying groups of species that characterize one or another group of assemblages.

To compare zooplankton communities, a non-metric cluster and multidimensional scaling (MDS) analyses were performed using the PRIMER-6 computer package. Species abundance data were log10(x + 1) transformed, and a station similarity matrix was generated using the Bray–Curtis metric. Cluster analysis was then applied using group average sorting. To test for significant numerical differences between identified clusters, ANOSIM was conducted on log-transformed abundance data. MDS was performed on the similarity matrix. A SIMPROF test was conducted ($\alpha = 0.01$) to determine the statistically significant differences between clusters [65]. Diversity was assessed using the Shannon–Weaver index (H').

2.2.5. Birds and Mammals

Onboard ship observations of marine mammals and birds were performed at the study site. The observations were carried out continuously during the daylight hours at winds (Beaufort scale < 5) and visibility of more than 50 m by two observers simultaneously from the port and starboard sides of the vessel from the direction-finding deck located at a height of 17 m above the sea level. The total observation time was 10 h 32 min. Relative abundance in units of birds per hour was calculated for the analysis of birds.

2.3. *Satellite Sea Surface Temperature and Chlorophyll-a*

We used satellite data on February 18, 2022, for additional analysis of the spatial fields of the chl-*a* concentration (Chl_{sat}) and sea surface temperature (*SST*) in the study area: Chl_{sat} were calculated from the OLCI radiometer Level-2 data (Sentinel-3B satellite) using the standard OC4ME global empirical algorithm with ~300 m resolution; *SST* values were obtained from the VIIRS radiometer Level-2 data (Suomi-NPP satellite) at 800 m resolution. Chl *a* data were downloaded through the Ocean Virtual Laboratory web portal (https://ovl.oceandatalab.com/ accessed on 25 November 2022), which uses CREODIAS Finder service (https://finder.creodias.eu/ accessed on 25 November 2022) and *SST* data from Ocean Color Web (https://oceancolor.gsfc.nasa.gov/ accessed on 25 November 2022), where the data are processed by standard algorithms with recommended quality flags set.

It was possible to qualitatively calculate the analyzed parameters in the absence of atmospheric cloudiness. The time difference between satellite and ship measurements was 3 days selecting the closest in time satellite images. This prevents a correct comparison of the measured values. However, it is acceptable for the analysis of the similarity of the observed spatial structures. Additionally, we used native resolution satellite data from a single satellite image without additional merging with other data and interpolation allowing the resolving of fine structure features. Application of satellite data is helpful to avoid misinterpretation of in-situ measurements along the section, as it becomes possible to analyze the instantaneous image of manifestations of oceanographic processes in the upper layer of the sea.

2.4. *Simulated Oceanographic Data and Bottom Topography*

Lagrangian analysis has been applied to study the origin of waters in the study site, in which geostrophic velocities calculated from the AVISO satellite altimetry data ($0.25° \times 0.25°$) were used to calculate particle trajectories. A large number of synthetic particles (the region is shown in Figure 1a) has been seeded daily with tracers on a grid of 700×700 points and their trajectories have been computed back in time for a fixed period of time [66]. The period of integration back-in-time in our study was specified from 21 February 2020, to 21 February 2022 (two-year period). One gets the so-called origin Lagrangian map (O-map) by marking the particles with different colors, which arrived from the northern, western, southern, and eastern boundaries of the study site in the past.

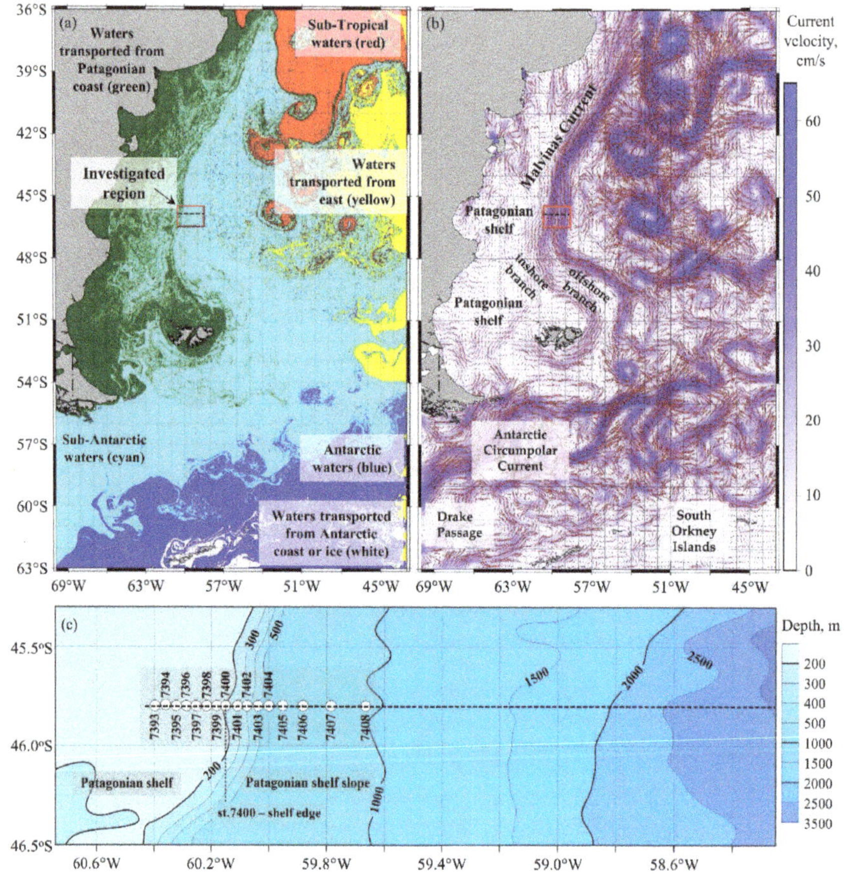

Figure 1. Map of the study site: Lagrangian map of the origin of waters (**a**); a diagram of currents in the southwestern part of the Atlantic on 21 February 2022 (**b**); map of the study site bathymetry and locations of stations (**c**). In Figure (**a**) blue color indicates the flow of water from the south, cyan is related to the flow from the west, red: from the north, yellow: from the east, green: from the Patagonian Shelf, white: from the Antarctic shelf or ice.

The data from the bathymetric database ETOPO1 (Earth topography 1 arc minute) with a resolution of 1 min (~1.85 km) were used to plot bottom topography maps.

3. Results

3.1. Common Oceanographic and Bio-Optical Characteristics of the Study Site

Results of the Lagrangian analysis of backward particle trajectories from 21 February 2022, with superimposed velocity vectors of the geostrophic currents calculated from the AVISO data, are shown in Figure 1a,b. The study site is marked with a rectangle. Figure 1c shows the bottom topography based on the ETOPO1 digital bathymetry data and the position of the oceanographic section marked with a black dotted line. The stations and track of continuous SADCP measurements are indicated.

The data can be divided into "shelf" (stations 7393–7400) at depths in the range of 130–150 m and "slope" stations (st. 7401–7408) at depths of 280–1000 m. St. 7400 and 7401 can also be classified as "shelf edge". It is seen from Figure 1 that the MC transports

Subantarctic waters from the Drake Passage to the study site. South of 43° S, the MC splits into two jets: inshore and offshore.

Figure 2 shows the measured SADCP northern current velocity component (V in cm/s, Figure 2a) and eastern current velocity component (U in cm/s, Figure 2b) over the section along 45.8° S with marked positions of vertical profiles at stations. One can see the northward propagation of two jets of the MC (Figure 2a) and anticyclonic circulation between the jets with intensification on the western side. In this work, the jet boundaries are chosen along the 40 cm/s isoline.

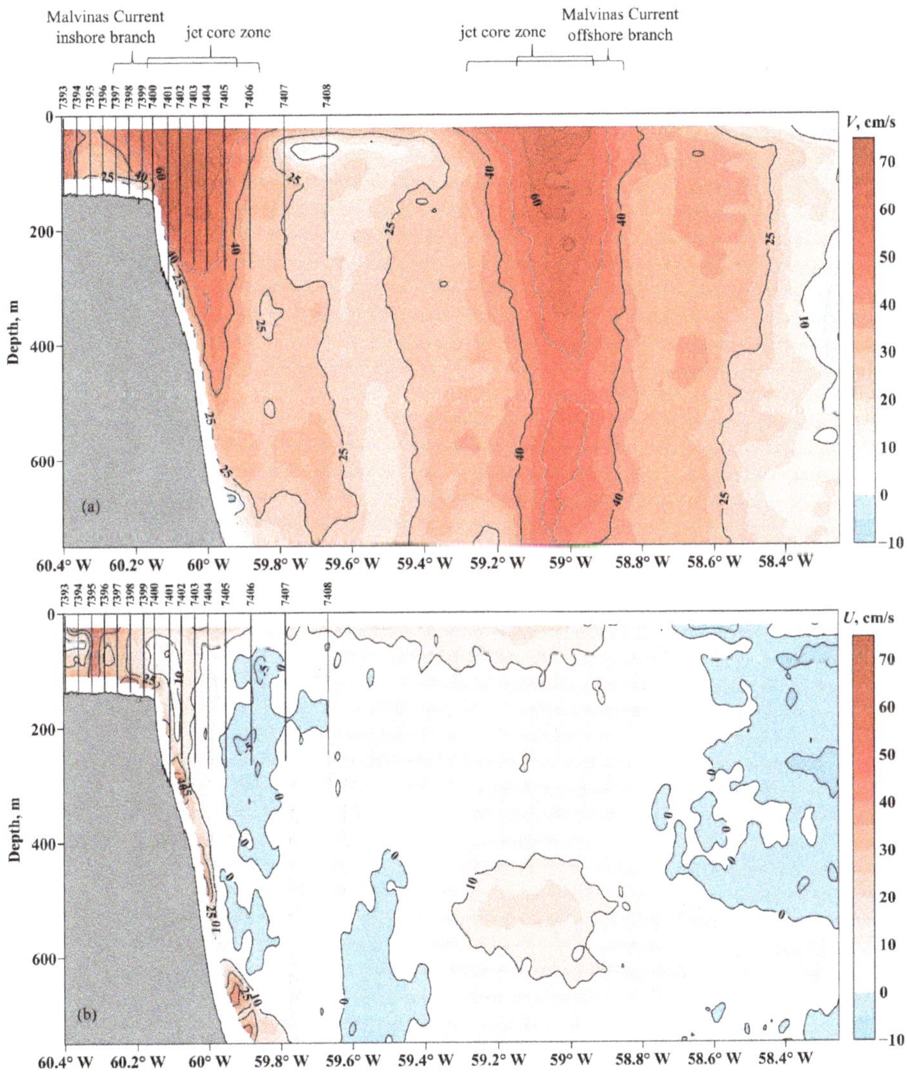

Figure 2. Meridional (**a**) and zonal (**b**) components of SADCP velocity measurements along the section of the study site.

The inshore branch is located close to the shelf and deepens to about 450 m. The western boundary of this branch in the surface layer is located near station 7397, and the eastern boundary is near station 7406. The width of the branch at the sea surface is about 40 km. The core of the current was recorded at stations 7400–7404, where the maximum velocities exceed 60 cm/s. The offshore branch of the current extends deeper than 750 m below the zone of correct SADCP measurements. The width of the offshore branch here at the sea surface is about 40 km similar to the inshore branch.

The satellite data on the concentration of chl-*a* (Chl_{sat}) and *SST* are presented in Figure 3, for an additional description of the study site. The section shows alternating high and low values of chl-*a* concentrations and *SST* values, which are associated with upwelling and downwelling manifestations in the surface layer of the sea. In general, both parameters decrease with distance from the shelf due to the influence of both jets of the MC. High Chl_{sat} values are observed not only over the shelf, but also approximately over the entire inshore jet of the MC, decreasing towards its eastern part.

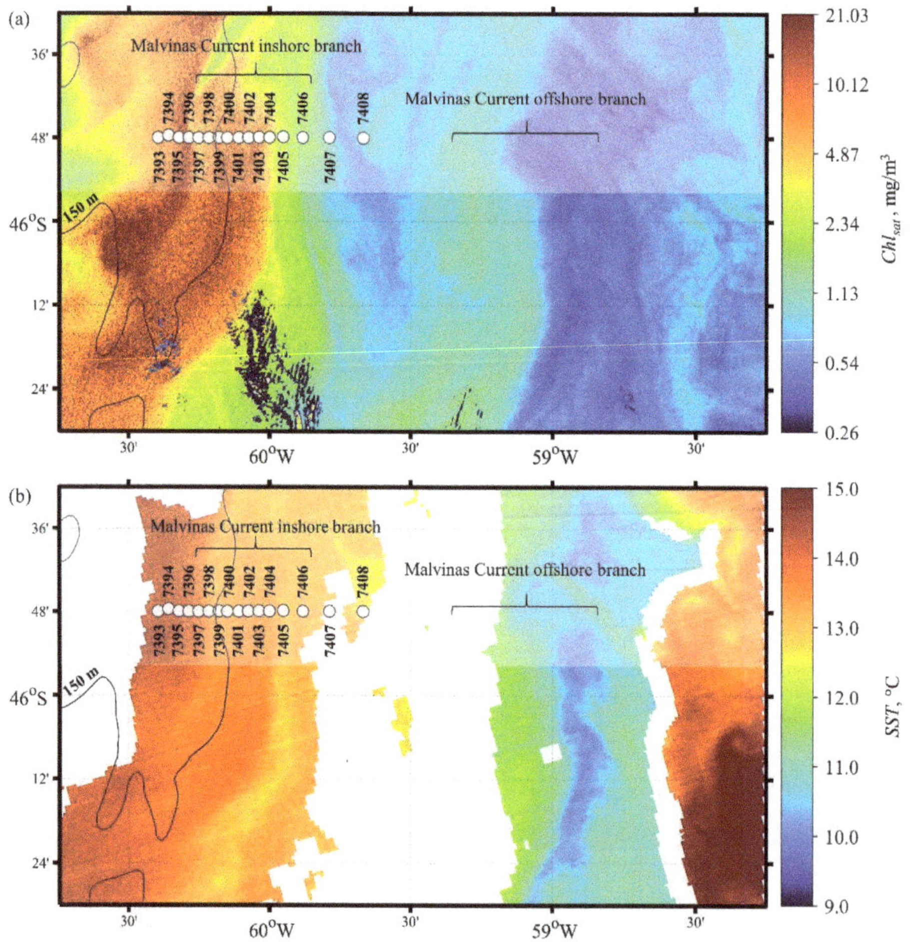

Figure 3. Satellite maps of chlorophyll-*a* concentration (Chl_{sat}) obtained by OLCI/Sentinel-3B (**a**) and sea surface temperature (*SST*) obtained by VIIRS/SNPP (**b**) on 18 February 2022.

3.2. Oceanographic Variables

Figure 4a–c show the vertical distributions of temperature, salinity, and density, respectively, and Figure 4d shows spatial variations of calculated UML thickness (Z_{UML}). The thickness of the UML varies within 25–55 m; seawater temperature in the UML is 11–13 °C. The thermocline coincides with the position of the pycnocline, and their depth changes in accordance with the variations between the upwelling and downwelling zones.

At stations 7400–7401, upwelling manifested itself most clearly in the UML as the ascent of the lower boundary of the mixed layer. Above the shelf break at a depth of about 80 m, this process is reflected by the elevation of the 6 °C isotherm at station 7400. An increase in the 6.5 °C isotherm at depths of 70–80 m at station 7397 is also clearly seen. Two additional upwellings at stations 7403–7405 and 7407–7408 can be distinguished from the upward deflection of the isopycnals. Downwelling was revealed at stations 7393–7396, and also at stations 7402 and 7406, due to the lowering of the thermocline and pycnocline and the corresponding increase in the thickness of the upper mixed layer.

In general, the differences in the UML thickness between upwelling and downwelling zones are significant. In the first case, they are in the range of ~25–35 m, and in the second case, in the range of 40–55 m. It is noteworthy that upwelling and downwelling zones change at the boundaries of the inshore jet of the MC. At the western boundary, upwelling in the current changes to downwelling next to the current. On the contrary, on the eastern boundary, downwelling in the current changes to upwelling. Another feature is downwelling at station 7402, located in the middle of the inshore jet core zone.

3.3. Biooptical Characteristics

Figure 5 shows sections of bio-optical characteristics, in which zones of upwelling and downwelling of UML waters were distinguished. Additionally, two upwelling zones in the middle layer (50–120 m) have been revealed: above the Patagonian Shelf and above the Patagonian continental slope. These zones correspond to the peculiarities revealed in the analysis of the oceanographic variables shown in Figure 4. The F_{pc} distribution is not shown here because it is very similar to the F_{chl} distribution, and the differences can only be seen in the scatterplots that will be shown in the following sections.

Increased values of F_{chl} and BAC_{660} are observed in the UML at stations 7393–7396 on the Patagonian Shelf and at stations 7400–7401 above the shelf edge. The local maximum of these values above the shelf edge is due to a decrease in the UML thickness during upwelling and a corresponding increase in the concentration of suspended phytoplankton and non-algal particles. Lower turbidity values are mostly observed in the upwelling zones in the UML rather than in adjacent downwelling zones. As the distance from the shelf increases in the eastern direction, the F_{CDOM} values decrease. A distinctive feature of the vertical distribution of CDOM from the other considered bio-optical characteristics is the absence of high F_{CDOM} values in the UML. In general, this vertical distribution is more uniform. At the same time, higher CDOM concentrations are observed under the pycnocline in the core of the inshore jet of the MC at stations 7400–7404, which may be associated with the intrusion of shelf waters into this layer.

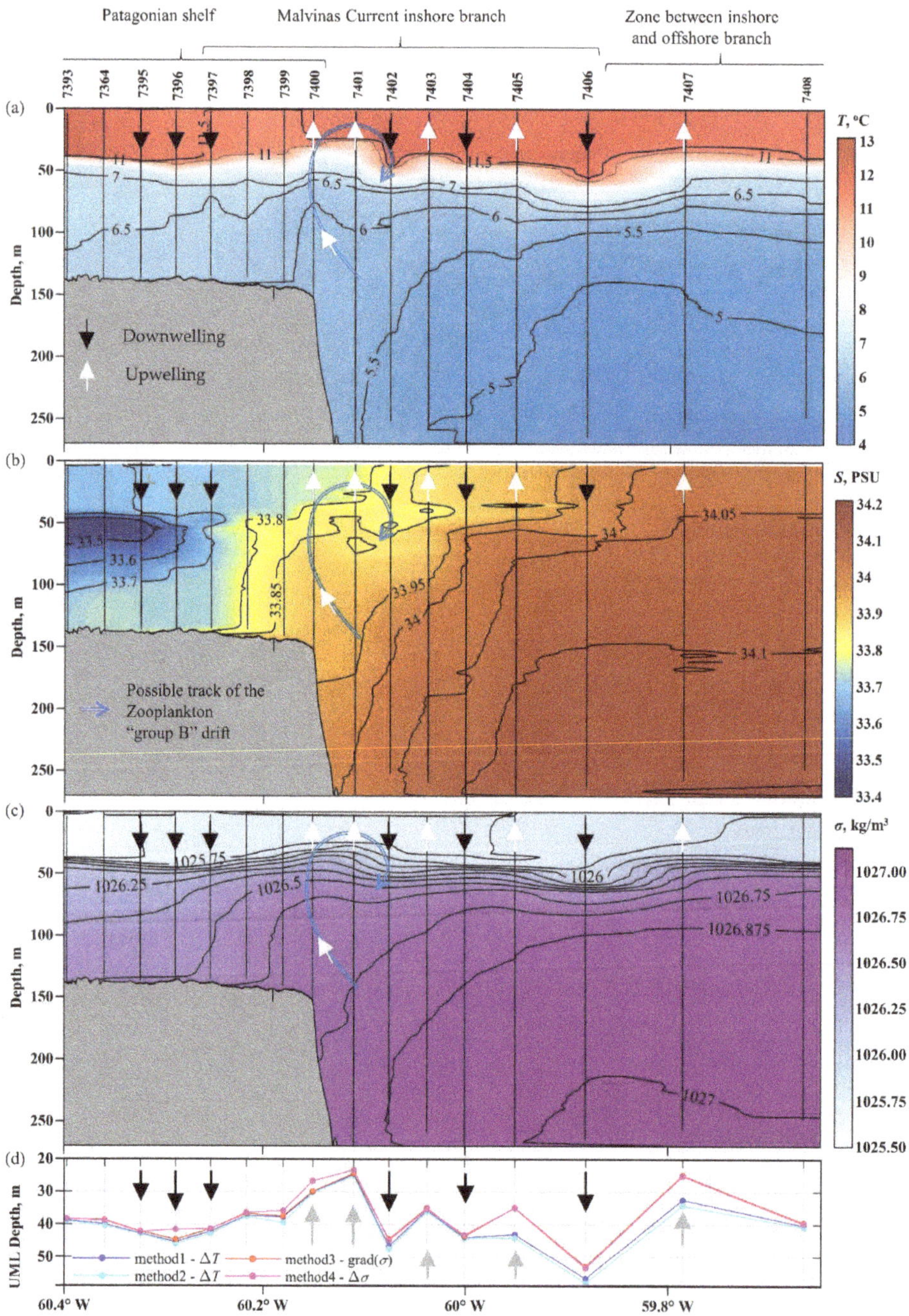

Figure 4. Zonal sections of: temperature, °C (**a**); salinity, PSU (**b**); density, kg/m3 (**c**); and spatial variations of upper mixed layer depth (UML), estimated by method 1 [33], 2 [34], 3 [35] or 4 [36] (**d**).

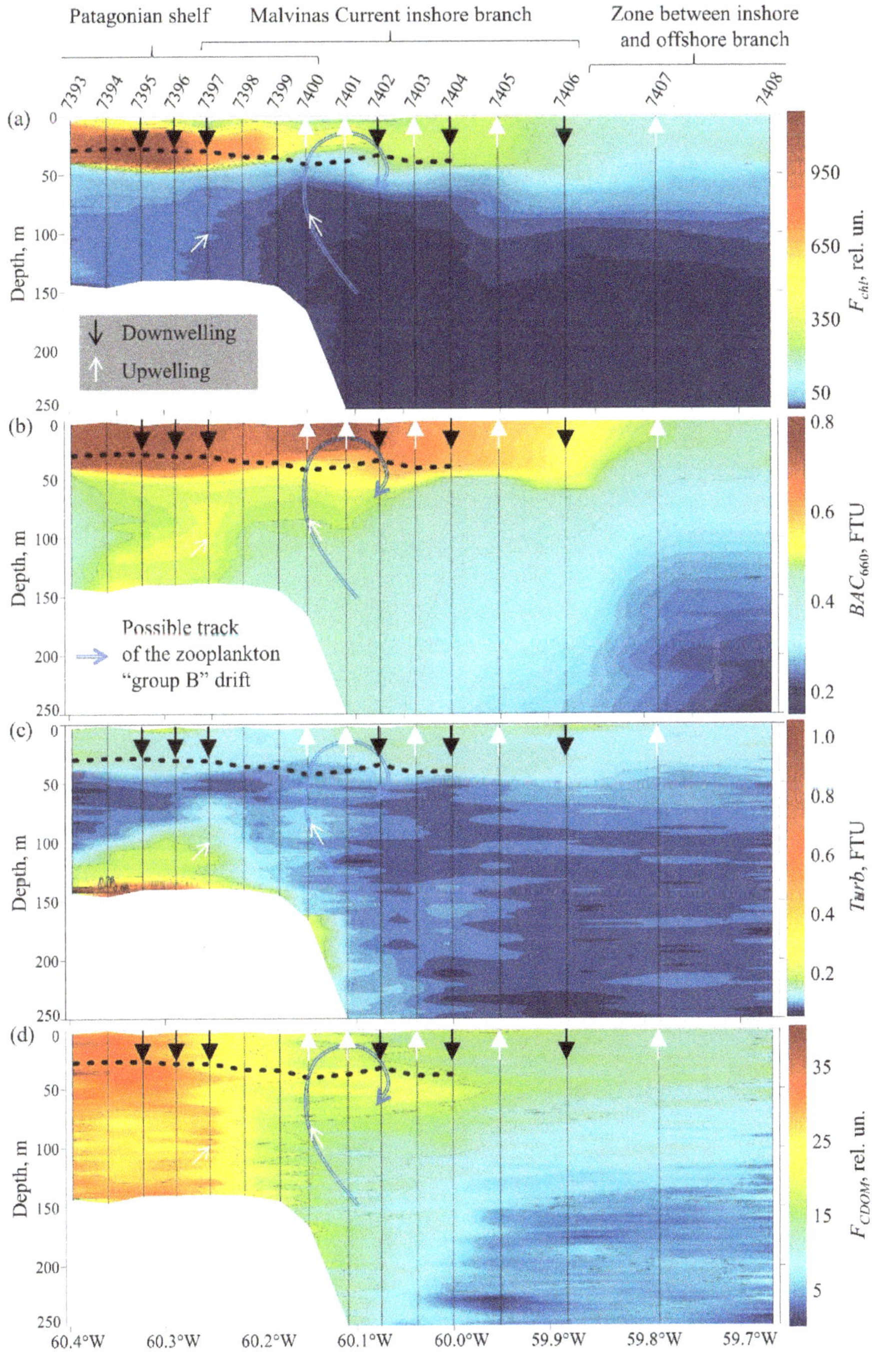

Figure 5. Sections of optical characteristics: intensity of chl-*a* fluorescence, F_{chl} (**a**); beam attenuation coefficient at a wavelength of 660 nm, BAC_{660} (**b**); turbidity, *Turb* (**c**); intensity of fluorescence of colored dissolved organic matter, F_{CDOM} (**d**). Black dashed line indicates the euphotic depth Z_{eu}.

3.4. Water Mass Classification

The joint analysis of oceanographic and bio-optical characteristics allows a more detailed classification of water masses. Figure 6 shows scatter diagrams for individual quantities and the corresponding distribution of the selected zones over the section. Figure 6a is a standard θ-S diagram, Figure 6b is a scatter plot of chlorophyll-a and phycoerythrin fluorescence intensities, the changes in which may indicate a different ratio of pigments in phytoplankton communities, Figure 6c shows CDOM fluorescence intensity versus salinity, which is also used to separate waters, especially in coastal areas [67,68], Figure 6d is a schematic representation of the section with highlighted water masses. Table 2 is a summary table for the identified types of water.

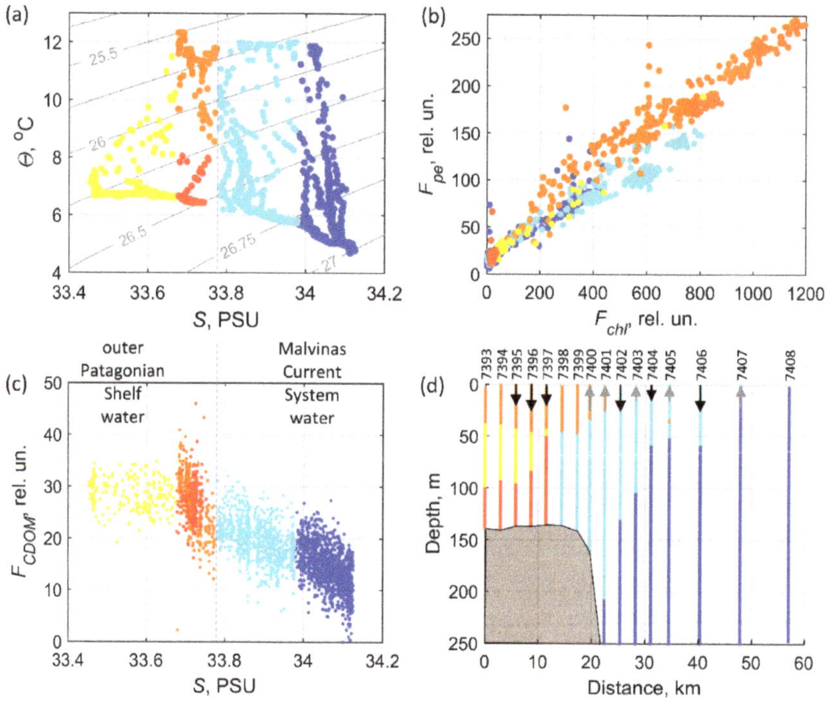

Figure 6. Scatter plots of "potential temperature—salinity" (**a**); "phycoerythrin—chl-*a* fluorescence intensity" (**b**); "CDOM fluorescence intensity—salinity" (**c**); and a scheme of the distribution of identified water masses along the section and vertical directions in the UML (**d**). Colors are described in Table 2.

The waters over the Patagonian Shelf (yellow, orange and red color in Figure 6) are generally fresher and less dense than those over the shelf slope part (cyan and blue color in Figure 6) of the section (Figures 4b,c and 6a). In general, the waters over the PS slope can be related to the MC system. The boundary between the PS and MC system waters is observed from the bottom (stations 7396–7397) to the surface (stations 7400–7401). This boundary passes through value S = 33.78 PSU, which allowed us to separate clusters in Figure 6a,c. Such separation is possible due to the ascent of bottom waters near stations 7396–7397 towards the shelf edge. In turn, there is the upwelling elevation of denser saline waters on the continental slope. These processes lead to the displacement of less dense waters in the upper mixed layer to the east.

Table 2. Identified water masses along the Section.

ID, Region, Water Mass Description	Color	Criteria	Characteristics
Y, Patagonian Shelf, intrusion from the distant part of the shelf	Yellow	$S < 33.675$	No correlation between salinity and CDOM fluorescence, relatively low salinity and high F_{CDOM} values
R, Patagonian Shelf, upwelled bottom waters of the outer shelf	Red	$S \geq 33.675$ & $S < 33.78$ & Depth $\leq$ 50 m	Low anticorrelation between salinity and CDOM fluorescence, relatively high F_{CDOM} and *Turb* values
O, Patagonian Shelf, high bio-productive water of the outer shelf	Orange	$S \geq 33.675$ & $S < 33.78$ & Depth > 50 m	High anticorrelation between salinity and CDOM fluorescence, specific phycoerythrin/chl-a fluorescence relationship, relatively high F_{CDOM} and F_{chl} values
C, Malvinas Current System, mainly in the inshore branch of the MC over continental slope	Cyan	$S \geq 33.78$ & $S < 33.95$	Low anticorrelation between salinity and CDOM fluorescence
B, Malvinas Current System, mainly between inshore and offshore branches of the MC	Blue	$S \geq 33.95$	Low anticorrelation between salinity and CDOM fluorescence, relatively low F_{CDOM} values and high salinity

The waters of the Patagonian shelf can be divided into three subtypes, according to Figure 6b,c. Waters highlighted by yellow color correspond to a decrease in salinity below 33.675 PSU, which is observed at stations 7393–7396 in the depth range of 50–100 m associated with the water intrusion from the distant part of the shelf. Waters marked in red are deeper than 50–100 m at stations 7393–7396, where suspended and dissolved organic matter ascends from the bottom. In the waters marked in orange, phytoplankton communities function, for which the ratio between the fluorescence intensities of phycoerythrin and chlorophyll-*a* changes and leads to the identification of cluster in Figure 6b, and in general, the highest F_{chl} values are observed in these waters.

Seawaters belonging to the Malvinas Current system can be divided into waters located above the boundary of the Patagonian Shelf, mainly in the zone of action of the coastal jet (cyan color), and into waters located above the deep part of the section between the coastal and offshore branches of the Malvinas Current (blue color). The separation boundary between them in salinity is 33.95 PSU, which made it possible to separate clusters in the "F_{CDOM}-S" scatterplot (Figure 6c).

3.5. Variations in the Euphotic Depth and the Upper Mixel Layer Depth

Vertical profiles of various variables at all stations are shown in Figure 7a. In addition, the purple horizontal line indicates the depth where 1% of the *PAR* incident on the surface is observed (euphotic depth). Figure 7b shows the distribution of Z_{UML} and the depth of the euphotic zone (Z_{eu}) along the section.

One can see in Figure 7 that at most stations where downwelling was identified, the euphotic depth is nearly in the middle of the phytoplankton layer (where increased F_{chl} values are observed). At all stations where upwelling was recorded, the euphotic depth was located below the phytoplankton layer. This feature is apparently related to the vertical displacement of the layer with phytoplankton and the change in the UML thickness.

In upwelling zones, the euphotic zone is generally deeper than in downwelling zones, despite the fact that in upwellings, a local increase in F_{chl} and BAC_{660} values can be observed in the surface layer (Figure 7a,b). As an example, this can be observed at nearby stations 7400, 7401, and 7402. The highest chl-*a* values at stations 7400 and 7401 are higher than at 7402. However, on the contrary, the integral concentration of chl-*a* in the layer 0–100 m, is higher at station 7402 than at 7400 and 7401. Therefore, the depth of the euphotic zone at station 7402 is lower than at 7400 and 7401.

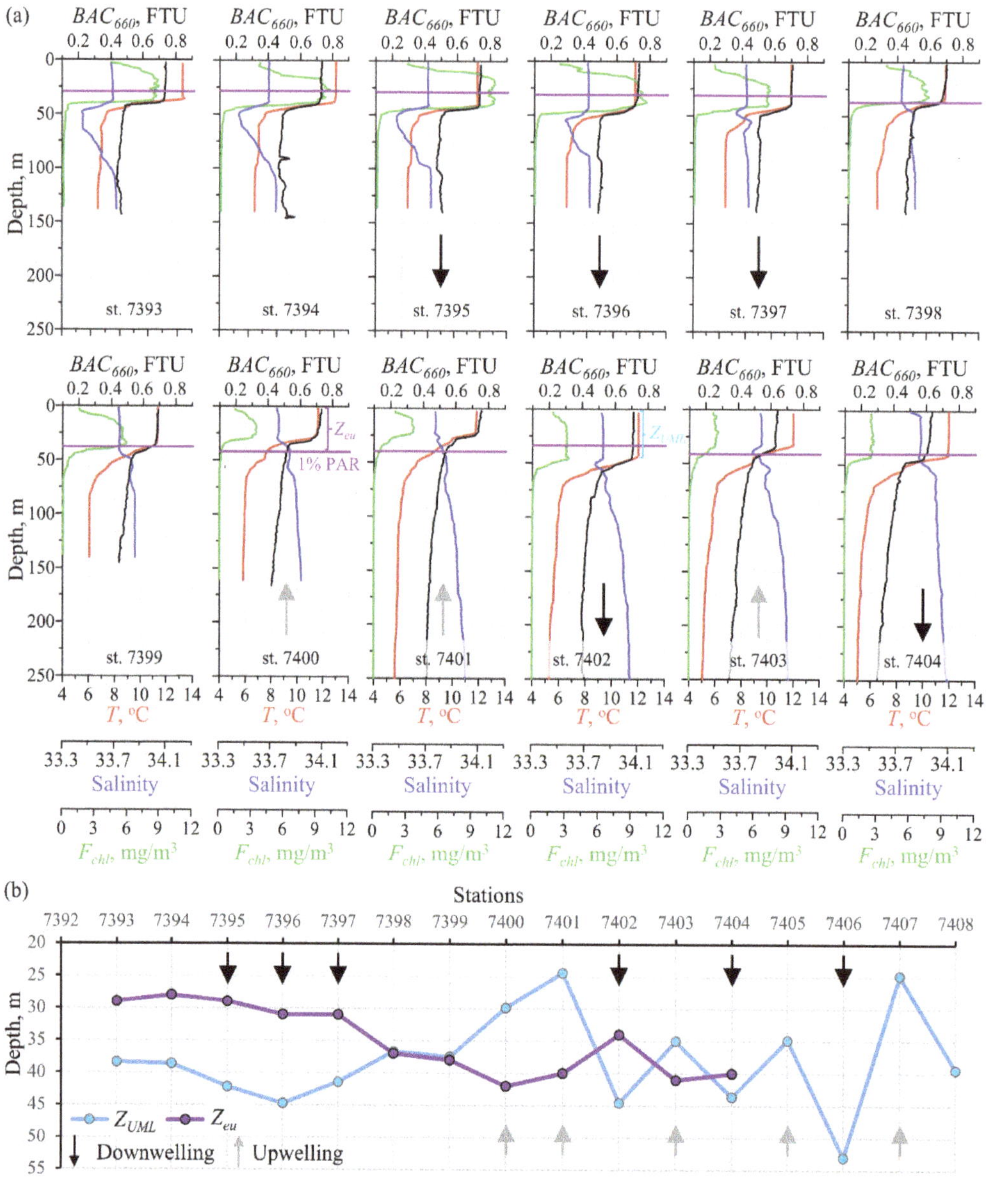

Figure 7. Vertical profiles of chl-*a* fluorescence (green), euphotic depth (purple), beam attenuation coefficient at 660 nm (black), temperature (red), and salinity (blue) at stations taken during daylight hours days (**a**). Variations in the euphotic depth (Z_{eu}) and UML depth (Z_{UML}) (**b**). Examples of Z_{eu} and Z_{UML} determination are presented for stations 7400 and 7402, respectively.

3.6. Phytoplankton Biological Activity

Primary production (*PP*) in the surface layer varied from 10.5 to 35.8 mgC/m^2 per day (on average 23.4 ± 8.6 mgC/m^2 per day) (Figure 8). The maximum values were revealed at the shelf station 7396 (35.8 mgC/m^2 per day) and in the inshore branch of the MC at stations 7400–7405 (23.5–27.9 mgC/m^2 per day). A notable decrease in *PP* (10.5–15.3 mgC/m^2 per day) was observed at the shelf edge (station 7499) and in the zone between the inshore and offshore branches (stations 7407–7408). The distribution of chl-*a* in the surface layer over the section was similar: an obvious trend towards a decrease from the shelf to the extreme oceanic station (from 3.57 to 1.06 μg/L). At the same time, unlike the pattern of distribution of *PP*, the concentration of chl-*a* in the inshore branch of the MC beyond the shelf edge remained at the same level (2.26–2.47 μg/L).

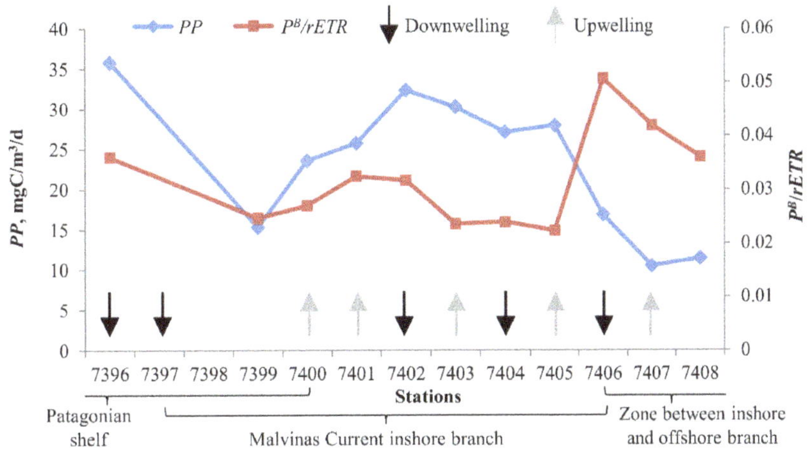

Figure 8. Phytoplankton photosynthetic efficiency ($P^B/rETR$) and primary production (*PP*) in the upper 5-m water layer.

The ratio of *rETR* and P^B values measured at the same light intensity for each subsample characterizes the photosynthetic efficiency of phytoplankton at different stations of the section, i.e., the degree of consuming the light energy for photosynthesis of organic matter. The photosynthetic efficiency ($P^B/rETR$) ranged from 0.022 to 0.051 (mean 0.032 ± 0.009) (Figure 8).

3.7. Phytoplankton Communities

A total of 43 species and subspecies of microalgae were noted at eight stations where phytoplankton has been studied (Table 1). Dinoflagellates (Myzozoa: Dinophyceae) predominated qualitatively among them (in qualitative terms according to the number of species): 27 species and subspecies (62.8% of the total list of phytoplankton). This was followed by diatoms (Ochrophyta: Bacillariophyceae) 12 (27.9%), pyramimonads (Chlorophyta: Pyramimonadophyceae) 2 (4.7%), and prasinophytes (Chlorophyta: Prasinophyceae), represented by the species *Pterosperma* cf. *polygonum*, and silicoflagellates (Ochrophyta: Dictyochophyceae) *Dictyocha fibula*.

Heterotrophs significantly prevailed in the list of dinoflagellates: they accounted for 19 species or 41.9% of the entire noted flora. In turn, at the level of genera, *Protoperidinium* was the most widely represented with 12 species and subspecies (27.9%, which is over a quarter of the entire flora). Thus, the phytoplankton species spectrum was represented mainly by dinoflagellates, and most of them had no adjustment to photosynthesis.

Only one species, the small-celled coccoid pyramimonad *Prasinoderma coloniale*, was found at all eight stations of the section. The other seven species are *Dinophysis rotundata*, *Tripos lineatus*, *T. fusus*, *Protoperidinium curvipes*, *P. pyriforme*, *P.* cf. *subsphaericum and Rhizosolenia hebetate*. These species had the status of widespread since they were noted at four of seven stations, but not everywhere.

The number of species at stations varied from four (in the core of the inshore branch) to 24 (at station 7395 in the downwelling zone above the shelf). On average, there were 12.5 species and subspecies per location. Structural diversity detected using the Shannon-

Weaver index (H') ranged from 0.725 (in the core of the inshore branch) to 2.038 (at the eastern periphery of the jet), with an average value of 1.467. The evenness of the species abundance, according to the Pielou index (J'), varied from 0.386 (at the westernmost point above the shelf in the downwelling zone) to 0.830 (at the eastern boundary of the jet), averaging 0.617. The probability of detecting two different species in one field of view ranged from 0.344 (at station 7400 in the upwelling zone above the shelf break) to 0.785 (in the downwelling zone at the eastern boundary of the jet), on average 0.582. In general, all the described indicators (species richness of communities, their diversity, and evenness of the species richness) were at relatively low levels.

Phytoplankton in the study site was not very rich in species, and in some places composition was poor (above the shelf edge and slope and especially in the core of the inshore jet), not particularly diverse, while at many locations the structure of communities was strongly influenced by the first most abundant species. At all stations, the pyramimonad *Prasinoderma coloniale* dominated in number. It was represented both among free-floating phytoplankton and as part of microaggregates. Small (up to 3–5 μm in diameter) spherical cells of this species were often found in living form and in the composition of spreading pellets of planktonic crustaceans. The cells proliferated again as part of aggregates built on the basis of such pellets. The contribution of this species to the total number of cenoses at the stations was 37.5–80.5%, 60.7% on average. The quantitative structure of communities at the stations is given in Table 3.

Table 3. Structure of phytoplankton communities at stations. The contributions (in % of the total number) formed by representatives of various large taxonomic groups are given. Specific species are indicated for the groups represented by one species.

Stations	7393	7395	7397	7400	7401	7402	7404	7406
Dinoflagellates	12.4	36.3	30.9	16.7	22.2	22.7	35.8	50.0
Diatoms	10.7	6.7	22.2	2.8	16.7	0.0	9.4	12.5
Pyramimonads	77.0	57.0	46.9	80.6	61.1	77.3	49.1	37.5
Dictyocha fibula	0.0	0.0	0.0	0.0	0.0	0.0	1.9	0.0
Pterosperma cf. *polygonum*	0.0	0.0	0.0	0.0	0.0	0.0	3.8	0.0

The total contribution of *Protoperidinium* species, widely represented at some stations, varied from 4.2% to 18.2% of the total abundance (9.3% on average). They played a particularly significant role in the total abundance at some locations over the shelf (in the downwelling zone and in the transition zone to upwelling), and in the core of the inshore branch. The number of species of this genus was especially high in the downwelling zone over the shelf (8–10), as was the total number of heterotrophic dinoflagellates (12–14). The total contribution of heterotrophic forms to the abundance of communities was especially high at station 7395 (downwelling over the shelf, 25.18%), as well as on the eastern periphery of the jet (22.64%) and, on its border (25%). On average over all samples, heterotrophs accounted for 17.72% of the total number of microalgae.

The average similarity of communities at the stations in terms of the composition of the microalgae flora was 26.07%. Assemblages of different locations, in general, differed significantly in composition. Structurally (based on the relative abundance of species), their average similarity was even lower, only 20.24%.

Two floristic groups (Figure 9a) can be distinguished based on the similarity of composition (total correlation $R = 0.85$, $p = 0.5\%$). Among them, the Af group was represented over the shelf and at the eastern boundary of the inshore branch; it was characterized by the presence of *Paralia* sp., *Dinophysis rotundata, Tripos lineatus, Rhizosolenia hebetata, Protoperidinium curvipes, P. pyriforme, P. cassum, P.* cf. *subsphaericum*, and *P. oviforme*. Group Bf was found above the shelf break and slope; it was represented by *Tripos lineatus, T. fusus*, and *Stephanopyxis nipponica*.

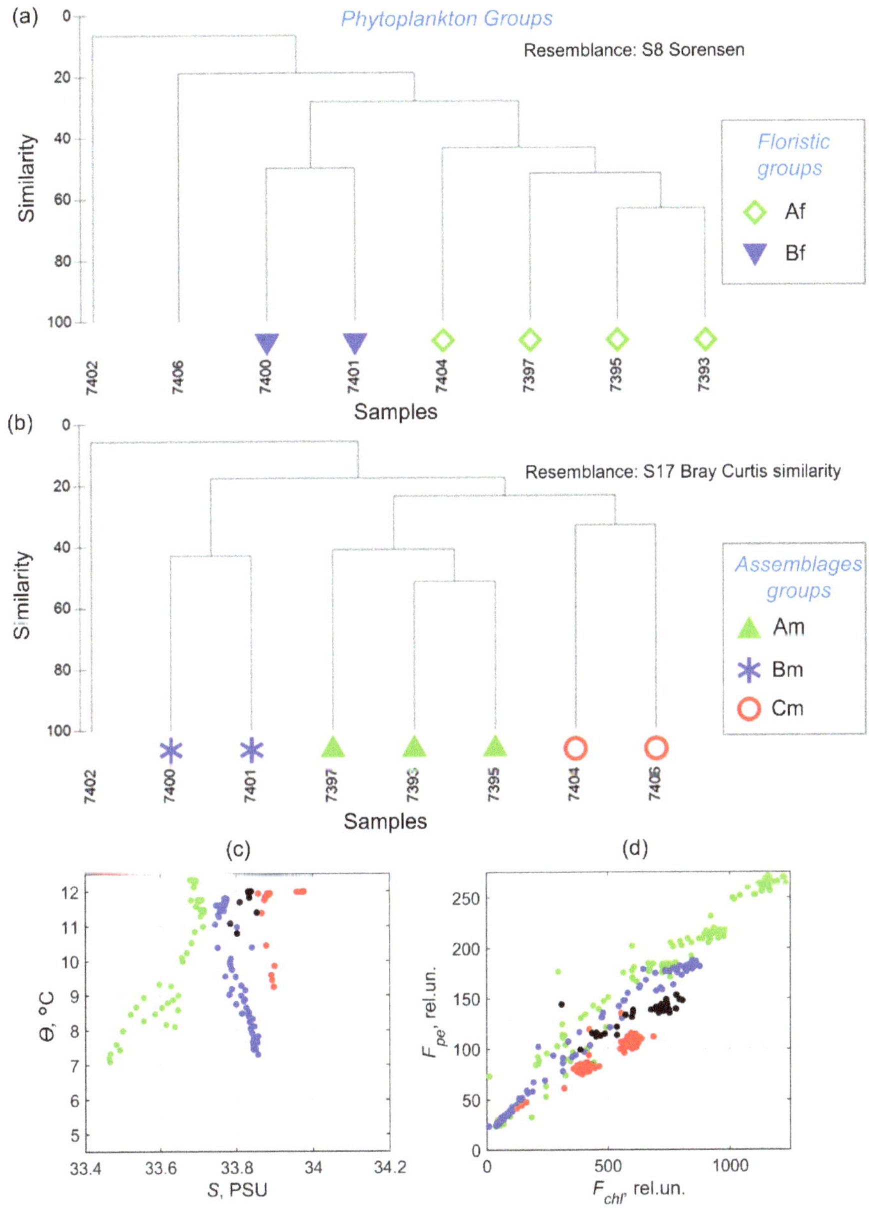

Figure 9. Dendrograms showing the grouping of phytoplankton communities according to qualitative features (floristic groups (**a**)) and quantitative structure (groups of assemblages (**b**)). Corresponding *θ-S* diagram (**c**) and "phycoerythrin—chl-*a* fluorescence intensity" scatter plot (**d**) for different phytoplankton assemblages where colors are associated with phytoplankton assemblage groups.

Within the section, it was possible to distinguish three groups of assemblages based on species ratios by relative abundance (total $R = 0.93$, $p = 0.1\%$) (Figure 9b). In particular, the Am group, which united the assemblages above the shelf (in the downwelling zone and in the transition zone to upwelling), was presented (in descending order of significance) by *Paralia* sp., *Tripos lineatus*, *Dinophysis rotundata*, *Protoperidinium curvipes*, *P.* cf. *subsphaericum*, *Rhizosolenia hebetata*, *P. pyriforme*, *Dinophysis acuminata*, *Protoperidinium pellucidum*, *P. solidicorne*, and *Oblea rotunda*. This group can be considered as characteristic of the Patagonian Shelf. Groups Am and Bm were distinguished in the zone of local maxima of F_{chl} and BAC_{660}.

In addition, Figure 9c shows scatterplots of oceanographic and bio-optical characteristics according to clustering by groups of assemblages where a significant separation is also seen, both in oceanographic conditions (T, S) and in the ratio of fluorescence intensity of various phytoplankton pigments. At the same time, such a clear separation did not happen for floristic groups, the same points belonged to different clusters, or different points were present in the same cluster.

Assemblages above the shelf break and slope were united in the Bm group characterized as equally significant by *Tripos lineatus*, *T. fusus*, and *Stephanopyxis nipponica*. Communities at the periphery of the inshore branch and in the region of its eastern boundary (in the deep-water region) belong to the Cm group of assemblages; this was presented in descending order of importance, by *Dinophysis rotundata*, *Neoceratium pentagonum*, *Protoperidinium curvipes*, and *Thalassiosira eccentrica*. Assemblage in the core of the inshore branch was not included in any of the groups due to extremely low number of species.

3.8. Zooplankton Species

A total of 30 zooplankton species/taxa were identified. The composition of zooplankton at stations varied from 9 (st. 7393) to 22 species/taxa (st. 7406 and 7408). The species diversity calculated using the Shannon-Weaver index (H') ranged from 0.984 (st. 7402) to 2.751 (st. 7406).

Zooplankton abundance ranged from 543 to 3067 ind/m^3 (Figure 10) with an average value of 1804 $\pm$ 951 ind/m^3. The maximum abundance values were located in the inshore branch of the MC in the area of the upwelling and downwelling zones, with high F_{chl} values (st. 7400–7402), while the minimum abundance values were observed in the center of the inshore branch of the MC (st. 7404). Copepods and euphausiids larvae were the dominant zooplankton components at all stations. Euphausiids larvae were found in the nearshore waters of the Patagonian Shelf. Other non-copepod groups were not as abundant (Figure 10).

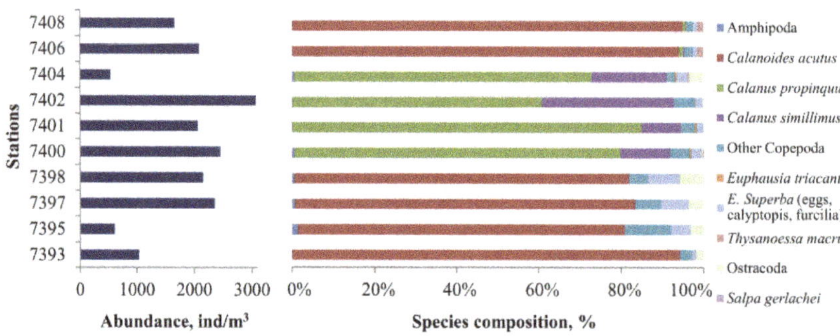

Figure 10. Contribution (%) of the major taxonomic groups to the total zooplankton abundance (ind/m^3).

Three (A–C) significant groups were identified (total R = 0.68, p = 0.05%) in the zooplankton community structure (Figure 11), which also clearly separated as seen in θ-S diagram (Figure 11c). Group A was detected in the nearshore waters of the Patagonian Shelf (st. 7393, 7395, 7397 and 7398) and mainly composed by copepod *Calanoides acutus*. Larvae of euphausiids, amphipods, and ostracods were present in this community; abundance was higher in the waters of the Patagonian Shelf.

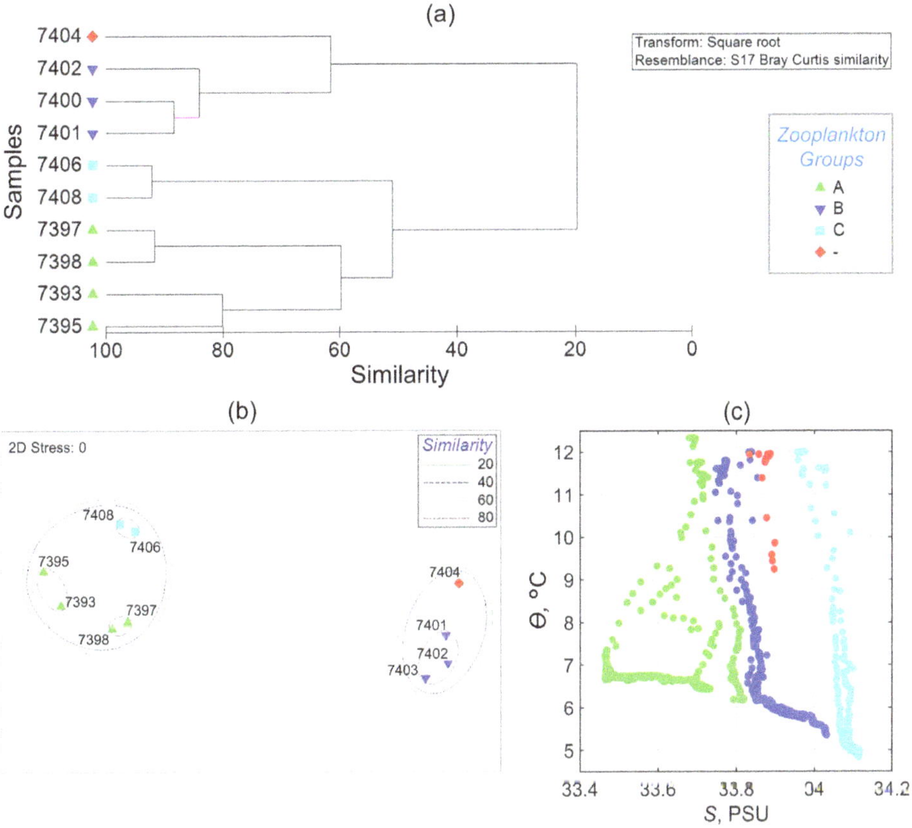

Figure 11. Cluster (**a**) and MDS (**b**) analyses of zooplankton collections (by species abundance), and θ-S scatter plot, where the colors of points correspond to the groups in (**a**–**c**).

Group B was restricted in the inshore branch of the MC in the upwelling and downwelling zones (st. 7400–7402). It was presented by Subantarctic species of the copepod genera *Calanus*, followed by euphausiids, salps, chaetognaths, appendicularians, amphipods, and polychaets. The deep-sea copepods of the genera *Paraeuchaeta* and *Rhincalanus* were abundant in the surface layer of 100-0 m.

Group C was identified in the zone between the inshore and offshore branches of the MC (st. 7406 and 7408) and presented by copepod *Calanoides acutus* and Antarctic oceanic species/taxa [69,70] copepod genera *Candacia*, *Euchirella*, *Haloptilus*, *Paraeuchaeta*, *Rhincalanus*, and euphausiids *Thysanoessa macrura*, Pteropod shells, siphonophores. One station (st. 7404) located in the center of the inshore branch of the MC was not included in any of the clusters and was characterized by mixed Subantarctic and Subtropical species/taxa with lower abundance and species richness (Figure 10).

3.9. Observations of Marine Mammals and Sea Birds

Summary results of observations of marine mammals and birds are shown in Figure 12. In the study site, we observed marine mammals belonging to the baleen whales Mysticeti and the toothed whales Odontoceti. Baleen whales were represented by one species: Antarctic minke whale *Balaenoptera bonaerensis* belonging to the family Balaenopteridae. Two species of recorded toothed whales belonged to two different families: the Peale's dolphin *Lagenorhynchus australis* in the dolphin family Delphinidae and the southern bottlenose whale *Hyperoodon planifrons* belonging to the family Ziphiidae.

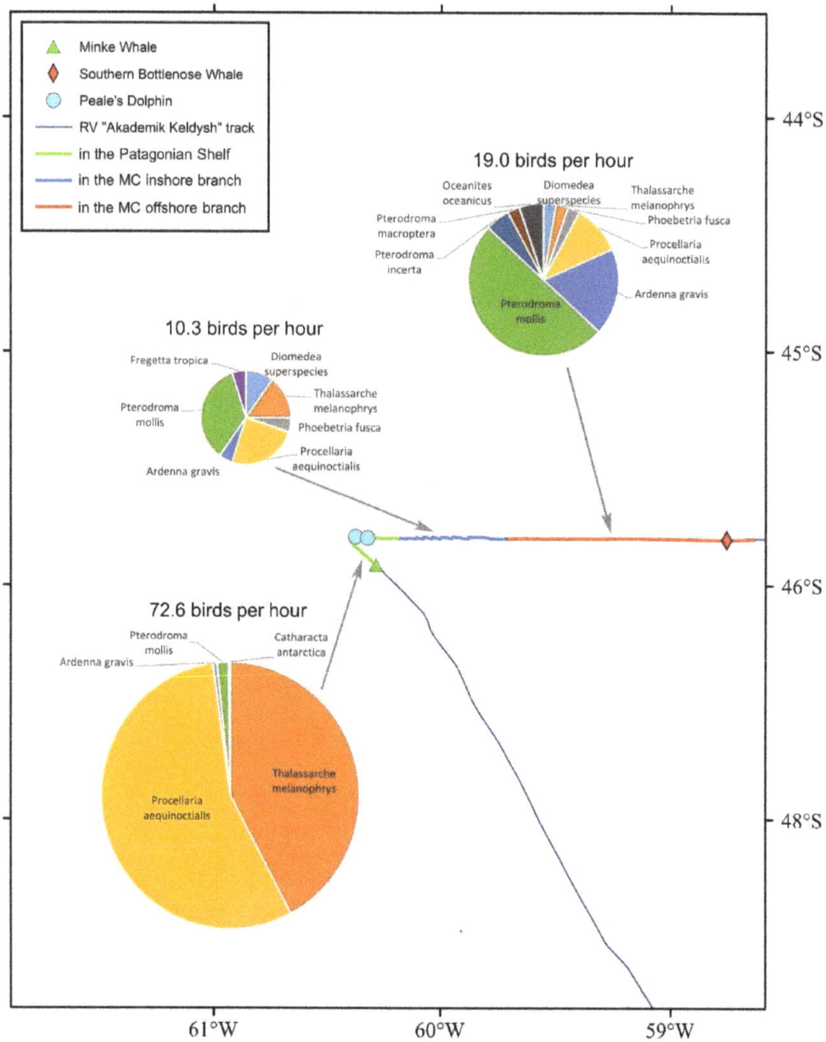

Figure 12. The ratio of the bird species at each site and sightings of marine mammals.

Two species, the Antarctic minke whale and the Peale's dolphin, were found in the shelf zone of the MC, one occurrence/one individual and two occurrences/five individuals, respectively. The southern bottlenose whale was noted in the offshore branch of the MC (one occurrence/one individual).

The birds encountered in the MC region were represented by two orders and four families. The order Procellariiformes included the following families: albatrosses Diomedeidae, petrels Procellariidae, and southern storm petrels Oceanitidae. The order Charadriiformes was represented by the skua family Stercorariidae. A total of 10 species of birds were recorded, with a total number of 422 individuals.

It was found that the petrel family was the most numerous and diverse in terms of the species composition. A total of 256 individuals belonging to five species of this family were recorded. However, there was a change in the occurrence of some species in different parts of the section across the MC. For example, the number of white-chinned petrel *Procellaria aequinoctialis*, the most abundant bird species, decreased by more than 40 times, from 202 individuals in the shelf zone to four and five individuals in the inshore and offshore branches of the MC, respectively. On the contrary, some species of this family appeared more often over one branch of the MC with respect to another; this is typical for the shelf zone and the offshore branch. An increase in the number of occurrences in these regions was noted for the soft-plumaged petrel *Pterodroma mollis* and the great shearwater *Ardenna gravis* from five to 19 individuals and from two to seven individuals, respectively. The Atlantic petrel *Pterodroma incerta* and the great-winged petrel *Pterodroma macroptera* were noted only by single occurrences over the offshore branch, namely two occurrences/one individual and one occurrence/one individual, respectively. In the upwelling zone, one individual of the petrel family was noted.

The albatross family, second in terms of the number of individuals encountered, was represented by two species. A change in the occurrence of the black-browed albatross *Thalassarche melanophris* has been established in the different sites of the MC, as discussed above for some species of the petrel family. This species is characterized by a decrease in the number of individuals in the transition from the shelf zone to the inshore branch of the MC from 153 to three individuals, and to one individual during the transition to the offshore branch of the current, respectively. Solitary specimens of the sooty albatross *Phoebetria fusca* have been recorded over the inshore and offshore branches. Over the inshore branch, two juveniles belonging to the so-called giant albatrosses were noted, but it was not possible to establish their species identity. The family Oceanitidae was not numerous. Among them, two specimens of Wilson's storm petrel *Oceanites oceanicus* were noted in the offshore branch and one specimen of the black-bellied storm-petrel *Fregetta tropica* in the inshore branch. The only member of the Stercorariidae family, the Antarctic Skua *Stercorarius antarcticus*, was recorded in the shelf zone of the MC.

The average number of birds encountered per hour was calculated for each site. The largest number of birds were found in the shelf zone: 363 individuals (72.6 birds per hour); in the coastal branch, 20 individuals (10.3 birds per hour); and in the offshore branch, 38 individuals (19 birds per hour). In the shelf zone, over a distance of several kilometers, fishing boats were seen, around which a large number of birds were circling. The ratio of the bird species at each site and sightings of marine mammals are presented in Figure 12.

4. Discussion

4.1. Types of Water Masses and Hydrodynamic Manifestations of Oceanographic Characteristics

The combined use of hydrophysical and bio-optical data, as well as model data from the Lagrangian analysis, allows better separation of water masses. The division of the waters over the section according to temperature, salinity, and optically active substances into a shelf and open-sea waters is consistent with the results reported in [71,72], where the boundary between the shelf waters and the waters of the Malvinas Current system is salinity isoline $S \approx 33.8$ PSU (in this work we assume 33.78 PSU). The ranges of temperature-salinity values for the Malvinas Current obtained in this work are consistent with the data in [6,71]. In addition, a similar spatial separation was also obtained in our work based on the Lagrangian analysis data (Figure 1). In this case, intrusions of water masses in both directions can be observed, both of the shelf waters into the system of the Malvinas Current, and of waters of the Malvinas Current into the shelf region. Both processes are seen on the Lagrangian maps in Figure 1. Intrusion of shelf waters into the open sea is presented in the results of our in situ observations (Figures 4–6). The strong intrusion of MC waters into the shelf is accompanied by the upward displacement of high saline, low temperature, and high nutrient waters reported in [73] indicating the upwelling of deep nutrients on a similar section near 47° S, where the slope of the continental margin is less abrupt.

The coastal branch of the Malvinas Current is well described in [4,5]. However, in this study, we found the unusual phenomenon that velocities in the coastal branch were higher than in the offshore branch of the Malvinas Current, and in addition, the coastal branch was located very close to the shelf edge (Figure 2). This can potentially lead to an increase in the vertical motion of water masses. In general, SADCP measurements in the Malvinas Current are rarely performed. Some results are reported in [4,5,74], but even there they were not accompanied by detailed CTD measurements, which can provide complete information about the vertical structure of waters in the zone of alternating upwellings/downwellings.

Satellite data of medium spatial resolution make it possible to see not just one band of elevated concentrations of chlorophyll-a, which belong to the well-known upwelling along the Patagonian Shelf break [16–18]. A complex pattern of several bands of varying chlorophyll-a concentrations and surface layer temperature is also seen, which indicates the action of complex dynamic processes in the study site leading to a series of successive upwellings and downwellings. In general, this suggests that the spatial resolution of the survey should be even better than 2–5 km (0.5–1 km), which can be achieved using submersible autonomous vehicles.

At the same time, satellite data still do not give a complete pattern of the ongoing processes, since they do not allow us to analyze the vertical variability of oceanographic and bio-optical characteristics. They also do not allow a correct assessment of the euphotic depth in this area, which will be discussed in more detail in the next section. Nevertheless, satellite data are very helpful as an additional source of data for analysis to highlight hydrophysical structures and corresponding fronts on the map, and to analyze the horizontal motion of water masses. Modern implementations of global reanalyses are also insufficient for a comprehensive study of thin vertical and horizontal oceanographic structures and processes in the study area since they lack spatial resolution and accurate bathymetry.

The detailed section made it possible to distinguish the submesoscale structure of alternating upwelling and downwelling zones, which was not the case in previous works, where the analysis focused on larger scales. Similar alternations are presented in measured in situ data [14,75], in model data [76], and similar manifestations in the satellite images are visible in [3,77]. They can be caused by the alternation of zones of divergence and convergence of the current velocity vector to the left and to the right of the MC inshore jet velocity maximum. The data presented here do not reveal erosion of the pycnocline caused by upwelling unlike the results presented in [14] at 43° S and in [76] at 51° S, which were performed on the Patagonian Shelf. It is important to note that in the cited works, erosion of the pycnocline occurred during upwelling over the shelf, and not over the outer boundary of the shelf.

Upwelling and downwelling zones in some cases are better manifested in the oceanographic variables (temperature, salinity, density), and in other cases in the optical characteristics presented here. For example, the BAC_{660} and turbidity sections show the ascent of turbid bottom waters from the Patagonian Shelf and their displacement towards its edge (from the bottom parts of stations 7396–7397 to the surface parts of stations 7400–7401). Downwelling between stations 7402 and 7403 in the surface layer of 50 m, which is displaced in the underlying layers at station 7401, is better pronounced in the BAC_{660} data. Downwelling at station 7406 is seen from all presented optical characteristics except F_{CDOM}. Therefore, in such works, it is important to have a large number of automatically measured vertical profiles of marine environment parameters in order to increase the reliability of the data.

4.2. Relation between the Depths of the Upper Mixed and Euphotic Zones

It is not obvious at first glance result (Figure 7) that the euphotic zone increases in upwelling zones, and decreases in downwelling zones, despite the fact that the maximum concentrations of chl-*a* are higher in upwelling zones; they appear on satellite data as bands of increased concentrations of chl-*a*.

The effect of the increased maximum of chl-*a* at stations with upwelling can be explained by a decrease in the thickness of the UML and the associated concentration of substances. The effect of shallower euphotic zone at the stations with downwelling may be associated with the spread of sufficiently high chl-*a* values to a greater depth within the UML; hence, the depth-integrated attenuation of light is more effective than at upwelling stations. During upwelling, the most transparent water of the intermediate layer of the sea, together with the seasonal pycnocline, ascends closer to the surface.

This leads to the paradoxical result that, on the one hand, during upwelling, the maximum chl-*a* values are higher, but the euphotic zone is deeper, due to the fact that the depth-integrated chl-*a* values are lower, since they are observed only within a thinner UML. This is also seen in Figure 7b, where local decreases in the thickness of the UML correspond to increases in the thickness of the euphotic zone at the stations with upwelling. Local increases in the UML thickness correspond to decreases of euphotic zone thickness at the stations with downwelling. This is not so obvious at stations 7403 and 7404, perhaps due to the fact that the upwelling and downwelling processes are not as strong here. In addition, at the first two stations 7393 and 7394 there are also all indications that downwelling may take place there based on the form of profiles in Figure 7a, and relationship between Z_{UML} and Z_{eu}.

It is important to understand that such a large deepening of the phytoplankton layer at stations 7393, 7394, 7395, 7396, 7397, and 7402 below the Z_{eu} is not normal from the point of view of the functioning of phytoplankton cells [78–81] and may be associated with dynamic effects, such as downwelling, which drive substances into deeper layers. A similar profile shape is presented in Figure 3 in [82] in the Drake Passage, where the ACC jet streams exist, the position of which changes in time [5], which can lead to the formation of powerful zones of divergence and convergence, and the corresponding ascends and descends of water masses. A similar analysis of Z_{eu} and Z_{UML} distribution was carried out in [73], however, with a lower spatial resolution, which did not allow us to connect the identified features in the spatial distribution of Z_{eu} and Z_{UML} with the fine structure of upwellings and downwellings.

In addition, the revealed phenomenon influences the satellite estimates of the euphotic layer depth in the study area, which is knowingly incorrect in the upwelling and downwelling zones. All modern algorithms for estimating Z_{eu} [83,84] use the feedback from satellite estimates of the chl-*a* concentration in the water layer seen from the space; the latter are higher in the case of upwelling zones and lower in downwelling zones. This leads to underestimation of Z_{eu} in the upwelling zone and to overestimation of Z_{eu} in the downwelling zone. It must be kept in mind that this conclusion is valid for the presented section and is not necessarily true for a different hydrophysical and bio-optical situation.

4.3. Analysis of Photosynthetic Activity of Phytoplankton

The results presented in 3.6 were approximately in the same range as those found for estuarine phytoplankton and microphytobenthos [85], where values of the *rETR* efficiency for C fixation (*EE*) varied between 0.04 and 0.16. These authors recalculated published data for cultures and presented *EE* values varying between 0.007 and 0.020 for different marine phytoplankton species.

The maximum values of the $P^B/rETR$ parameter were found in the outer part of the inshore branch of the MC (stations 7406, 0.051) and in the zone between the inshore and offshore branches (stations 7407–7408). When analyzing the spatial distribution of photosynthetic efficiency values, shown in Figure 8, it should be noted that stations with high $P^B/rETR$ were located at the boundaries of upwelling and downwelling zones throughout the entire section from the shelf to the oceanic part. Specific hydrodynamic conditions at the boundaries of zones likely contribute to the increase in the photosynthetic activity of phytoplankton in these regions.

Earlier, one of the authors of this paper reported the results in the Kara Sea, where an increase in $P^B/rETR$ in the local areas with strong gradients of oceanographic conditions was caused by various factors: the area of the outer shelf in the zone of mixing of river and sea waters in the estuary of the Yenisei River, and a strong frontal zone at the edge of the shelf that appeared as a result of the interaction of two currents in the upwelling zone above the continental slope [39]. Despite the geographical remoteness of these two regions and the different nature of the gradients of oceanographic variables, an increase in photosynthetic efficiency precisely in the zones of gradients of variables is manifested in both cases. This allows us to assume the general regularity of this phenomenon.

4.4. Phytoplankton Communities

The work by Antacli et al. [86] reporting detailed studies of phyto- and protozooplankton of the southern part of the Patagonian Shelf in the summer of 2004, provides very detailed data on the species composition of these communities (319 species), their size structure (90% of the total number accounted for picoplankton 2–5 µm in size), as well as the significant spatial heterogeneity of the communities in this vast region. These surveys were performed on four transects across the shelf located between 47° and 55° S. In turn, our observations covered the marginal and close-to-marginal regions of the Patagonian Shelf, its slope part, as well as the coastal jet of the Malvinas Current up to its eastern periphery. This expands the geographic scope of the results.

Our studies were carried out 18 years later, which gives us a certain opportunity to assess changes in communities over time. It should be noted that almost 20 years later, in the more eastern region, which covers a significantly greater diversity of water masses, the smallest size fraction of phytoplankton also dominated, "headed by a coccal ultraphytoeukaryotic cell (3 µm) (probably chlorophyte/prasinophyte), which was the most important morpho-species in the region (frequency of occurrence FO = 35%, relative abundance RA = 45%)". Based on our data and descriptions of colleagues, we can assume that in the summer of 2004, *Prasinoderma coloniale* developed just as abundantly in the water area studied in [86]. Thus, this species still plays the most significant role in the abundance of phytoplankton in the entire region under consideration.

Against the background of the widespread significant abundance of this species, the phytoplankton size fraction >10 µm made a relatively small total contribution to the community structure, both in 2004 and in 2022 (except for the eastern boundary of the MC). However, these relatively larger forms accounted for the largest part of the species richness of communities in both seasons and both areas. In general, the total number of species noted by us (43) seems significantly less than the sum of dinoflagellates (148) + diatoms (73) encountered by our colleagues in 2004.

It should be taken into account that our observations show an increase in the species richness of communities during the transition to the shelf zone, while in the summer of 2004, four vast areas were studied just above the shelf. This could already be the reason for the higher floristic diversity. In terms of the significant heterogeneity of the community structure in the section studied by us and in the sections made in [86], there is also a certain correlation because assemblages both above the shelf and in the zone of influence of the coastal jet of the Malvinas Current were and remain significantly heterogeneous in space.

The work [87] is also devoted to the study of phytoplankton including the summer one in the southern part of the Patagonian Shelf. However, according to the observations made by the authors in the summer of 2008, in a vast area northwest of the Falkland Islands, the main components of phytoplankton were the coccolithophores Emiliania huxleyi and the primesiophytes Phaeocystis antarctica. The authors attribute their flowering to a notable decrease in the level of nutrients from spring to summer due to the abundant development of diatoms in spring. At the same time, their observations were made in the middle of summer (4–7 January 2008), while our studies on the section from the shelf and through the Malvinas Current were made at the end of summer (21–22 February 2022). The results in [86] were also obtained at the end of summer (18 March–4 April 2004). The observed differences in mass species may well be explained, among other issues, by different stages of seasonal succession of phytoplankton.

4.5. Zooplankton Communities

The majority of identified zooplankton species (80%) were resident species, which are shared with the Subantarctic zone, and other taxa can be considered advected species common to the Antarctic zone [88]. Previous observations [76] showed "zooplankton hot spot—high biomass and abundance" associated with frontal upwelling. Additionally, all identified groups of zooplankton are well separated in accordance with oceanographic characteristics (Figure 11). All identified groups of zooplankton are well separated in accordance with oceanographic characteristics (Figure 11).

Copepods of the genera *Calanoides* from group A observed in the Patagonian Shelf, and from group C in the zone between inshore and offshore branches of the MC, were the most abundant members of zooplankton communities in the nearshore waters throughout the Antarctic zone and in the south of the Subantarctic zone as described in [88,89], and were also one of the main dominant taxa by biomass in the Drake Passage [90]. They were found at depths of 0–200 m in austral summer and 500–800 m in winter [91].

Copepods of the genera *Calanus* related to Group B in the inshore branch of the MC in the zone of most significant upwelling and downwelling, were found previously in the deep-sea layers (300–500 m) in austral summer [92]. Possibly, deep-sea copepods in the surface waters (0–100 m) can be associated with the local zones of upwelling. The path of possible vertical motions of this group of zooplankton is shown in all sections of oceanographic (Figure 4) and bio-optical (Figure 5) characteristics.

In the study region, the Antarctic oceanic species/taxa were found in the communities of group C, while previously (January 2022) in the same expedition, similar species composition was detected in the water northeast of the South Orkney Islands. This area was influenced by the wide Antarctic Circumpolar Current (ACC) system where water flows are generally directed to the east [93]. Possibly, the eastward currents could transport these species both to the northeast of the South Orkney Islands and to the Patagonian Shelf. Generally, a correlation exists between the faunal and floristic relations in the Southeast Pacific and Southwest Atlantic within the biogeographic Magellanic Province [94]. This is provided by the eastward water mass transport through the Magellan Strait and the Drake Passage [11,95]. The zone of sterile migration of Antarctic species is usually insignificant, but in some cases, it reaches a large extent [88]. The application of mathematical modeling is required for the confirmation of these ideas.

4.6. Observations of Marine Mammals and Sea Birds

The study site is a very important food base for animals of higher trophic levels. Unfortunately, due to the different efforts of observations at each site and limited data, it was not possible to identify a statistically significant dependence in the distribution of marine mammals and birds on oceanographic and biological parameters. However, we can make a few qualitative conclusions.

All of the found mammal species belong to different families and have different biological traits, including different trophic ecology. Antarctic minke whales feed mainly on zooplankton, while the Peale's dolphins feed on demersal fish, some species of cephalopods and shrimp. The southern bottlenose whale is known for feeding at great depths, deeper than 1000 m, mainly on squid, but also fish, such as the Patagonian toothfish *Dissostichus eleginoides*; therefore, it is not surprising that this species was observed in the offshore branch of the Malvinas Current, where depths reach 3000 m. The distribution of observed species also differs, although in some areas their ranges overlap [96].

Antarctic minke whale is widely distributed in the Southern Hemisphere. The Peale's dolphin mainly lives near the coast and on the shelf of South America and the Malvinas (Falkland) Islands. The southern bottlenose whale has a circumpolar distribution in the Southern Hemisphere in the deep regions [96]. No marine mammals were encountered in the shallow upwelling zone and the inshore branch of the MC.

All species of birds we met are typical for this region of the South Atlantic Ocean [97]. The highest density of concentration of birds encountered was recorded in the shelf part of the route; a large amount of bioproduction was observed in the same area. At the same time, it is interesting that the minimum density of birds was recorded in the zone of maximum upwelling, where there were also high concentrations of chlorophyll-a and zooplankton. This is probably due to the limited observation time in this relatively small but very dynamic area with high horizontal and vertical velocities of water masses. It is likely that birds prefer less dynamic, but no less productive zones for feeding farther from the shelf slope. In addition, the change in the species diversity of the food supply, characteristic of this area, can also affect the distribution of birds. These hypotheses require further testing and longer observations in this and other areas with similar conditions. Similar results with large variability in the species and quantitative composition of marine mammals and birds within a few tens of kilometers in a highly dynamic area of the Southwest Atlantic across an eddy dipole in the interaction zone between Subtropical and Subantarctic waters are presented in [98].

4.7. Summary Information about Hydro-Physical and Biological Charactristics

Table 4 presents summary information about identified water masses, seawater vertical velocity directions, groups of assemblages of phytoplankton communities and groups of zooplankton communities, and also the density of observed birds. One can see that the division into different biological groups occurred similarly in accordance with the zones of the Patagonian Shelf, the edge of the shelf, and the zone between the inshore and offshore branches of the MC. In addition, the separation of biological communities is influenced by different water masses, as well as upwelling and downwelling zones.

Table 4. Summary information about hydro-physical and biological characteristics of investigated waters. The background colors correspond to the colors that mark the planktonic groups in Figures 9b–d and 11, and ship tracks in Figure 12.

Station	Bottom Depth, m	UML Depth, m (Figure 4)	Euphotic Depth, m (Figure 7)	Up or Down Vert. Dir. * (Figure 4)	Region, (Figures 1 and 2)	Water Mass, (Figure 6, Table 2)	Phytoplankton Group ** (0–50 m), (Figure 9)	Zooplankton Group (0–200 m), (Figure 11)	Birds' Density, ind/h (Figure 12)
7393	137	38	29		PS	R, Y, O	Am	A	
7394	140	39	28		PS	R, Y, O			
7395	136	42	29	D	PS	R, Y, O	Am	A	72.6
7396	134	45	31	D	PS	R, Y, O			
7397	134	41	31	D	PS	R, O	Am	A	
7398	139	37	37		PS, MCi	C, O		A	
7399	139	37	38		PS, Mci	C, O			
7400	155	30	42	U	Pse, Mci	C, O	Bm	B	
7401	293	24	40	U	Pse, Mci	B, C, O	Bm	B	10.3
7402	365	45	34	D	Mci	B, C	no cluster	B	
7403	462	35	41	U	Mci	B, C			
7404	631	44	40	D	Mci	B, C	Cm	no cluster	
7405	742	35		U	Mci	B, C			
7406	790	53		D	Mci	B	Cm	C	
7407	853	25		U	ZB	B			
7408	942	39			ZB	B		C	
-	>1000			?	ZB & MCo	-			19.0

Note(s): * Upwelling (U) or downwelling (D) in the upper mixed layer. ** Groups of assemblages of phytoplankton.

Alternating zones of upwelling and downwelling are formed over the slope of the Patagonian Shelf under the influence of the MC and wind. This is especially pronounced in the region of the inshore jet of the MC. The detected zones significantly influenced all the characteristics analyzed in the work. It is shown that at the boundaries of the inshore jet, upwelling and downwelling alternate in space. At the western boundary of the jet, upwelling in the current changes to downwelling beyond the current, and on the contrary, at the eastern boundary of the jet, downwelling within the current changes to upwelling at the open sea side.

It was found that the euphotic zone is larger in the upwelling zones and smaller in the downwelling zones, which is associated with variations in mixed layer depth and depth-integrated abundance of phytoplankton cells and other optically active components of seawater within the UML, which are higher in downwelling zones and lower in upwelling zones. At the same time, on the contrary, the local maxima of the chl-*a* content and the light attenuation coefficient can be higher in the upwelling region than in downwelling due to the concentration of phytoplankton cells and non-living particles in the reduced upper mixed layer. It was found from the collected dataset that all spatially local maxima of phytoplankton photosynthetic efficiency are observed in the highly dynamic zones where upwelling changes to downwelling. A study of photosynthetic efficiency based on parallel measurements of primary production and electron transport rate in photosystem II is new for the Malvinas upwelling area.

Generally, all considered biological communities of different trophic levels were similarly distributed in space in accordance with the zones of the Patagonian Shelf and branches of the MC, as well as in accordance with the corresponding upwelling and downwelling fine structure.

Phytoplankton, studied for the first time in summer both in the marginal regions of the shelf and in the inshore branch of the MC, included more than 40 species. Qualitatively, dinoflagellates predominated, and heterotrophic forms were among them. Overall, a-diversity, structural diversity, and heterogeneity scores were low especially in the shelf edge and in the core of the MC inshore branch. Above the Patagonian Shelf, the richness of the flora was high. The picoplankton photosynthetic pyramidomonads *Prasinoderma coloniale* quantitatively dominated everywhere. Two floristic groups have been identified: Af, on the Patagonian Shelf, and Bf, above the shelf slope. Among the communities, against the background of a common dominant, three groups of complexes were distinguished: Am—above the shelf; Bm—above the slope of the shelf; Cm—on the eastern margin of the MC coastal jet.

Zooplankton was mostly dominated by copepods. The average abundance and species diversity of zooplankton increased from the inshore branch towards the offshore branch of the MC. Three major zooplankton communities were identified, which are associated with the physical properties in the inshore and offshore branches of the Malvinas Current. Copepods and euphausiids larvae are concentrated in the water of the Patagonian Shelf. The upwelling and downwelling zones were characterized by the presence of species of Subantarctic communities. In the zone between the inshore and offshore branches of the Malvinas Current zooplankton was mainly composed of the species belonging to the Antarctic communities. Deep-sea species that were found in the surface water of the MC can be associated with the local zones of upwelling.

Marine mammals and birds were more common on the Patagonian Shelf. At the same time, the greatest species diversity of birds was noted in the region of the offshore branch of the MC. The smallest relative abundance of birds was recorded in the inshore branch despite the fact that the strongest upwelling and highly bio-productive waters were observed in this region.

5. Conclusions

The main findings of the work are the following:

- Sub-mesoscale structure of alternating upwellings and downwellings in the upper mixed layer has been identified in the region of the outer Patagonian Shelf and inshore branch of the Malvinas Current along ~46° S section.
- In general, the differences in the upper mixed layer thickness between identified upwelling and downwelling zones are significant. In the first case, they are in the range of ~25–35 m, and in the second case, in the range of 40–55 m.
- The euphotic zone is larger in upwelling zones and smaller in downwelling zones. The difference can be as high as 6–10 m in adjacent upwelling and downwelling zones.

- Phytoplankton photosynthetic efficiency increases during the changeover between upwelling and downwelling.
- Distribution of all considered biological communities of different trophic levels is similar in accordance with the identified oceanographic features.
- Most of the phytoplankton species list was formed by dinoflagellates, among which heterotrophic forms prevailed. In quantitative terms, the pyramimonad *Prasinoderma coloniale*, a coccoid representative of picoplankton, dominated everywhere. Two floristic and three assemblage groups were distinguished among the communities. Biodiversity was high above the Patagonian Shelf and low above the shelf edge and in the core of Malvinas Current inshore branch.
- Zooplankton was mostly dominated by copepods. Subantarctic species/taxa of zooplankton concentrated in the nearshore waters of the Patagonian Shelf, while Antarctic species/taxa were most abundant in the zone between the inshore and offshore branches of the MC.
- The distribution of birds and mammals is different in the waters of shelf and Malvinas Current; the relative abundance of birds is minimal in the inshore branch of the Malvinas Current, where the strongest upwelling was identified.

Author Contributions: Conceptualization, P.A.S. and D.I.F.; methodology, S.A.M., V.V.K., P.V.S., A.Y.M., P.A.S. and V.I.P.; software, M.Y.U.; validation, V.A.K. and N.A.S.; formal analysis, P.A.S., D.I.F., V.I.P., P.V.S., S.A.O., N.A.L., M.V.B., P.V.C. and P.A.F.; investigation, P.A.S., S.A.M., O.Y.K., P.V.S., A.A.L., I.V.M. and A.D.C.; resources, A.Y.M. and N.A.S.; data curation, P.A.S.; writing—original draft, P.A.S., S.A.M., V.V.K., P.V.S. and P.V.C.; writing—review and editing, P.A.S., V.I.P., E.G.M., S.P.S. and V.V.K.; visualization, P.A.S., V.V.K., A.A.L., S.A.O., N.A.L. and M.V.B.; supervision, P.A.S.; project administration, E.G.M.; funding acquisition, E.G.M. All authors have read and agreed to the published version of the manuscript.

Funding: This research was supported by the Russian State Tasks FWMM-2022-0033, 0211-2021-008 and 0211-2021-0007 (POI FEB RAS), FMWE-2022-0001 (IO RAS), 122072000067-9 (NSCMB FEB RAS), FNNN-2022-0001, FNNN-2021-0003 (MHI RAS), FFER-2019-0021 (IEE RAS), 0202-2021-0001 (IACP FEB RAS) and grant of the Russian Science Foundation 21-77-20004 (ship operations). Resources of the computing cluster of the POI FEB RAS were also involved.

Institutional Review Board Statement: Not applicable.

Informed Consent Statement: No studies involving humans were performed. Birds and mammals were observed using binoculars and photo cameras. No experiments with animals have been carried out.

Data Availability Statement: Data are available upon request.

Conflicts of Interest: The authors declare no conflict of interest.

References

1. Hoffmeyer, M.; Sabatini, M.; Brandini, F.; Calliari, D.; Santinelli, N. (Eds.) *Plankton Ecology of the Southwestern Atlantic: From the Subtropical to the Subantarctic Realm*; Springer: Cham, Switzerland, 2018; p. 586. [CrossRef]
2. Peterson, R.G.; Whitworth, T. The Subantarctic and Polar Fronts in Relation to Deep Water Masses through the Southwestern Atlantic. *J. Geophys. Res.* **1989**, *94*, 10817. [CrossRef]
3. Franco, B.C.; Piola, A.R.; Rivas, A.L.; Baldoni, A.; Pisoni, J.P. Multiple Thermal Fronts near the Patagonian Shelf Break. *Geophys. Res. Lett.* **2008**, *35*, L02607. [CrossRef]
4. Piola, A.R.; Franco, B.C.; Palma, E.D.; Saraceno, M. Multiple Jets in the Malvinas Current. *J. Geophys. Res. Ocean.* **2013**, *118*, 2107–2117. [CrossRef]
5. Frey, D.I.; Piola, A.R.; Krechik, V.A.; Fofanov, D.V.; Morozov, E.G.; Silvestrova, K.P.; Tarakanov, R.Y.; Gladyshev, S.V. Direct Measurements of the Malvinas Current Velocity Structure. *J. Geophys. Res. Ocean.* **2021**, *126*, e2020JC016727. [CrossRef]
6. Artana, C.; Provost, C.; Poli, L.; Ferrari, R.; Lellouche, J. Revisiting the Malvinas Current Upper Circulation and Water Masses Using a High-Resolution Ocean Reanalysis. *J. Geophys. Res. Ocean.* **2021**, *126*, e2021JC017271. [CrossRef]
7. Paniagua, G.F.; Saraceno, M.; Piola, A.R.; Charo, M.; Ferrari, R.; Artana, C.; Provost, C. Malvinas Current at 44.7° S: First Assessment of Velocity Temporal Variability from in Situ Data. *Prog. Oceanogr.* **2021**, *195*, 102592. [CrossRef]

8. Palma, E.D.; Matano, R.P.; Piola, A.R. A Numerical Study of the Southwestern Atlantic Shelf Circulation: Stratified Ocean Response to Local and Offshore Forcing. *J. Geophys. Res.* **2008**, *113*, C11010. [CrossRef]
9. Piola, A.R.; Avellaneda, N.M.; Guerrero, R.A.; Jardón, F.P.; Palma, E.D.; Romero, S.I. Malvinas-Slope Water Intrusions on the Northern Patagonia Continental Shelf. *Ocean. Sci.* **2010**, *6*, 345–359. [CrossRef]
10. Matano, R.P.; Palma, E.D.; Piola, A.R. The Influence of the Brazil and Malvinas Currents on the Southwestern Atlantic Shelf Circulation. *Ocean. Sci.* **2010**, *6*, 983–995. [CrossRef]
11. Acha, E.M.; Mianzan, H.W.; Guerrero, R.A.; Favero, M.; Bava, J. Marine Fronts at the Continental Shelves of Austral South America. *J. Mar. Syst.* **2004**, *44*, 83–105. [CrossRef]
12. Miloslavich, P.; Klein, E.; Díaz, J.M.; Hernández, C.E.; Bigatti, G.; Campos, L.; Artigas, F.; Castillo, J.; Penchaszadeh, P.E.; Neill, P.E.; et al. Marine Biodiversity in the Atlantic and Pacific Coasts of South America: Knowledge and Gaps. *PLoS ONE* **2011**, *6*, e14631. [CrossRef] [PubMed]
13. Matano, R.P.; Palma, E.D. On the Upwelling of Downwelling Currents. *J. Phys. Oceanogr.* **2008**, *38*, 2482–2500. [CrossRef]
14. Valla, D.; Piola, A.R. Evidence of Upwelling Events at the Northern Patagonian Shelf Break. *J. Geophys. Res. Ocean.* **2015**, *120*, 7635–7656. [CrossRef]
15. Carranza, M.M.; Gille, S.T.; Piola, A.R.; Charo, M.; Romero, S.I. Wind Modulation of Upwelling at the Shelf-Break Front off Patagonia: Observational Evidence. *J. Geophys. Res. Ocean.* **2017**, *122*, 2401–2421. [CrossRef]
16. Garcia, V.M.T.; Garcia, C.A.E.; Mata, M.M.; Pollery, R.C.; Piola, A.R.; Signorini, S.R.; McClain, C.R.; Iglesias-Rodriguez, M.D. Environmental Factors Controlling the Phytoplankton Blooms at the Patagonia Shelf-Break in Spring. *Deep. Sea Res. Part I Oceanogr. Res. Pap.* **2008**, *55*, 1150–1166. [CrossRef]
17. Rivas, A.L.; Dogliotti, A.I.; Gagliardini, D.A. Seasonal Variability in Satellite-Measured Surface Chlorophyll in the Patagonian Shelf. *Cont. Shelf Res.* **2006**, *26*, 703–720. [CrossRef]
18. Romero, S.I.; Piola, A.R.; Charo, M.; Garcia, C.A.E. Chlorophyll- a Variability off Patagonia Based on SeaWiFS Data. *J. Geophys. Res.* **2006**, *111*, C05021. [CrossRef]
19. Salyuk, P.A.; Glukhovets, D.I.; Lipinskaya, N.A.; Moiseeva, N.A.; Churilova, T.Y.; Ponomarev, V.I.; Aglova, E.A.; Artemiev, V.A.; Latushkin, A.A.; Major, A.Y. Variability of the Sea Surface Bio-Optical Characteristics in the Region of Falkland Current and Patagonian Shelf. *Sovrem. Probl. Distantsionnogo Zondirovaniya Zemli Iz Kosm.* **2021**, *18*, 200–213. [CrossRef]
20. Pakhomov, E.; Pshenichnov, L.; Krot, A.; Paramonov, V.; Slypko, I.; Zabroda, P. Zooplankton distribution and community structure in the Pacific and Atlantic Sectors of the Southern Ocean during austral summer 2017–2018: A Pilot Study Conducted from Ukrainian Long–Liners. *J. Mar. Sci. Eng.* **2020**, *8*, 488. [CrossRef]
21. Turner, J.T. The importance of small planktonic copepods and their roles in pelagic marine food webs. *Zool. Stud.* **2004**, *43*, 255–266.
22. Von Dassow, P.; Montresor, M. Unveiling the Mysteries of Phytoplankton Life Cycles: Patterns and Opportunities behind Complexity. *J. Plankton Res.* **2011**, *33*, 3–12. [CrossRef]
23. Otto, S.A.; Niiranen, S.; Blenckner, T.; Tomczak, M.T.; Müller-Karulis, B.; Rubene, G.; Möllmann, C. Life Cycle Dynamics of a Key Marine Species Under Multiple Stressors. *Front. Mar. Sci.* **2020**, *7*. [CrossRef]
24. Barton, A.D.; Irwin, A.J.; Finkel, Z.V.; Stock, C.A. Anthropogenic Climate Change Drives Shift and Shuffle in North Atlantic Phytoplankton Communities. *Proc. Natl. Acad. Sci. USA* **2016**, *113*, 2964–2969. [CrossRef] [PubMed]
25. Franco, B.C.; Defeo, O.; Piola, A.R.; Barreiro, M.; Yang, H.; Ortega, L.; Gianelli, I.; Castello, J.P.; Vera, C.; Buratti, C.; et al. Climate Change Impacts on the Atmospheric Circulation, Ocean, and Fisheries in the Southwest South Atlantic Ocean: A Review. *Clim. Chang.* **2020**, *162*, 2359–2377. [CrossRef]
26. Nievas El Makte, M.L.; Polifroni, R.; Sepúlveda, M.A.; Fazio, A. Petroleum Hydrocarbons in Atlantic Coastal Patagonia. In *Anthropogenic Pollution of Aquatic Ecosystems;* Häder, D.P., Helbling, E.W., Villafañe, V.E., Eds.; Springer: Cham, Switzerland, 2021; pp. 325–352. [CrossRef]
27. Goni, G.J.; Bingas, F.; DiNezio, P.N. Observed low frequency variability of the Brazil Current front. *J. Geophys. Res. Ocean.* **2011**, *116*, C10037. [CrossRef]
28. Combes, V.; Matano, R.P. Trends in the Brazil/Malvinas confluence region. *Geophys. Res. Lett.* **2014**, *41*, 8971–8977. [CrossRef]
29. Franco, B.; Ruiz-Etcheverry, L.; Marrari, M.; Piola, A.; Matano, R. Climate Change Impacts on the Patagonian Shelf Break Front. *Geophys. Res. Lett.* **2022**, *49*, e2021GL096513. [CrossRef]
30. Helbling, E.W.; Narvarte, M.A.; González, R.A.; Villafañe, V.E. (Eds.) *Global Change in Atlantic Coastal Patagonian Ecosystems;* Springer: Cham, Switzerland, 2022; p. 463. [CrossRef]
31. Chereskin, T.K.; Harris, C.L. *Shipboard Acoustic Doppler Current Profiling during the WOCE Indian Ocean Expedition: I10;* Scripps Institution of Oceanography, University of California: San Diego, CA, USA, 1997.
32. Egbert, G.D.; Erofeeva, S.Y. Efficient Inverse Modeling of Barotropic Ocean Tides. *J. Atmos. Ocean. Technol.* **2002**, *19*, 183–204. [CrossRef]
33. Monterey, G.I.; Levitus, S. Seasonal variability of mixed layer depth for the world ocean. *NOAA Atlas NESDIS* **1997**, *14*, 96.
34. Kara, A.B.; Rochford, P.A.; Hurlburt, H.E. An optimal definition for ocean mixed layer depth. *J. Geophys. Res.* **2000**, *105*, 16803–16821. [CrossRef]
35. Brainerd, K.E.; Gregg, M.C. Surface mixed and mixing layer depths. *Deep. Sea Res. Part I Oceanogr. Res. Pap.* **1995**, *42*, 1521–1543. [CrossRef]

36. deBoyer Montégut, C.; Madec, G.; Fischer, A.S.; Lazar, A.; Ludicone, D. Mixed layer depth over the global ocean: An examination of profile data and a profile-based climatology. *J. Geophys. Res.* **2004**, 109. [CrossRef]
37. Nielsen, E.S. The Use of Radio-Active Carbon (C14) for Measuring Organic Production in the Sea. *ICES J. Mar. Sci.* **1952**, *18*, 117–140. [CrossRef]
38. Intergovernmental Oceanographic Commission. Protocols for the Joint Global Ocean Flux Study (JGOFS) Core Measurements. In *Intergovernmental Oceanographic Commission Manuals and Guides, 29*; UNESCO-IOC: Paris, France, 1994; p. 170. [CrossRef]
39. Mosharov, S.A.; Sergeeva, V.M.; Kremenetskiy, V.V.; Sazhin, A.F.; Stepanova, S.V. Assessment of Phytoplankton Photosynthetic Efficiency Based on Measurement of Fluorescence Parameters and Radiocarbon Uptake in the Kara Sea. *Estuar. Coast. Shelf Sci.* **2019**, *218*, 59–69. [CrossRef]
40. Hoppe, C.J.M.; Klaas, C.; Osserbaar, S.; Soppa, M.A.; Cheah, W.; Laglera, L.M.; Santos-Echeandia, J.; Rost, B.; Wolf-Gladrow, D.A.; Bracher, A.; et al. Controls of primary production in two phytoplankton blooms in the Antarctic Circumpolar Current. *Deep.-Sea Res. II* **2017**, *138*, 63–73. [CrossRef] [PubMed]
41. Genty, B.; Briantais, J.-M.; Baker, N.R. The Relationship between the Quantum Yield of Photosynthetic Electron Transport and Quenching of Chlorophyll Fluorescence. *Biochim. Et Biophys. Acta (BBA)-Gen. Subj.* **1989**, *990*, 87–92. [CrossRef]
42. Schreiber, U. Pulse-Amplitude-Modulation (PAM) Fluorometry and Saturation Pulse Method: An Overview. In *Chlorophyll a Fluorescence. Advances in Photosynthesis and Respiration, vol 19*; Papageorgiou, G.C., Govindjee, Eds.; Springer: Dordrecht, The Netherlands, 2004; pp. 279–319. [CrossRef]
43. Napoleon, C.; Raimbault, V.; Claquin, P. Influence of nutrient stress on the relationships between PAM measurements and carbon incorporation in four phytoplankton species. *PLoS ONE* **2013**, *8*, e66423. [CrossRef] [PubMed]
44. Hannam, G. Relative abundance and distribution of phytoplankton in Glacier Bay, Alaska. In *Proceedings from the University of Washington School of Oceanography Undergraduate Thesis Research Expedition to Glacier Bay Alaska, March 2008*; Keil, R., Kelley, D., D'Asaro, E., Krembs, C., Collins, R.E., Eds.; School of Oceanography, University of Washington: Seattle, WA, USA, 2008; p. 14.
45. Lim, H.C.; Teng, S.T.; Leaw, C.P.; Wataki, M.; Lim, P. Phytoplankton assemblage of the Merambong Shoal, Tebrau Straits with note on potentially harmful species. *Malay Nat. J.* **2014**, *66*, 198–221.
46. Horner, R. *A Taxonomic Guide to Some Common Marine Phytoplankton*; Biopress Limited: Bristol, England, 2002; pp. 1–110.
47. Dodge, J.; Hermes, H. A Revision of the Diplopsalis Group of Dinoflagellates (Dinophyceae) Based on Material from the British Isles. *Bot. J. Linn. Soc.* **1981**, *83*, 15–26. [CrossRef]
48. Scott, F.J.; Marchant, H.J. (Eds.) *Antarctic Marine Protists. Australian Biological Resources Study, Canberra and Australian Antarctic Division*; Australian Antarctic Division: Hobart, Australia, 2005; p. 563.
49. Okolodkov, Y.B. Protoperidinium Bergh (Dinoflagellata) in the Southeastern Mexican Pacific Ocean: Part I. *Bot. Mar.* **2005**, *48*, 284–296. [CrossRef]
50. Okolodkov, Y.B. Protoperidinium Bergh (Dinophyceae) of the National Park Sistema Arrecifal Veracruzano, Gulf of Mexico, with a Key for Identification. *Acta Bot. Mex.* **2008**, *84*, 93–149. [CrossRef]
51. Pienaar, R.N.; Sakai, H.; Horiguchi, T. Description of a New Dinoflagellate with a Diatom Endosymbiont, *Durinskia capensis* Sp. Nov. (Peridiniales, Dinophyceae) from South Africa. *J. Plant Res.* **2007**, *120*, 247. [CrossRef] [PubMed]
52. Al-Kandari, M.; Al-Yamani, F.; Al-Rifaie, K. *Marine Phytoplankton Atlas of Kuwait's Waters*; Kuwait Institute for Scientific Research: Safat, Kuwait, 2009; p. 351.
53. Myat, S.; Thaw, M.S.H.; Matsuoka, K.; Lay, K.K.; Koike, K. Phytoplankton surveys off the southern Myanmar coast of the Andaman Sea: An emphasis on dinoflagellates including potentially harmful species. *Fish. Sci.* **2012**, *78*, 1091–1106. [CrossRef]
54. Gul, S.; Nawaz, M.F. The Dinoflagellate Genera Protoperidinium and Podolampas from Pakistan's shelf and deep sea vicinity (North Arabian Sea). *Turk. J. Fish. Aquat. Sci.* **2014**, *14*, 91–100. [CrossRef] [PubMed]
55. Estrada, M.; Delgado, M.; Blasco, D.; Latasa, M.; Cabello, A.M.; Benítez-Barrios, V.; Fraile-Nuez, E.; Mozetič, P.; Vidal, M. Phytoplankton across Tropical and Subtropical Regions of the Atlantic, Indian and Pacific Oceans. *PLoS ONE* **2016**, *11*, e0151699. [CrossRef]
56. Kretschmann, J.; Čalasan, A.Ž.; Gottschling, M. Molecular Phylogenetics of Dinophytes Harboring Diatoms as Endosymbionts (Kryptoperidiniaceae, Peridiniales), with Evolutionary Interpretations and a Focus on the Identity of Durinskia Oculata from Prague. *Mol. Phylogenetics Evol.* **2018**, *118*, 392–402. [CrossRef]
57. Phan-Tan, L.; Nguyen-Ngoc, L.; Doan-Nhu, H.; Raine, R.; Larsen, J. Species diversity of Protoperidinium sect. Oceanica (Dinophyceae, Peridiniales) in Vietnamese waters, with description of the new species P. larsenii sp. nov. *Nord. J. Bot.* **2017**, *35*, 129–146. [CrossRef]
58. Furuya, K.; Mitsunori, I.; Po Teen, L.; Songhui, L.; Chui-Pin, L.; Azanza, R.V.; Kim, H.-G.; Fukuyo, Y. Overview of harmful algal blooms in Asia. In *Global Ecology and Oceanography of Harmful Algal Blooms. Ecological Studies*; Glibert, P., Berdalet, E., Burford, M., Pitcher, G., Zhou, M., Eds.; Springer: Cham, Switzerland, 2018; Volume 232, pp. 289–308. [CrossRef]
59. Weikert, H.; John, H.-C. Experiences with a Modified Bé Multiple Opening-Closing Plankton Net. *J. Plankton Res.* **1981**, *3*, 167–176. [CrossRef]
60. Bouchard, S.; Cote, S.; St-Jacques, J.; Robillard, G.; Renaud, P. Effectiveness of Virtual Reality Exposure in the Treatment of Arachnophobia Using 3D Games. *Technol. Health Care* **2006**, *14*, 19–27. [CrossRef]

61. Razouls, C.; Desreumaux, N.; Kouwenberg, J.; de Bovee, F. *Biodiversity of Marine Planktonic Copepods (Morphology, Geographical Distribution and Biological Data)*; Sorbonne University, CNRS: Paris, France, 2022. Available online: http://copepodes.obs-banyuls.fr/en (accessed on 23 May 2022).
62. Boltovskoy, D. Radiolaria Polycystina. *South Atl. Zooplankton* **1999**, *1*, 149–212.
63. WoRMS Editorial Board. World Register of Marine Species. 2015. Available online: http://www.marinespecies.org (accessed on 22 May 2022).
64. Goswami, S.C. *Zooplankton Methodology Collection and Identification-a Field Manual: National Institute of Oceanography*; Dhargalkar, V.K., Verlecar, X.N., Eds.; National Institute of Oceanography: Panaji, India, 2004; pp. 5–8.
65. Clarke, K.R.; Warwick, R.M.; Marine, P. *Change in Marine Communities: An Approach to Statistical Analysis and Interpretation*, 2nd ed.; PRIMER-E: Plymouth, UK, 2001.
66. Prants, S.V. Backward-in-Time Methods to Simulate Large-Scale Transport and Mixing in the Ocean. *Phys. Scr.* **2015**, *90*, 074054. [CrossRef]
67. Bowers, D.G.; Brett, H.L. The relationship between CDOM and salinity in estuaries: An analytical and graphical solution. *J. Mar. Syst.* **2008**, *73*, 1–7. [CrossRef]
68. Stedmon, C.A.; Osburn, C.L.; Kragh, T. Tracing water mass mixing in the Baltic–North Sea transition zone using the optical properties of coloured dissolved organic matter. *Estuar. Coast. Shelf Sci.* **2010**, *87*, 156–162. [CrossRef]
69. Fransz, H.G. Vernal Abundance, Structure and Development of Epipelagic Copepod Populations of the Eastern Weddell Sea (Antarctica). *Polar Biol.* **1988**, *9*, 107–114. [CrossRef]
70. Foster, B.A. Time and Depth Comparisons of Sub-Ice Zooplankton in McMurdo Sound, Antarctica. *Polar Biol.* **1989**, *9*, 431–435. [CrossRef]
71. Piola, A.R.; Matano, R.P.; Steele, J.H.; Thorpe, S.A.; Turekian, K.K. Brazil and Falklands (Malvinas) currents. In *Ocean Currents, Encyclopedia Ocean Sciences*; Steele, J.H., Thorpe, S.A., Turekian, K.K., Eds.; Academia Press: Cambridge, MA, USA, 2010; pp. 35–43. [CrossRef]
72. Ehrlich, M.D.; Sánchez, R.P.; De Ciechomski, J.D.; Machinandiarena, L.; Pájaro, M. Ichthyoplankton composition, distribution and abundance on the southern patagonian shelf and adjacent waters. *INIDEP Doc. Científico* **1999**, *5*, 37–65. Available online: http://hdl.handle.net/1834/2574 (accessed on 19 November 2022).
73. Carreto, J.I.; Montoya, N.G.; Carignan, M.O.; Akselman, R.; Acha, E.M.; Derisio, C. Environmental and Biological Factors Controlling the Spring Phytoplankton Bloom at the Patagonian Shelf-Break Front–Degraded Fucoxanthin Pigments and the Importance of Microzooplankton Grazing. *Prog. Oceanogr.* **2016**, *146*, 1–21. [CrossRef]
74. Morozov, E.G.; Tarakanov, R.Y.; Demidova, T.A.; Frey, D.I.; Makarenko, N.I.; Remeslo, A.V.; Gritsenko, A.M. Velocity and transport of the Falkland Current at 46°S. *Russ. J. Earth Sci.* **2016**, *16*, ES6005. [CrossRef]
75. Artana, C.; Provost, C.; Lellouche, J.M.; Rio, M.H.; Ferrari, R.; Sennéchael, N. The Malvinas current at the confluence with the Brazil current: Inferences from 25 years of Mercator ocean reanalysis. *J. Geophys. Res. Ocean.* **2019**, *124*, 7178–7200. [CrossRef]
76. Sabatini, M.; Reta, R.; Matano, R. Circulation and Zooplankton Biomass Distribution over the Southern Patagonian Shelf during Late Summer. *Cont. Shelf Res.* **2004**, *24*, 1359–1373. [CrossRef]
77. Karabashev, G.S.; Evdoshenko, M.A. Narrowband shortwave minima of multispectral reflectance as indication of algal blooms associated with the mesoscale variability in the Brazil-Malvinas Confluence. *Oceanologia* **2018**, *60*, 527–543. [CrossRef]
78. Baldry, K.; Strutton, P.G.; Hill, N.A.; Boyd, P.W. Subsurface chlorophyll-a maxima in the Southern Ocean. *Front. Mar. Sci.* **2020**, *7*, 671. [CrossRef]
79. Ardyna, M.; Babin, M.; Gosselin, M.; Devred, E.; Bélanger, S.; Matsuoka, A.; Tremblay, J.É. Parameterization of vertical chlorophyll a in the Arctic Ocean: Impact of the subsurface chlorophyll maximum on regional, seasonal, and annual primary production estimates. *Biogeosciences* **2013**, *10*, 4383–4404. [CrossRef]
80. Lavigne, H.; D'ortenzio, F.; Ribera D'Alcalà, M.; Claustre, H.; Sauzède, R.; Gacic, M. On the vertical distribution of the chlorophyll a concentration in the Mediterranean Sea: A basin-scale and seasonal approach. *Biogeosciences* **2015**, *12*, 5021–5039. [CrossRef]
81. Cornec, M.; Claustre, H.; Mignot, A.; Guidi, L.; Lacour, L.; Poteau, A.; d'Ortenzio, F.; Gentili, B.; Schmechtig, C. Deep chlorophyll maxima in the global ocean: Occurrences, drivers and characteristics. *Glob. Biogeochem. Cycles* **2021**, *35*, e2020GB006759. [CrossRef]
82. Shtraikhert, E.A.; Zakharkov, S.P.; Salyuk, P.A.; Ponomarev, V.I.; Artemiev, V.A.; Glukhovets, D.I.; Latushkin, A.A. The Chlorophyll-a Content Distribution in the Atlantic Ocean in December 2019–January 2020 according to Ship Measurements at the Different Hydrometeorological Conditions. *Fundam. Appl. Hydrophys.* **2022**, *15*, 97–113. (In Russian) [CrossRef]
83. Lee, Z.; Weidemann, A.; Kindle, J.; Arnone, R.; Carder, K.L.; Davis, C. Euphotic zone depth: Its derivation and implication to ocean-color remote sensing. *J. Geophys. Res.* **2007**, *112*, C03009. [CrossRef]
84. Soppa, M.A.; Dinter, T.; Taylor, B.B.; Bracher, A. Satellite derived euphotic depth in the Southern Ocean: Implications for primary production modelling. *Remote Sens. Environ.* **2013**, *137*, 198–211. [CrossRef]
85. Barranguet, C.; Kromkamp, J. Estimating Primary Production Rates from Photosynthetic Electron Transport in Estuarine Microphytobenthos. *Mar. Ecol. Prog. Ser.* **2000**, *204*, 39–52. [CrossRef]
86. Antacli, J.C.; Silva, R.I.; Jaureguizar, A.J.; Hernández, D.R.; Mendiolar, M.; Sabatini, M.E.; Akselman, R. Phytoplankton and protozooplankton on the southern Patagonian shelf (Argentina, 47°–55°S) in late summer: Potentially toxic species and community assemblage structure linked to environmental features. *J. Sea Res.* **2018**, *140*, 63–80. [CrossRef]

87. Gonçalves-Araujo, R.; de Souza, M.S.; Borges Mendes, C.R.; Tavano, V.M.; Garcia, C.A.E. Seasonal change of phytoplankton (spring vs. summer) in the southern Patagonian shelf. *Cont. Shelf Res.* **2016**, *124*, 142–152. [CrossRef]
88. Voronina, N.M. *Ecosystems of Pelagial of the Southern Ocean*; Nauka: Moscow, Russia, 1984; Volume 1, p. 206.
89. Lee, W.-C.; Kim, S.-A.; Kang, S.-H.; Bang, H.-W.; Lee, K.-H.; Kwak, I.-S. Distribution and Abundance of Zooplankton in the Bransfield Strait and the Western Weddell Sea during Austral Summer. *Ocean. Polar Res.* **2004**, *26*, 607–618. [CrossRef]
90. Vedenin, A.A.; Musaeva, E.I.; Zasko, D.N.; Vereshchaka, A.L. Zooplankton communities in the Drake Passage through environmental boundaries: A snapshot of 2010, early spring. *PeerJ* **2019**, *7*, e7994. [CrossRef] [PubMed]
91. Delgado, L.E.; Jaña, R.; Marin, V.H. Testing Hypotheses on Life-Cycle Models for Antarctic Calanoid Copepods, Using Qualitative, Winter, Zooplankton Samples. *Polar Biol.* **1998**, *20*, 74–76. [CrossRef]
92. Michels, J.; Schnack-Schiel, S.B.; Pasternak, A.; Mizdalski, E.; Isla, E.; Gerdes, D. Abundance, Population Structure and Vertical Distribution of Dominant Calanoid Copepods on the Eastern Weddell Sea Shelf during a Spring Phytoplankton Bloom. *Polar Biol.* **2012**, *35*, 369–386. [CrossRef]
93. Orsi, A.H.; Whitworth III, T.; Nowlin Jr, W.D. On the Meridional Extent and Fronts of the Antarctic Circumpolar Current. *Deep. Sea Res. Part I Oceanogr. Res. Pap.* **1995**, *42*, 641–673. [CrossRef]
94. Boschi, E.E. Species of decapod crustaceans and their distribution in the American marine zoogeographic provinces. *Rev. Invest. Desarro. Pesq.* **2000**, *13*, 7–136.
95. Brun, A.A.; Ramirez, N.; Pizarro, O.; Piola, A.R. The role of the Magellan Strait on the southwest South Atlantic shelf. *Estuar. Coast. Shelf Sci.* **2020**, *237*, 106661. [CrossRef]
96. Jefferson, T.A.; Webber, M.A.; Pitman, R.L. *Marine Mammals of the World: A comprehensive Guide to Their Identification*; Elsevier: San Diego, CA, USA, 2011; p. 608.
97. Harrison, P. *Seabirds: An Indetification Guide*; Croom Helm: Beckenham, England, 1985; p. 448.
98. Morozov, E.G.; Frey, D.I.; Krechik, V.A.; Latushkin, A.A.; Salyuk, P.A.; Seliverstova, A.M.; Mosharov, S.A.; Orlov, A.M.; Murzina, S.A.; Mishin, A.V.; et al. Multidisciplinary Observations across an Eddy Dipole in the Interaction Zone between Subtropical and Subantarctic Waters in the Southwest Atlantic. *Water* **2022**, *14*, 2701. [CrossRef]

water

MDPI

Article

Hydraulically Controlled Bottom Flow in the Orkney Passage

Eugene G. Morozov [1,*], Dmitry I. Frey [1], Oleg A. Zuev [1], Manuel G. Velarde [2,3], Viktor A. Krechik [1] and Rinat Z. Mukhametianov [1,4]

1 Shirshov Institute of Oceanology, Nakhimovsky 36, 117997 Moscow, Russia
2 Instituto Pluridisciplinar, Universidad Complutense, Paseo Juan XXIII, 1, 28040 Madrid, Spain
3 Escuela de Arquitectura, Ingenieria y Diseno, Universidad Europea, 28670 Villaviciosa de Odon, Spain
4 Moscow Institute of Physics and Technology, Institutsky per. 9, Dolgoprudny, 141700 Moscow, Russia
* Correspondence: egmorozov@mail.ru

Abstract: Supercritical hydraulically controlled overflow of Antarctic Bottom Water from the Weddell Sea has been observed in the Orkney Passage during field measurements in February 2022. The Orkney Passage is the main pathway for the densest layer of Antarctic Bottom Water flow from the Weddell Sea to the Scotia Sea. The bottom current overflows the sill across the passage and flows down from the crest of the sill at 3600 m deeper than 4000 m. The descending flow accelerates because of the difference in the height of the sill and its foot. An estimate of the Froude number of this flow was greater than unity. Near the foot of the slope the kinetic energy of the flow becomes insufficient to continue moving in this regime. The flow slows down, and strong mixing and warming of the bottom water occurs due to the exchange with the surrounding waters. This hydrodynamic phenomenon is called supercritical hydraulically controlled flow. However, the flow of bottom water continues further and eventually fills the abyssal depths of the Atlantic.

Keywords: Orkney Passage; hydraulic control; subcritical; supercritical flows

Citation: Morozov, E.G.; Frey, D.I., Zuev, O.A.; Velarde, M.G.; Krechik, V.A.; Mukhametianov, R.Z. Hydraulically Controlled Bottom Flow in the Orkney Passage. *Water* **2022**, *14*, 3088. https://doi.org/10.3390/w14193088

Academic Editor: Fangxin Fang

Received: 22 August 2022
Accepted: 27 September 2022
Published: 1 October 2022

1. Introduction

The study site is located in the Weddell Gyre, which is a region in the Weddell Sea with specific properties that controls the deep and bottom ocean circulation and affects the global climate with the capability of influencing the global climate on time scales of hundreds to thousands of years [1]. The waters of the Weddell Sea deeper than 3000 m are connected with the Scotia Sea by the passages in the South Scotia Ridge.

Antarctic Bottom Water (AABW) that occupies the bottom layer of the Atlantic Ocean is generally formed in the Weddell Sea. Sinking of this dense water plays a key role in driving the lower limb of the global overturning circulation [2–4]. A major part of the AABW formation occurs in the Weddell Sea [5]. Formation of AABW generally occurs in the southern and western parts of the Weddell Sea over the ice shelves. Low temperatures and saline shelf waters are formed as a result of surface water cooling during ice freezing and brine rejection in coastal polynyas. Formation of highly dense water is maintained by cyclonic circulation in the Weddell Sea, which is driven by western winds in the north and eastern winds in the south of the Weddell Sea. The flow over the South Scotia Ridge represents a significant component of the AABW export from the Weddell Sea [6].

Weddell Sea Deep Water (WSDW) is the coldest part of AABW. Weddell Sea Deep Water is defined as water with potential temperatures between ~0°C and −0.7 °C. This water is formed due to the air–sea–ice interaction at the periphery of the Weddell Sea combined with the upwelling of Weddell Sea Bottom Water [7,8]. The main pathway for the export of WSDW from the Weddell Sea is the Orkney Passage [9]. Part of the bottom water is exported from the Weddell Sea through the South Sandwich Trench and other fractures of the South Scotia Ridge [2].

2. Hydraulic Control

We shall explain the physical phenomenon of hydraulic control following [10,11]. Fluids can transmit a signal by advection or wave propagation. We consider a water flow over an underwater ridge or dam. The flow from a deep reservoir overflows the crest of a dam or underwater ridge and spills down the spillway. A hydraulic jump is formed at the base of the spillway, which is an abrupt increase in the fluid depth (or depth of isopycnals) accompanied by turbulence, wave generation and mixing.

If perturbations of the flow appear upstream of the ridge (dam) crest, the waves (surface or internal) are generated, which can propagate in either direction. The flow downstream from the crest is so rapid that waves cannot propagate upstream. In other words, the velocity of the waves is smaller than the velocity of the descending flow. This type of flow is called supercritical. As the flow reaches the hydraulic jump, its velocity decreases and it returns to a subcritical state. In addition, the regime of the flow at the crest is critical, i.e., changing from subcritical to supercritical. Thus, the outflow from the reservoir is said to be choked or hydraulically controlled by the ridge (dam). The size of the dam, water depth and stratification are parameters that control the flow. A scheme of the phenomenon is shown in Figure 1a.

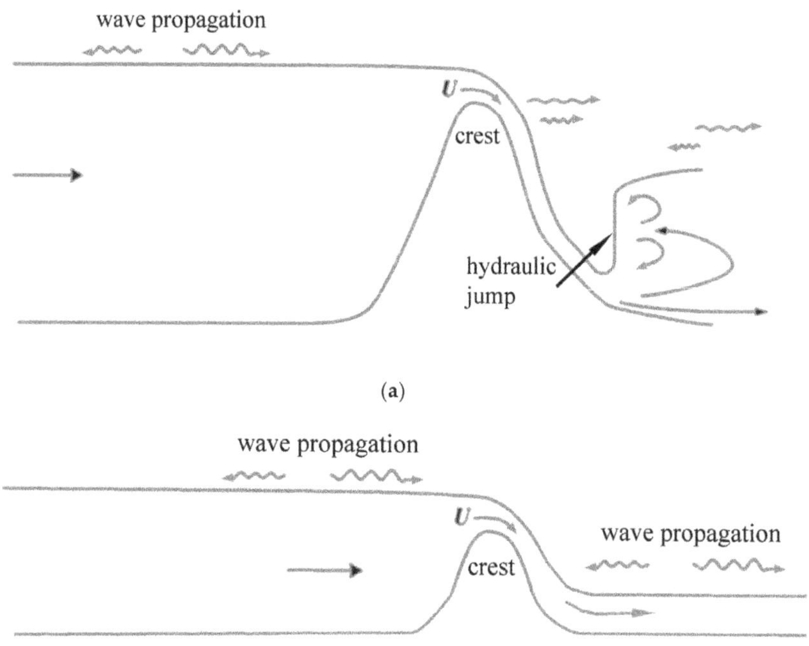

(b)

Figure 1. (**a**). Scheme of hydraulic control. Modified from [11]. (**b**). Scheme of laminar current overflowing an obstacle.

If the dam is low and the flow is slow the current just overflows the dam (Figure 1b).

Whitehead et al. [12] and Pratt and Whitehead [11] gave examples of hydraulic control in the passage between the Virgin Islands and in the Strait of Denmark; other examples were found in the Strait of Gibraltar [13,14].

This phenomenon has also been studied in laboratory experiments [15]. Even if the flow is subcritical, the flow entrains surrounding waters depending not only on the Froude number but also on the Reynolds number, as shown in [16]. The authors demonstrated this on the example of laboratory experiments and overflow in the Denmark Strait. Field measurements of a bolus of cold water overflowing the sill in the Denmark Strait and properties of the flow with variable velocity of the overflowing water were reported in [17]. Overflow in the Faroe Bank Channel was studied in [18].

We will show some more illustrative examples. First is the well-known Niagara Falls, which is a surface flow (Figure 2).

Figure 2. Subcritical and supercritical regimes in the Niagara Falls.

Another example is a flow over an underwater ridge in a less known region in the Kara Gates Strait between the Barents and Kara seas [19]. The phenomenon of the flow is very similar to that in the Strait of Gibraltar. A surface flow from the Barents Sea to the Kara Sea forms a hydraulic jump, whereas strong tidal currents over a sill generate large-amplitude internal tides [19–21] (Figure 3).

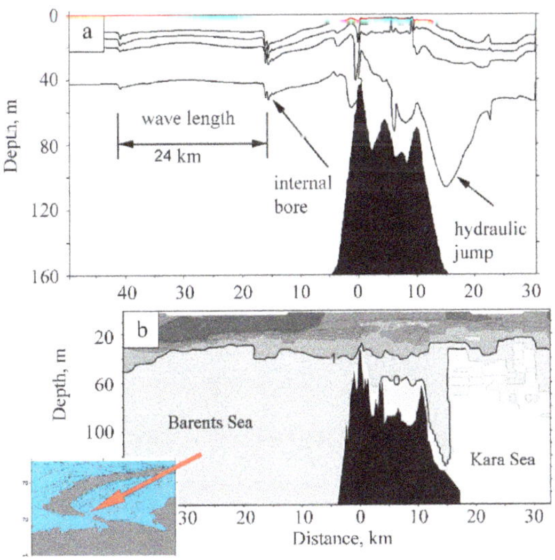

Figure 3. Field of isopycnals 23.5, 24.5, 25.5, and 26.5 perturbed by the flow in the Kara Gates Strait and internal waves based on numerical calculations using the model in [21,22]. The black color shows the bottom topography as specified in the model (**a**). Distribution of temperature over the towed section (**b**). Isotherms are shown with an interval of 1 °C. Thicker lines show isotherms of 1 and 0 °C to emphasize the observed effect. The black color shows the bottom profile. Location of the region on a larger scale is shown in the inset.

The flow of AABW through the Orkney Passage has been previously studied based on CTD/LADCP casts. A map of stations in different years is shown in Figure 4.

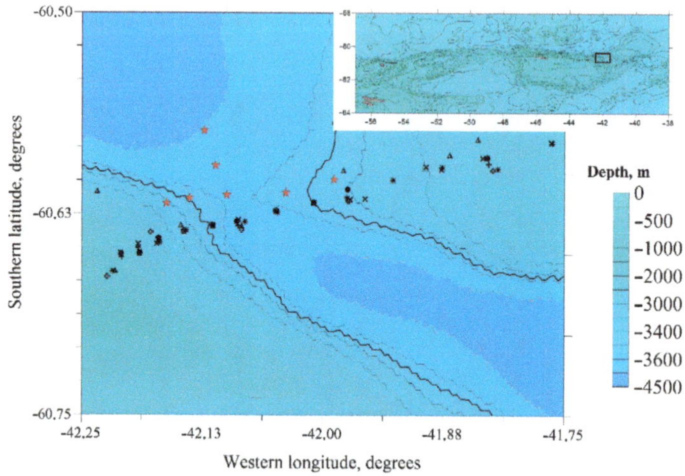

Figure 4. CTD/LADCP casts in the Orkney Passage in different years. Red stars indicate our stations in 2022. A chart of the region is shown in the inset. Black symbols indicate previous stations of CTD-casts in different years.

3. Data and Methods

Seven CTD casts with velocity profiling (LADCP) were made in the cruise in 2022 in the Orkney Passage region. Five stations were made across the passage and two stations down into the Laurie Depression from the central station. The depth of the ridge across the passage is about 3600 m. The sounding stations were displaced to the north relative to the repeated section in order to exclude possible interference with moorings that could be placed in this area.

The stations were performed using a lowered acoustic doppler current profiler (LADCP) and conductivity, temperature, and depth (CTD) profilers mounted on a General Oceanics GO1018 rosette water sampler. The water sampler was equipped with a Valeport VA500 altimeter allowing measurements close to the ocean bottom (3–7 m above the seafloor). An Idronaut Ocean Seven 320plus CTD probe was used for the measurements together with an MKplus Deck Unit. The CTD data were collected using the standard package REDAS5. The declared accuracy of CTD measurements is 0.001 °C for temperature and 0.001 mS/cm for conductivity sensors. The LADCP data measured by the TRDI WorkHorse Monitor 300 kHz profiler were processed using the programming package LDEO Software version IX.10 [23]. The accuracy of velocity measurements estimated by the processing program is usually 3–4 cm/s. In the bottom layers due to the bottom track signals, the errors decrease to 1–2 cm/s.

The temperature section is shown in Figure 5.

One can see in Figures 5 and 6 two characteristic features that are common to all flows in underwater channels. The cold core is displaced to the right side of the channel along the flow. This occurs due to the Ekman circulation, which arises from the bottom friction when the current flows in a narrow channel. A second core appears in the western (left) part of the passage above the bottom, which is displaced to the left of the flow and is located over the western wall of the channel [24]. The minimum potential temperature is −0.54 °C, which is close to the previous estimates [6].

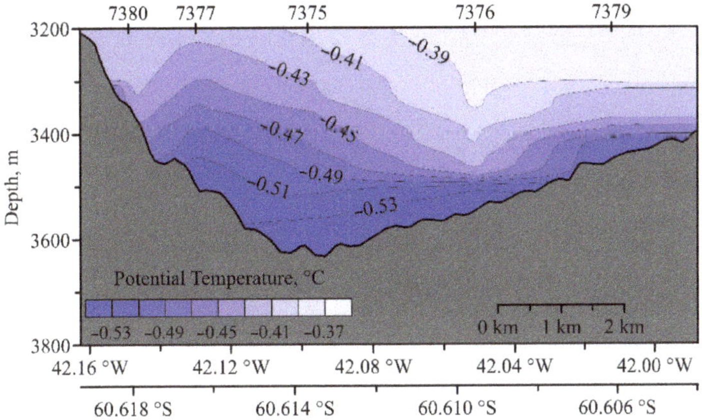

Figure 5. Section of potential temperature across the Orkney Passage.

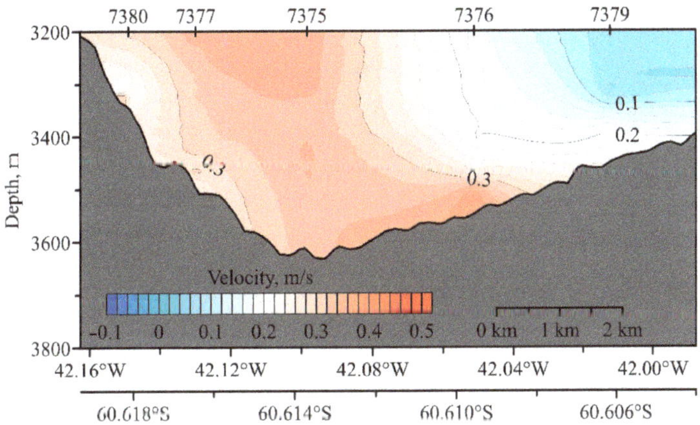

Figure 6. Longitudinal velocity component (normal to the section) across the Orkney Passage.

The velocities of the flow in the passage correspond to similar flows in the southern hemisphere and in particular the flow in the Vema Channel. The core of the maximum velocity is displaced to the left side relative to the flow.

Let us now consider the downflow of bottom water after overflowing the ridge. Figure 7 shows the flow velocities down the slope. The flow accelerates due to the transition of potential energy on the crest of the ridge into kinetic energy. Due to the expansion of the passage, the flow becomes wider and after flowing down by more than 200 m at a distance of about 3500 m, the flow slows down, descending into the deep layers of the ocean and not having enough energy to continue the fast flow. The maximum flow velocities are close to 50 cm/s. The potential temperature behaves similarly, forming a tongue of cold water flowing down the slope (Figure 8). In the tongue of cold water, which rolls down the slope, the temperature gradually increases from −0.54 to −0.45 °C. Salinity distribution downslope in the direction of the descending flow is shown in Figure 9. Contour lines of salinity that show the further propagation of the flow downslope are more illustrative than the isotherms because bottom water is warming due to the contact with the overlying waters. The salinity increase is due only to mixing.

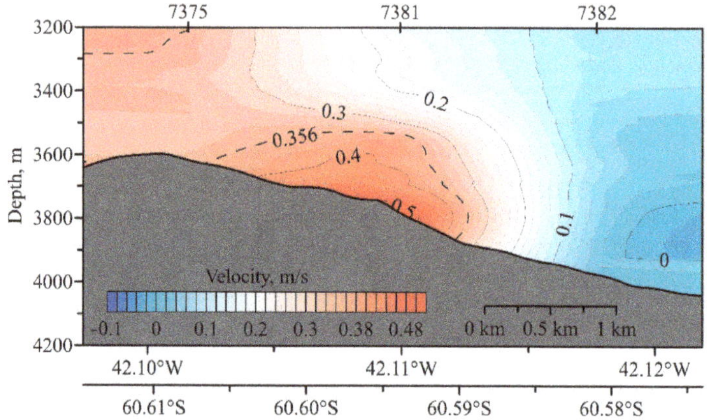

Figure 7. Longitudinal velocity (through flow) section downslope the Orkney Passage. Contour line of 0.356 m/s is additionally shown.

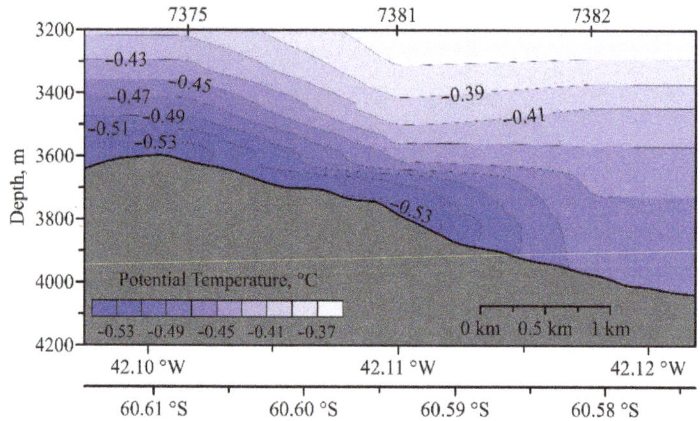

Figure 8. Potential temperature distribution over the downslope section of the Orkney Passage.

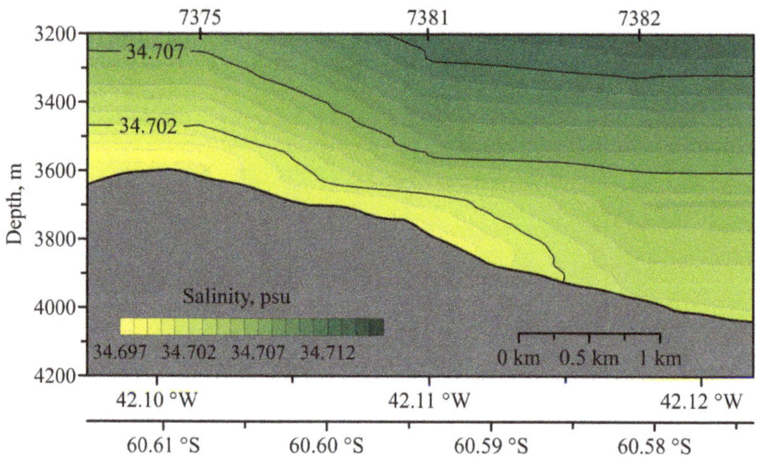

Figure 9. Salinity distribution over the downslope section of the Orkney Passage.

It was reported in [25] that hydraulically controlled flow over the sills is responsible for bottom mixing above the slope. A local overturning circulation cell is formed over the slope with increasing vertical velocity that mixes the bottom flow [25,26]. The flow regime changes as the velocity increases and the Froude number exceeds unity. The critical regime appears when the Froude number is $Fr > 1$. When the current overflows the sill of the Orkney Passage the Froude number is calculated as:

$$Fr = v/\sqrt{g'H}$$

where v is the meridional component of velocity, $\sqrt{g'H}$ is the velocity of gravity waves in the shallow flow, H = 100 m is the thickness of the flow, $g' = g\Delta\rho/\rho$ is reduced acceleration due to gravity, ρ = 1 g/cm^3 is the in situ density, g = 9.8 m s^{-1}, and $\Delta\rho$ = 0.00013 g/cm^3 in the overflowing current. The critical regime of overflowing appears when the bottom current exceeds v = 0.356 m s^{-1}:

$$Fr = \frac{0.356}{\sqrt{\frac{9.8\times100\times0.00013}{1}}} \approx 1.0$$

4. Conclusions

We studied the flow of Antarctic Bottom Water from the Weddell Sea to the Scotia Sea through the Orkney Passage. This passage is considered the main pathway for the outflow of Antarctic Bottom Water from the Weddell Sea; hence, it is the main source of Antarctic Bottom Water in the Atlantic. The transport of bottom water through this passage is 5 Sv based on reports in many publications. We made five CTD/LADCP casts to the bottom over the sill of the passage and three casts along the slope downstream. During the overflow of Antarctic Bottom Water from the Weddell Sea over the ridge in the Orkney Passage, a supercritical regime with a Froude number greater than unity was noted. The flow accelerates when descending the slope. Near the foot of the slope there is not enough kinetic energy of the flow to continue moving in this regime. The flow slows down; strong mixing and warming of the bottom water occur due to the exchange with the surrounding waters. This mode is called hydraulically controlled flow.

Author Contributions: E.C.M., D.I.F., M.G.V.: Conceptualization, original draft preparation, writing; V.A.K., O.A.Z.: original draft preparation; R.Z.M.: data procerssing. All authors have read and agreed to the published version of the manuscript.

Funding: The work was supported by the Russian Science Foundation grant 21-77-20004 (ship measurements). Field data processing was supported by the Russian Foundation for Basic Research grant no 20-08-00246.

Institutional Review Board Statement: Not applicable.

Informed Consent Statement: No studies involving humans were performed. Birds and mammals were observed using binoculars and photo cameras. No experiments with animals have been carried out.

Data Availability Statement: Data are available upon request.

Conflicts of Interest: The authors declare no conflict of interest.

Abbreviations

AABW	Antarctic bottom water
WSDW	Weddell Sea deep water
LADCP	Lowered acoustic Doppler current profiler
CTD	Conductivity, temperature, and depth profiler
Fr	Froude number
v	Meridional component of velocity
$g' = g\Delta\rho/\rho$	Reduced acceleration due to gravity

H	Thickness of the flow
$\rho = 1\ g/cm^3$	In situ density
$g = 9.8\ m\ s^{-1}$	Acceleration due to gravity
$\Delta\rho$	Density difference between layers

References

1. Vernet, M.; Geibert, W.; Hoppema, M.; Brown, P.J.; Haas, C.; Hellmer, H.H.; Jokat, W.; Jullion, L.; Mazloff, M.; Bakker, D.C.E.; et al. The Weddell Gyre, Southern Ocean: Present Knowledge and Future Challenges. *Rev. Geophys.* **2019**, *57*, 623–708. [CrossRef]
2. Abrahamsen, E.P.; Meijers, A.J.S.; Polzin, K.L.; Garabato, A.C.N.; King, B.; Firing, Y.L.; Sallée, J.-B.; Sheen, K.L.; Gordon, A.L.; Huber, B.A.; et al. Stabilization of dense Antarctic water supply to the Atlantic Ocean overturning circulation. *Nat. Clim. Chang.* **2019**, *9*, 742–746. [CrossRef]
3. Meredith, M.P.; Gordon, A.L.; Garabato, A.C.N.; Abrahamsen, E.P.; Huber, B.A.; Jullion, L.; Venables, H.J. Synchronous intensification and warming of Antarctic Bottom Water outflow from the Weddell Gyre. *Geophys. Res. Lett.* **2011**, *38*, L03603. [CrossRef]
4. Sloyan, B.; Rintoul, S. The Southern Ocean Limb of the Global Deep Overturning Circulation*. *J. Phys. Oceanogr.* **2001**, *31*, 143–173. [CrossRef]
5. Orsi, A.H.; Johnson, G.C.; Bullister, J.L. Circulation, mixing and production of Antarctic Bottom Water. *Prog. Oceanogr.* **1999**, *43*, 55–109. [CrossRef]
6. Garabato, A.C.N.; McDonagh, E.L.; Stevens, D.P.; Heywood, K.J.; Sanders, R.J. On the export of Antarctic Bottom Water from the Weddell Sea. *Deep Sea Res. Part II: Top. Stud. Oceanogr.* **2002**, *49*, 4715–4742. [CrossRef]
7. Gordon, A.L.; Visbeck, M.; Huber, B. Export of Weddell Sea deep and bottom water. *J. Geophys. Res. Earth Surf.* **2001**, *106*, 9005–9017. [CrossRef]
8. Meredith, M.P.; Locarnini, R.A.; Van Scoy, K.A.; Watson, A.; Heywood, K.; King, B.A. On the sources of Weddell Gyre Antarctic Bottom Water. *J. Geophys. Res. Earth Surf.* **2000**, *105*, 1093–1104. [CrossRef]
9. Schodlok, M.P.; Hellmer, H.H.; Beckmann, A. On the transport, variability and origin of dense water masses crossing the South Scotia Ridge. *Deep Sea Res. Part II: Top. Stud. Oceanogr.* **2002**, *49*, 4807–4825. [CrossRef]
10. Pratt, L.J.; Lundberg, P.A. Hydraulics of rotating strait and sill flow. *Annu. Rev. Fluid Mech.* **1991**, *23*, 81–106. [CrossRef]
11. Pratt, L.J.; Whitehead, J.A. *Rotating Hydraulics: Nonlinear Topographic Effects in the Ocean and Atmosphere*; Springer: New York, NY, USA, 2007; p. 550. [CrossRef]
12. Whitehead, J.A.; Leetmaa, A.; Knox, R.A. Rotating hydraulics of strait and sill flows. *Geophys. Fluid Dyn.* **1974**, *6*, 101–125. [CrossRef]
13. Fratantoni, D.M.; Zantopp, W.E.; Johns, E.; Miller, J.L. Updated bathymetry of the Anegada-Jungfern passage complex and implications for Atlantic inflow to the abyssal Caribbean Sea. *J. Mar. Res.* **1997**, *55*, 847–860. [CrossRef]
14. Armi, L.; Farmer, D.M. The flow of Mediterranean water through the Strait of Gibraltar. *Prog. Oceanogr.* **1988**, *21*, 1–105.
15. Cenedese, C.; Whitehead, J.A.; Ascarelli, T.A.; Ohiwa, M. A Dense Current Flowing down a Sloping Bottom in a Rotating Fluid. *J. Phys. Oceanogr.* **2004**, *34*, 188–203. [CrossRef]
16. Cenedese, C.; Adduce, C. A New Parameterization for Entrainment in Overflows. *J. Phys. Oceanogr.* **2010**, *40*, 1835–1850. [CrossRef]
17. Morozov, E.G.; Frey, D.I.; Gladyshev, S.V.; Gladyshev, V.S. Hydrodynamics of the Bottom-Water Flow from the Arctic to the Atlantic through the Strait of Denmark. *Izv. Atmospheric Ocean. Phys.* **2020**, *56*, 479–487. [CrossRef]
18. Fer, I.; Voet, G.; Seim, K.S.; Rudels, B.; Latarius, K. Intense mixing of the Faroe Bank Channel overflow. *Geophys. Res. Lett.* **2010**, *37*, L026042. [CrossRef]
19. Morozov, E.G.; Parrilla-Barrera, G.; Velarde, M.G.; Scherbinin, A.D. The Straits of Gibraltar and Kara Gates: A Comparison of internal tides. *Oceanol. Acta* **2003**, *26*, 231–241. [CrossRef]
20. Morozov, E.; Trulsen, K.; Velarde, M.G.; Vlasenko, V.I. Internal Tides in the Strait of Gibraltar. *J. Phys. Oceanogr.* **2002**, *32*, 3193–3206. [CrossRef]
21. Morozov, E.G.; Paka, V.T.; Bakhanov, V.V. Strong internal tides in the Kara Gates Strait. *Geophys. Res. Lett.* **2008**, *35*, L16603. [CrossRef]
22. Vlasenko, V.I. Nonlinear model for the generation of baroclinic tides over extensive inhomogeneities of bottom topography. *Phys. Oceanogr.* **1992**, *3*, 417–424.
23. Visbeck, M. Deep Velocity Profiling Using Lowered Acoustic Doppler Current Profilers: Bottom Track and Inverse Solutions*. *J. Atmospheric Ocean. Technol.* **2002**, *19*, 794–807. [CrossRef]
24. Morozov, E.G.; Tarakanov, R.Y.; Frey, D.I. *Bottom Gravity Currents and Overflows in Deep Channels of the Atlantic. Observations, Analysis, and Modeling*; Springer Nature: Berlin/Heidelberg, Germany, 2021; 483p.
25. Thurnherr, A.M.; Speer, K.G. Boundary Mixing and Topographic Blocking on the Mid-Atlantic Ridge in the South Atlantic*. *J. Phys. Oceanogr.* **2003**, *33*, 848–862. [CrossRef]
26. Thurnherr, A.M.; Clément, L.; Laurent, L.S.; Ferrari, R.; Ijichi, T. Transformation and Upwelling of Bottom Water in Fracture Zone Valleys. *J. Phys. Oceanogr.* **2020**, *50*, 715–726. [CrossRef]

water

MDPI

Article

Multidisciplinary Observations across an Eddy Dipole in the Interaction Zone between Subtropical and Subantarctic Waters in the Southwest Atlantic

Eugene G. Morozov [1,2,*], Dmitry I. Frey [1,2], Victor A. Krechik [1], Aleksandr A. Latushkin [2], Pavel A. Salyuk [3], Anna M. Seliverstova [1], Sergey A. Mosharov [1], Alexei M. Orlov [1,4], Svetlana A. Murzina [1,5], Alexej V. Mishin [1], Pavel V. Chukmasov [4], Arseny A. Kubryakov [2], Maxim V. Budyansky [3], Oleg A. Zuev [1], Olga S. Mekhova [1], Vladimir I. Ponomarev [3], Anna L. Chultsova [1], Anna V. Masevich [2], Nadezhda I. Torgunova [1], Andrey O. Kholmogorov [3], Elena A. Shtraikhert [3], Irina V. Mosharova [1], Nikolay Yu. Neretin [1,6], Glafira D. Kolbasova [1,6], Vitaly L. Syomin [1], Andrey V. Tretiakov [4], Larisa G. Tretiakova [4] and Anton D. Chernetsky [1]

1 Shirshov Institute of Oceanology, Russian Academy of Science, 36 Nakhimovsky pr., 117997 Moscow, Russia
2 Marine Hydrophysical Institute, Russian Academy of Science, 2 Kapitanskaya ul., 299011 Sevastopol, Russia
3 V.I. Ilichev Pacific Oceanological Institute, Far Eastern Branch of the Russian Academy of Science, 43 Baltiiskaya ul., 690041 Vladivostok, Russia
4 Severtsov Institute of Ecology and Evolution, Russian Academy of Sciences, 33 Leninsky Prospekt, 119071 Moscow, Russia
5 Institute of Biology of the Karelian Research Centre of the Russian Academy of Sciences, 11 Pushkinskaya Street, 185910 Petrozavodsk, Russia
6 N.A. Pertsov White Sea Biological Station, Faculty of Biology, Moscow State University, 1 Leninskiye Gory, 119991 Moscow, Russia
* Correspondence: egmorozov@mail.ru

Citation: Morozov, E.G.; Frey, D.I.; Krechik, V.A.; Latushkin, A.A.; Salyuk, P.A.; Seliverstova, A.M.; Mosharov, S.A.; Orlov, A.M.; Murzina, S.A.; Mishin, A.V.; et al. Multidisciplinary Observations across an Eddy Dipole in the Interaction Zone between Subtropical and Subantarctic Waters in the Southwest Atlantic. *Water* **2022**, *14*, 2701. https://doi.org/10.3390/w14172701

Academic Editor: Bruno Charrière

Received: 6 June 2022
Accepted: 22 August 2022
Published: 30 August 2022

Abstract: Seawater properties in two intense rings in the South Atlantic are considered. One ring separated from the Brazil Current and the other from the Malvinas Current. The analysis is based on the CTD casts and SADCP measurements from the onboard velocity profiler. The optical properties, chemical parameters, methane concentration, and biological properties such as primary production, plankton, and fish were also analyzed. Analysis of strong differences between the eddies is supplemented by observations of whales and birds in the region.

Keywords: subtropical rings; Subantarctic rings; CTD; ADCP; optical and hydrochemical parameters; methane; primary production; plankton; fish; birds and whales

1. Introduction

Mesoscale eddies are very important elements of the global ocean circulation. Mesoscale eddies transport mass, momentum, heat, and freshwater over the basins of the ocean. They strongly influence the mean sate of the ocean and together with internal waves transport nutrients to the upper layer of the ocean. The energy of eddies exceeds that of the mean circulation by an order of magnitude, which was first reported during the Atlantic Polygon-70 experiment [1]. Eddies are generally formed due to the instabilities of the mean flow [2]. The most energetic frontal rings are formed due to meandering of the jet currents. Mesoscale rings are generated by finite-amplitude instabilities of currents, which result in the spreading of a volume of fluid into a region with different environmental characteristics [3]. It is clear that temperature and salt anomalies associated with these rings are important sources for the transformation of properties of water masses [3].

Two intense currents in the South Atlantic, the warm Brazil Current and cold Malvinas Current, form a confluence region at approximately 38–40° S [4,5]. An intense oceanic front exists between these two currents; rings of opposite sign are generated at the front with intermediate water masses between rings. Thus, one of the most intense oceanic fronts appears that generates rings of opposite sign and forms intermediate water masses. The exchange of mass, heat, and salt between Subantarctic and subtropical waters leads to intense primary production and transformation of water properties [6]. The Malvinas Current is related to the Subantarctic Front (SAF) and the Brazil Current forms the Brazil Current Front (BCF). Some studies indicate that the SAF is organized in two branches [7,8] that merge further downstream at approximately 44° S [9]. In the confluence region the SAF and BCF converge to a single front, the Brazil–Malvinas Confluence Front (BMCF) [10]. Confluence of the Malvinas and Brazil currents forms a strong front that generates many rings and eddies. After the collision, the Malvinas Current turns back to the south as the Malvinas Return Flow [11]. The retroflection of the Malvinas Current and continuation of the Brazil Current results in meandering and formations of rings, eddies, and filaments [12]. It is noteworthy that eddy advection has an important effect on the variability of the Atlantic meridional overturning circulation (AMOC) on different time scales [13–15]. Some publications report a strong influence of eddies on the cross-shelf transport of fresh river water from the coast of Brazil in the BMCF region, which can cause sharp changes in the salinity in the study region [11,16].

The studies of the BMCF included CTD surveys and moorings [17], satellite data [18], and modeling [19,20]. A new approach to studying the propagation of eddies was suggested in [21], which involved applying observations of the motion of loggerhead sea turtles to trace mesoscale eddies. Unfortunately, there are very few in situ measurements of the vertical structure of eddies in the study region. One of the rare exceptions is the publication by Gordon [11], in which the data are presented on the thermohaline structure of two anticyclones and one cyclone in the BMCF. The authors showed that a significant modification of the characteristics of water masses occurs in eddies. In particular, it was shown that intense winter mixing in the cores of anticyclones leads to the penetration of salty and cold waters into their lower layers.

Lentini et al., [22] analyzed the statistics of warm-core rings formed in the Brazil–Malvinas Confluence region based on a continuous time series of satellite data from January 1993 to December 1998. On average, seven rings per year were released into Subantarctic waters. The lifetime of the rings ranged from 11 to 95 days, with a mean value of 35 days. These rings were mostly elliptical anticyclonic rings with a mean major radius of 126 ± 50 km and a minor radius of 65 ± 22 km, and with translation speeds ranging from 4.2 to 27.2 km per day (4.8–31.5 cm/s). None of them seemed to remain more than 4 months in the confluence region.

Lentini et al., [23] analyzed rings of the Brazil Current. They report that warm-core anticyclonic rings were shed by the meander of the current after poleward excursions of the Brazil Current. Approximately one ring is formed by the current in 2 months. The observed lifetime ranges between 1 and 4 months, with a mean value of approximately 2 months. Two or three anticyclonic rings of the Brazil Current were observed to coexist in the confluence region. Most of the rings were moving to the south. The mean horizontal scale of the eddies was 55 km; and mean translation speed was 10 km per day [23].

Mason et al., [24] report on the basis of altimetry data that motions associated with mesoscale eddies and meanders are responsible for significant transports of mass and heat. Mesoscale vertical fluxes also influence upper ocean biological productivity by transporting nutrients into the euphotic layer. They estimated that the maximum eddy lifetime in the region varies from 78 to 198 days.

Mason et al., [24] distinguish the following eddy properties: amplitudes, radii, intensity, and nonlinearity. Eddy intensity is the ratio between the eddy amplitude and its radius, A/Ls [25]. Eddy nonlinearity is the ratio between eddy swirl speed and translation velocity, U/c [26]. Eddy intensities generally range from zero to 0.5, while eddy nonlinearity varies from zero to 15. Eddy intensity and nonlinearity are greater in cold cyclones. We emphasize that strong currents exist at the boundary between cyclones and anticyclones. The eddy radii of cyclones based on altimetry data in Mason et al., [24] varies from 69 km in the region at 45° S to 75 km in the northern region (~35° S). The radii of anticyclones vary from 90 to 62 km. The amplitudes do not have a clearly pronounced regularity and vary from 38 to 9 cm; the amplitudes of anticyclones are generally smaller. The swirl speed is generally greater in cyclones and varies from 58 to 17 cm/s. Propagation velocity varies from 0.5 to 8.8 cm/s.

Studies of variations in biooptical characteristics in the eddy zone of the Southwest Atlantic have been carried out mainly using satellite ocean color imaging [12,21,27,28]. In these works, the surface distribution of remotely determined chlorophyll-a in anticyclonic and cyclonic eddies was mainly considered, and there are almost no works that analyze a wider number of biooptical characteristics, and also do not consider the features of their vertical distribution. At the same time, with the help of multidisciplinary analysis of biooptical and other oceanographic characteristics, a more detailed study of physical and biological processes in the eddy structures is possible.

Despite numerous above-mentioned papers dedicated to studies of eddies in the Southwest Atlantic, there is still no available data on synchronous measurements of physical, optical, chemical, and biological properties of waters within eddies of different polarity in this region. The main goal of this multidisciplinary study is to trace changes in water properties between the waters of the subtropical and Subantarctic origin, which are observed in anticyclonic and cyclonic eddies, respectively. With this in mind, we selected an eddy dipole based on satellite altimetry data and performed measurements at seven stations along a section across the rings connecting their centers (Figure 1).

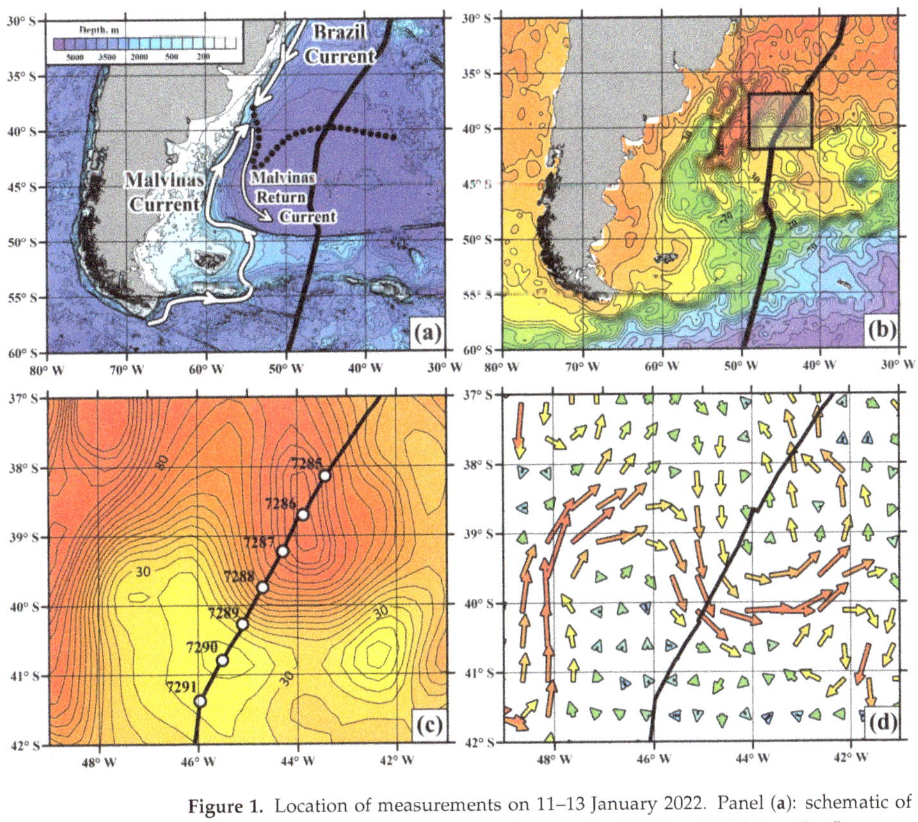

Figure 1. Location of measurements on 11–13 January 2022. Panel (**a**): schematic of the surface circulation in the Southwest Atlantic. The location of the Brazil–Malvinas Confluence zone is shown

by dotted line; the main surface currents are indicated by white arrows. Bottom topography is shown based on the GEBCO database [https://www.gebco.net/data_and_products/gridded_bathymetry_data/gebco_2021] (access date 21 August 2022); the shoreline is shown based on the GSHHS data [29]. Panel (**b**): Absolute dynamic topography (ADT) map at the time of in situ measurements (10 January 2022). Gray rectangle indicates the limits of maps (**c**,**d**). Panel (**c**): detailed ADT map in the region of rings. White dots indicate the location of CTD stations. Panel (**d**): sea surface altimetry-derived geostrophic circulation on 10 January 2022. Solid black line denotes the ship route on all panels.

2. Data and Methods

This study is based on the multidisciplinary data linking physical and biological processes in a pair of cyclonic and anticyclonic rings near the BMC zone in the Southwest Atlantic. In this section, we describe our approach for the selection of rings for our measurements based on satellite altimetry data (Section 2.1), present Lagrangian simulation methods for studies of water origin (Section 2.2), discuss measurements of thermohaline eddy structure together with its velocity field based on CTD and SADCP measurements (Section 2.3), describe the measurements of optical water properties (Section 2.4), and the biological studies of waters within selected eddies (Sections 2.5–2.7).

The measurements over the section across two rings were carried out on 11–13 January 2022; seven stations of CTD profiling were occupied with an AML probe up to 500 m. water sampling with Niskin bottles was performed up to 200 m. According to the altimetry data and the data of the survey, the section crossed two rings: with warm waters from the warm Brazil and cold Malvinas currents. Their centers were at stations 7286 (Brazil Current ring) and 7290 (Malvinas Current ring). Stations 7285, 7287, 7289, and 7291 were peripheral, and station 7288 was in the mixing zone.

2.1. Satellite Altimetry Data

In this work, we used satellite altimetry data for selection of a section between the centers of the cyclonic and anticyclonic eddies and locations of stations along the section. We used the Data Unification and Altimeter Combination System (DUACS) near real time altimeter gridded product of 0.25° [30] available from Copernicus Marine Environment Monitoring Service (CMEMS, http://marine.copernicus.eu/ (accessed on 15 February 2022)). This product includes the data from all available altimeters and consists of the absolute dynamic topography (ADT), sea level anomalies (SLA), surface-geostrophic velocities, and surface-geostrophic velocity anomalies, which are sampled daily on a Mercator regular grid. Thus, we can estimate the location of eddy centers with an accuracy of 0.25°, which corresponds to 28 km in the meridional direction and 21 km in the zonal direction at this latitude (40° S).

2.2. Lagrangian Analysis

Lagrangian analysis has been applied to study the origin of waters in the study area, in which geostrophic current velocities calculated from the AVISO satellite altimetry data ($0.25 \times 0.25°$) were used to calculate particle trajectories. A large number of synthetic particles (the area shown in Figure 2 has been seeded daily with tracers on a grid of 700×700 points) and their trajectories have been computed back in time for a fixed period of time [31]. The back-in-time integration period in our study is two years. One obtains an origin Lagrangian map (O-map) by marking the particles with different colors, which arrived from the northern, western, southern, and eastern boundaries of the study area in the past. The origin Lagrangian maps were computed daily on 11–13 January 2022.

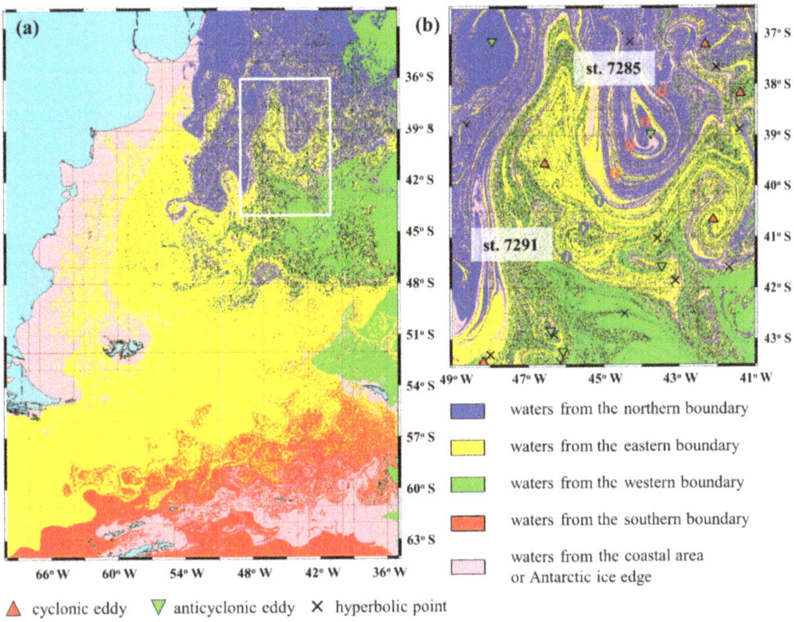

Figure 2. (**a**,**b**) shows the backward-in-time Lagrangian origin map on 12 January 2022. Blue color denotes water particles from the northern edge of the map from the subtropical region, yellow color denotes waters from the western edge of the map from the Drake Passage, red color denotes water from the southern edge of the map (Antarctica), green color denotes water from the eastern border of the map. Pink color denotes water particles that "touched" the shore or Antarctic ice edge and spread from the shore or the edge.

2.3. SADCP, CTD, and Hydrochemistry Data

Direct velocity measurements were carried out from the R/V *Akademik Mstislav Keldysh* (R/V AMK) equipped with an SADCP system Teledyne RD Instruments Ocean Surveyor (TRDI OS) with a frequency of 76.8 kHz. During the survey the profiler was set in the narrowband mode, which increases the profiling range up to 700 m depth. We set 60 vertical bins 16 m each with an 8 m blank distance immediately below the transducer. The draught of the ship is 6 m, which gives 22 m depth for the center of the first bin (the depth of the uppermost layer of velocity measurements). Time averaging of the raw data was made over 120 s intervals. Since the ship speed varied between 8 and 10 knots, this time average represents an along-track averaging of roughly 500 m. Measurement errors in the amplitude of the horizontal velocities were small, approximately 1–2 cm/s [32]. The TPXO9 model [33] was used to subtract the barotropic tidal velocities at the moment of measurements. Typical tidal velocities in the region are less than 2–3 cm/s which is insignificant for the high-velocity jets at the periphery of the rings.

The CTD data were collected at seven stations performed from the sea surface to 500 m depth (Table 1). An AML Oceanographic BaseX CTD probe was used for these measurements. Sampling rate was set to 4 Hz, the CTD vertical speed was 1 m/s, which gives the initial vertical resolution equal to 0.25 m. Only downcast data were used as the CTD profile data. The standard SeaCast software (Version 4.4.0) (software was supplied with the purchased instrument from SBE-electronics Inc. Bellevue, USA) was used for the data collection.

Table 1. Location, time, and type of measurements at seven stations between centers of cyclonic and anticyclonic rings in the Southwest Atlantic performed on 11–13 January 2022.

No. st.	Date/Time (UTC)	Coordinates	CTD Depth, m	Type of Measurements
7285	11 January 2022, 12:22	38°08.3′ S, 43°26.2′ W	500	CTD, Optics, Niskin bottles (0, 15, 70, 110, 200, 500 m)
7286	11 January 2022, 18:42	38°41.7′ S, 43°53.2′ W	500	CTD, Optics, Double square net (DSN), Niskin bottles (0, 15, 40, 80, 130, 150, 200, 500 m)
7287	12 January 2022, 02:16	39°13.3′ S, 44°17.8′ W	500	CTD, Optics, Niskin bottles (0, 20, 60, 83, 130, 200, 500 m)
7288	12 January 2022, 08:40	39°44.7′ S, 44°42.3′ W	500	CTD, Optics, Niskin bottles (0, 30, 45, 73, 110, 200, 500 m)
7289	12 January 2022, 15:23	40°16.4′ S, 45°06.4′ W	500	CTD, Optics, Niskin bottles (0, 18, 40, 70, 110, 200, 500 m)
7290	12 January 2022, 21:55	40°47.6′ S, 45°31.1′ W	500	CTD, Optics, DSN, Niskin bottles (0, 15, 40, 60, 100, 200, 500 m)
7291	13 January 2022, 06:05	41°22.8′ S, 45°58.3′ W	500	CTD, Optics, Pleiston net, Niskin bottles (0, 18, 45, 70, 200, 500 m)

Water samples for chemical analysis and methane were taken from the Niskin bottles to 200 m. The contents of the main forms of nutrients were determined: inorganic forms of silicate, mineral phosphorus, nitrate, nitrite and ammonium nitrogen, as well as dissolved oxygen, total alkalinity (*Ta*) and pH. Further processing of samples for the study of all these parameters was carried out in the onboard laboratory immediately after sampling.

The "HeadSpace" equilibrium concentration method [34] was used to analyze the methane content. The gas chromatograph Crystal Lux 4000 was used. Calculation of dissolved methane concentration in seawater was performed using the method described in [35] and modified in [36] using the calculated methane solubility constants.

2.4. Biooptical Measurements Using a Hydro-Optical Profiler

Biooptical measurements were carried out to the depths of 250 m using a complex of profiling hydrooptical equipment, consisting of Turner C6P and Kondor meters (manufactured by Akvastandart-Yug, Sevastopol, Russia, http://ecodevice.com.ru/ecodevice-catalogue/multiturbidimeter-kondor, accessed on 1 December 2021). Chlorophyll-a (Chl-a) and colored dissolved organic matter (CDOM) fluorescence intensity measurements were performed using a Turner C6P instrument. The Kondor probe determined the attenuation index of directional light at 660 nm (BAC660) and photosynthetically active radiation (PAR). BAC660 in the red band of the spectrum was determined from the absorbing and scattering properties of the total suspended matter (TSM); it depends on the CDOM absorption [37]. The PAR measurements were made only during daylight hours.

The fluorescence intensities CDOM (FCDOM) were calibrated to the quinine sulfate units (QSU) under laboratory conditions just before the expedition. Chl-a fluorescence intensities were recalculated into mass concentrations of Chl-a ($C_{chl\text{-}a}$, μg/L) based on the results of standard measurements by the extract method in water samples [38] taken synchronously with hydrooptical profiling. Calibration of the BAC660 was carried out in the laboratory before the expedition based on the results of measurements in suspension solutions of formazin with a given concentration (FTU units).

2.5. Primary Production

Water samples were taken from five depths within the euphotic layer (0–80 m), where the levels of illumination allow photosynthesis. To assess the primary productivity of phytoplankton, the following parameters were determined: the rate of primary production, potential photosynthetic activity, and the concentration of the main phytopigment chlorophyll "a".

The rate of primary production was determined experimentally using the radiocarbon method [39] by simulating light and temperature conditions using the original laboratory phyto-incubator developed by the authors with adjustable LED illumination. Subsamples were exposed for 3 h, then filtered through Vladipor membrane filters (0.45 μm). The radioactivity of the stock solution and filters was determined using a Triathler liquid scintillation counter (Hidex, Finland). The assimilation number (AN, specific primary production, mg C/mg chl/h) was calculated by normalizing the value of primary production at individual depths based on the corresponding concentration of chlorophyll "a". Chlorophyll "a" concentration in water was measured by the acetone extract fluorescence method [38].

Active fluorescence of chlorophyll "a" was measured using an ultrasensitive PAM fluorometer WATER-PAM (Walz, Germany). To estimate the current state of phytoplankton and its adjustment to light, the "fast light curves" method was used, according to which the effective quantum yield of phytoplankton photosystem II (PPSII) is measured and the relative electron transport rate in photosystem II (rETR) is calculated as a function of illumination [40,41]. The rETR value reflects the rate of conversion of light energy captured by chlorophyll "a" into chemically bound energy in phytoplankton cells, which provides the processes of biosynthesis of organic matter by phytoplankton. Photosynthetic efficiency reflecting the extent of use of the light energy in organic matter synthesis can be expressed through the ratio between AN and rETR values.

2.6. Fish and Invertebrates

Materials on fish and invertebrates were collected using a pelagic double square micronekton net (DSN) with an inlet area of 1 m^2 and a 6 m long filter cone made of gas with a mesh size of 0.5 mm [42], equipped with a water flow counter (Hydrobios, Germany) and a pterygoid deepener weighing 24 kg (Hydrobios, Germany). Oblique catches were carried out in the 600–0 m layer at a vessel speed of 2 knots. The hauling depth was determined based on the pressure sensor of the Senti DT probe (StarOddi, Iceland) readings. Samples were fixed in 2% formaldehyde solution, followed by transfer in 24 h to 96% ethanol. The species identification of ichthyoplankton was based on available data in the literature [43–48].

2.7. Sea Birds and Mammals

On-board ship observations of marine mammals and birds were performed in the study site. The observations were carried out continuously during the daylight hours at winds (Beaufort scale < 5) and visibility more than 50 m by two observers simultaneously from the port and starboard sides of the vessel from the direction-finding deck located at a height of 17 m above the sea level. The birds were detectable approximately 300 m from the ship.

3. Results

The data collected during the survey on 11–13 January 2022 in the Southwest Atlantic revealed the existence of a strong front between the relatively warm and saline subtropical waters within the anticyclonic eddy and cold and fresh Subantarctic waters within the cyclonic eddy. The physical, optical, chemical, and biological properties of these waters were studied along the transect between centers of the eddies. The results of these multidisciplinary measurements are presented separately below for each subject.

3.1. Origin of Waters within the Eddy Dipole

Figure 2 shows the backward-in-time Lagrangian origin map on 12 January 2022, which highlights the considered pair of interacting rings of opposite vorticity (Figure 1b). Similar methods of Lagrangian analysis were used in [49,50]. It can be seen that waters of various origins are involved in the rings, namely, subtropical (blue), Subantarctic (yellow), as well as waters from the South American shelf (pink). The anticyclonic ring (in the figure, the centers of anticyclones are marked with green triangles) contains mainly waters of subtropical origin, and the cyclonic eddy (the centers of cyclones are marked with red triangles) contains Subantarctic waters.

3.2. SADCP and CTD Measurements and Chemistry

Thermohaline structure of the eddy dipole was studied based on CTD measurements at seven stations along a section connecting centers of the cyclonic and anticyclonic eddies (Figure 3; the locations of stations are shown in Figure 1 and Table 1). The second (station number 7286) and sixth (station number 7290) stations were made exactly at the centers of the eddies; the rest stations were made at a constant distance of 70 km (38 nm) from each other. All CTD measurements were performed from the sea surface to 500 m depth.

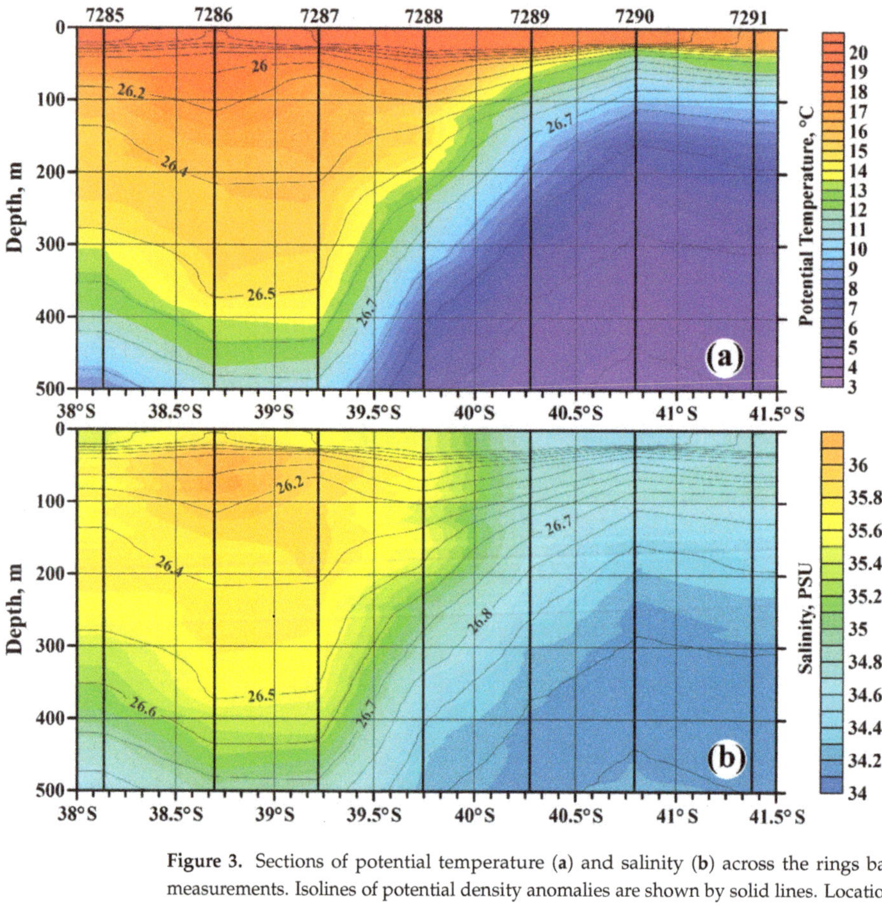

Figure 3. Sections of potential temperature (**a**) and salinity (**b**) across the rings based on CTD measurements. Isolines of potential density anomalies are shown by solid lines. Locations of stations are shown by thick black lines.

Distributions of potential temperature and salinity (Figure 3) show significant difference in properties of subtropical and Subantarctic waters. Thus, the potential temperature at 100 m depth changes from 17.6 °C (station 7286, subtropical waters, center of anticyclonic eddy) to 9.6 °C (station 7290, Subantarctic waters, center of cyclonic eddy) while the salinity changes from 36.02 PSU to 34.67 PSU. Corresponding differences are observed at all measured depths; for example, the changes at a depth of 500 m are from 11.1° to

4.3 °C and from 34.97 PSU to 34.23 PSU. The depth of 12 °C potential temperature isotherm changes from 468 m in subtropical waters to 37 m in Subantarctic waters. Note that the thermohaline structure of the upper 30 m ocean layer is not affected by the influence of subtropical and Subantarctic waters. This layer is warm and relatively fresh in comparison with deep layers of the ocean. The stratification is defined by temperature distribution; the pycnocline is observed at 25–35 m depending on a station. The thickness of the mixed ocean layer is almost the same within the cyclonic and anticyclonic eddies.

The θ,S-diagrams were analyzed from the data of seven stations along the section through the eddy dipole (Figure 4). Intense transformation of water masses takes place in the eddies. Saline surface tropical waters (TW) accumulate in anticyclones. They are differently grouped in the upper and lower parts of the water column. As a result, waters of almost uniform salinity (S = 35.5–36.0 PSU) with a temperature of 15–20 °C appear. Anticyclones contribute to the transfer of salt to the lower layers of water, as a result of which salty and relatively cold waters form in their lower part. Such conditions are favorable for the formation of salt fingers [11], which transport saline waters to the deep part and the salinization of South Atlantic Central Water (SACW).

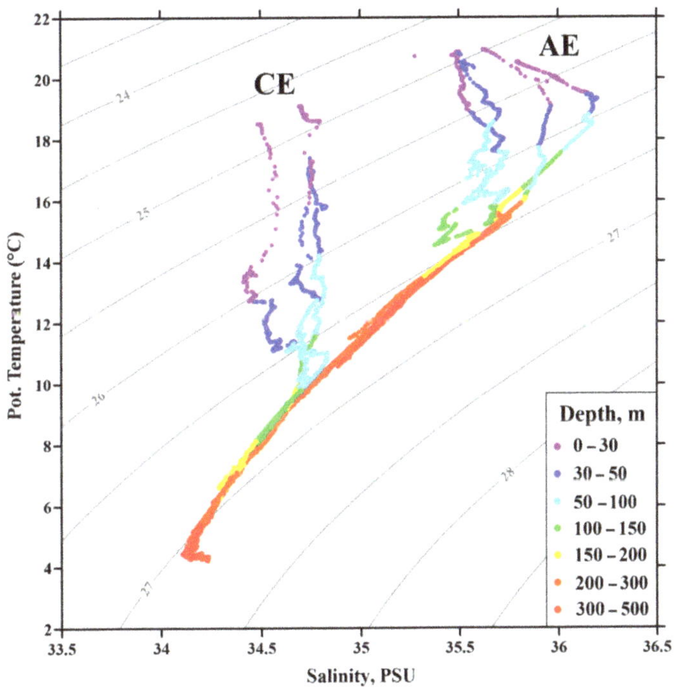

Figure 4. θ,S-diagram based on data from seven stations through an eddy dipole. The depth of measurements is shown with colors. CE is cyclonic eddy; AE is anticyclonic eddy.

Fresher SACW waters in cyclones ascend to the surface. They become warmer and mix with the tropical waters under the influence of the wind. As a result, relatively fresh waters with salinity (S = 34.5–34.7 PSU) are formed; their temperature varies from 10 °C to 18 °C, which is similar to the data reported in [11] for the Cleopatra eddy. In the upper part of the cyclones, due to the warming of fresh waters, waters of very low density appear and stratification increases, strongly blocking the vertical exchange (Figure 4).

The CTD stations were accompanied by along-track SADCP measurements (Figure 5). Measured velocities were projected to a direction of 120° (East-south-east), which corresponds to the direction of maximum velocity at the periphery of the eddies. This periphery is organized in a form of two main high-velocity jets. They are located at a distance of 73 km from each other; maximum velocity of 84 cm/s in the first jet (40.2° S) is located at the sea surface while the second jet (39.4° S) with maximum velocity of 71 cm/s is located at a depth of 165 m. Velocities near the centers of the eddies are much lower and do not exceed 20 cm/s.

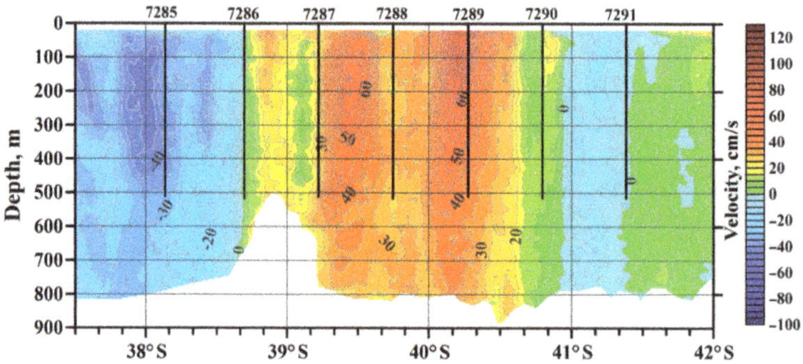

Figure 5. Velocity section across the rings based on SADCP measurements. Locations of stations are shown by thick black lines.

Three types of water can be clearly distinguished over the section: waters of two rings with different hydrochemical parameters and pronounced centers and a zone of their mixing. The warm ring of the Brazil Current is characterized by low concentrations of oxygen and nutrients as well as an increased total alkalinity. The cold ring of the Malvinas Current is characterized by high oxygen content and nutrients such as dissolved silicates, phosphorus and nitrates (Figures 6–9).

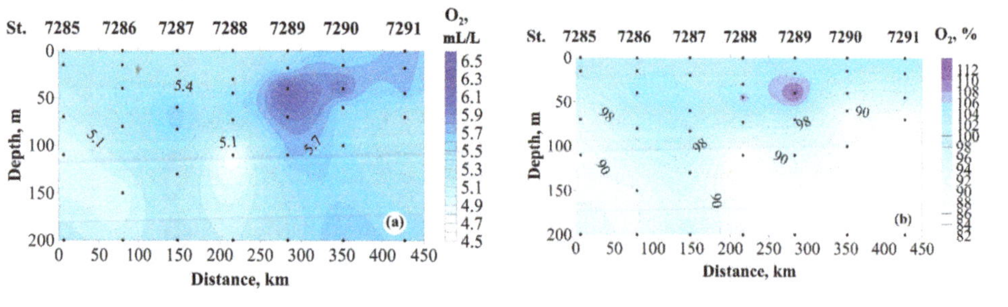

Figure 6. Concentration of dissolved oxygen (O_2, mL/L) (**a**) and degree of oxygen saturation (**b**) over the section.

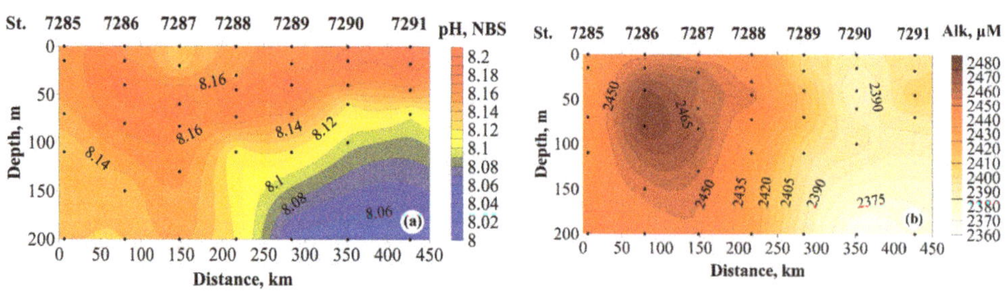

Figure 7. Distribution of pH (pH, NBS units) (**a**) and total alkalinity (Alk, µM) (**b**) over the section.

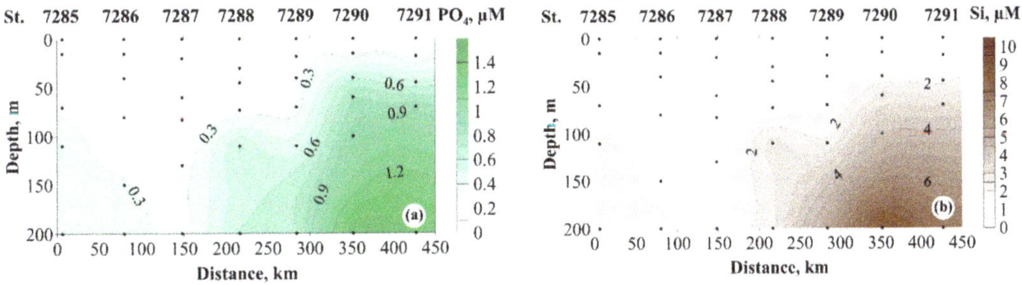

Figure 8. Distribution of dissolved phosphorus (PO_4, µM) (**a**) and silicates (Si, µM) (**b**) over the section.

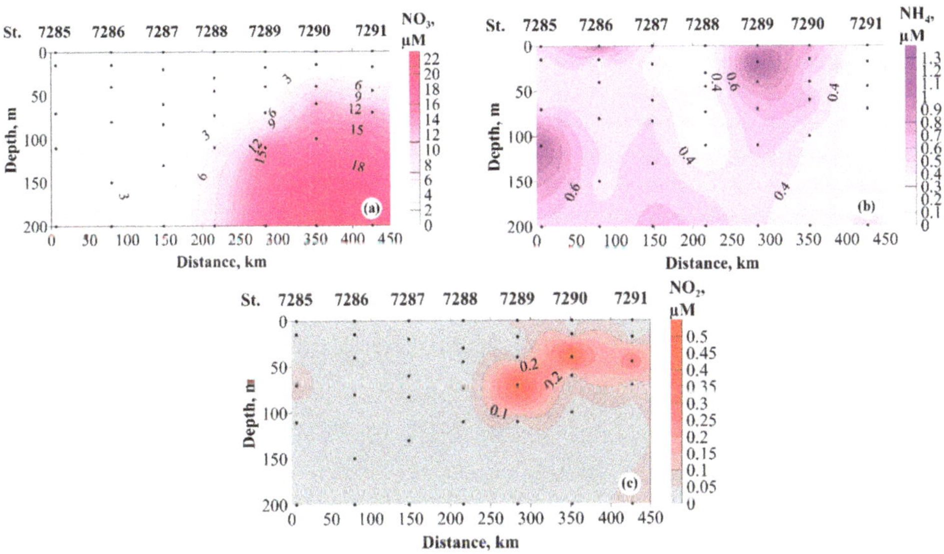

Figure 9. Distribution of nitrates (NO_3, µM) (**a**), nitrites (NO_2, µM) (**b**), and ammonium nitrogen (NH_4, µM) (**c**) over the section.

3.3. Methane Concentration in the Rings

We detected significantly different methane concentrations in the anticyclonic and cyclonic rings, as well as between them. In the anticyclonic ring, the average concentrations of methane are lower than in the cyclonic ring.

Medium (4.5–6 nM/L) and high (6–12.9 nM/L) methane concentrations were observed in the layer 2–200 m, and low (2.8–4.5 nM/L) between 200 and 500 m (Figure 10). Lower methane concentrations were observed in the central parts of the cyclonic (up to 3.8 nM/L at station 7290) and anticyclonic (up to 2.8 nM/L at station 7286) rings at a depth of 200 m and 500 m, respectively. The maximum concentration of methane (12.9 nM/L) was observed at the central station 7288 of the section between the rings at a depth of 110 m.

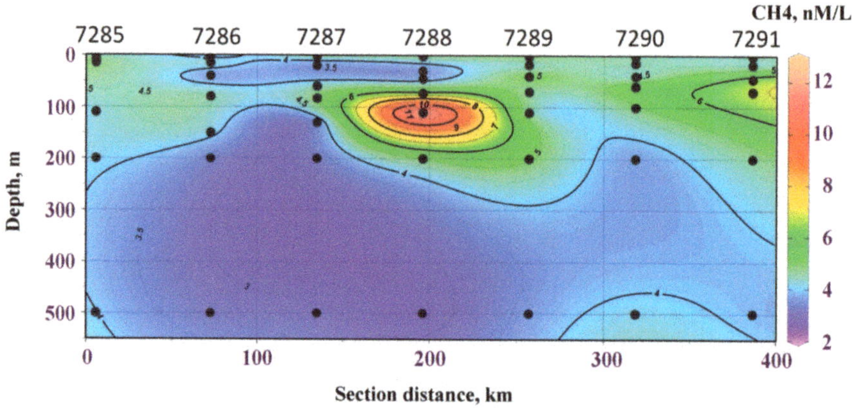

Figure 10. Distribution of methane concentration over the section.

3.4. Biooptics

Studies over a section in the rings made it possible to reveal some features of the vertical distribution of biooptical parameters (Figure 11). In the anticyclonic eddy, the depth of the maximum concentration of chlorophyll-a (deep chlorophyll maximum, DCM) was 70–80 m, which is deeper compared to the cyclonic eddy, where DCM varied in the range of 45–60 m (Figure 11a). At stations located closer to the centers of the rings, the deepest DCMs were found (station 7286, ~80 m) in the anticyclonic eddy, and the least deep ones were in the cyclone (station 7290, ~45 m). This was caused by velocities directed downward in the central part of the anticyclonic eddy and upward in the central part of the cyclonic eddy. In both cases, the vertical velocities increase towards the center of the rings.

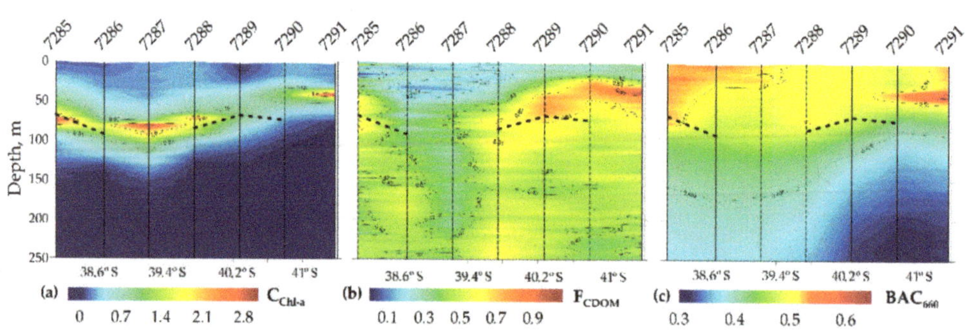

Figure 11. Section of chlorophyll "a" concentration (C_{chl-a}, μg/L) (**a**), fluorescence intensity of colored dissolved organic matter (FCDOM, QSU) (**b**) and directional light attenuation index (BAC660, FTU) (**c**); Dashed black lines on the sections mark the depth of the photic zone (Zeu).

One can see on the C_{chl-a} section (Figure 11a) that in the anticyclonic eddy the width of the peak on the vertical profile of C_{chl-a} is greater than in the cyclonic eddy (50–70 m and 30–50 m, respectively). This is due to large vertical density gradients in the seasonal pycnocline in the central part of the cyclonic ring compared to the anticyclonic ring. High values of vertical density gradients are a barrier to the vertical motion of phytoplankton cells, nutrients, and other suspended and dissolved substances [51–53], which leads to an increase in the concentration of substances in the seasonal pycnocline. This is the reason for the appearance of pronounced layers of increased FCDOM and BAC660 values, and a decrease in the width of the C_{chl-a} peak in the pycnocline of the central part of the cyclonic ring, which can be seen in the vertical sections of these parameters (Figure 11).

Despite the previously shown differences in the vertical distributions of C_{chl-a}, the maximum values of C_{chl-a} in the anticyclonic and cyclonic rings were approximately the same and reached 3 μg/L. However, the maximum values of other biooptical characteristics (FCDOM and BAC660) were different.

The vertical distribution of FCDOM (Figure 11b) differs significantly in the two rings. In the upper mixed layer (UML) of the anticyclonic ring, the mean FCDOM values are lower than those in the underlying layers. At the same time, in the center of the anticyclonic ring, there is a zone of water deepening up to 220 m with reduced FCDOM values (St. 7286, 7287). In the cyclonic ring, within the UML, the minimum values of FCDOM were also observed, and the maximum values, similar to the C_{chl-a} distributions, were found in the pycnocline. In the layer below the pycnocline up to the maximum sounding depth, FCDOM almost does not change with depth (0.4–0.5 QSU). In the cyclonic ring, the average FCDOM values are higher in each of the layers compared to the values in the anticyclonic ring.

The vertical distribution of BAC660 (Figure 11c) revealed common patterns with the distribution of C_{chl-a} but some differences were found. In the anticyclonic ring, waters with elevated BAC660 values, such as Cchl-a, deepen in the central part of the ring. However, in the vertical distribution of BAC660 there is no pronounced deepened local maximum, as in the case of Cchl-a. In the anticyclonic ring, the BAC660 values decreased monotonously with depth from 0.55 FTU at the surface to 0.4 FTU at a depth of 250 m. In the cyclonic ring, the vertical profile of BAC_{660} is similar to that of C_{chl-a}. Namely, BAC660 slightly changes in the UML; it has a maximum in the seasonal pycnocline of about 0.7 FTU, decreases below the pycnocline to 0.4 FTU at depths of 90–110 m and further to a minimum value of 0.3 FTU at a depth of 250 m. The absolute maximum the BAC660 value was observed in the pycnocline of the central part of the cyclonic ring, similar to the FCDOM distribution.

Based on underwater PAR measurements, the depth of the photic zone (Zeu) was determined as 1% of the PAR incident on the sea surface. The Zeu estimates in the section ranged from 69 to 93 m and are marked as dotted black lines in Figure 11a,c. At night stations 7287 and 7291, Zeu data are not available. In the presented data, Zeu was determined from the total effect on light attenuation of TSM (Figure 11c) and CDOM (Figure 11b).

These features of the vertical distribution of biooptical characteristics indicate that anticyclonic eddies manifest themselves on the satellite images on the color of the ocean as areas with low chlorophyll-a concentrations, and cyclonic eddies appear as areas with increased chlorophyll-a concentrations [21,27]. This is due not only to the chlorophyll-a concentration, but also to the DCM value, and the content of additional TSM or CDOM [54].

3.5. Primary Production

The primary productivity of phytoplankton (i.e., photosynthetic ability) is determined by three main factors: the amount of the main photopigment chlorophyll "a", the amount of available light energy, and the amount of nutrients (the material basis of photosynthesis). Analysis of the distribution of these factors over the section revealed certain differences.

The total content of chlorophyll "a" (extract method) in the euphotic layer showed a uniform distribution over water masses (45.4–83.9 mg/m^2) (Figure 12a) with a maximum content at the northern periphery of the anticyclonic ring (station 7285). The integral primary production in the cold cyclonic ring was more than twice as high as in the subtropical one (on average 155.8 and 72.0 mgC/m^2 per day, respectively) (Figures 12 and 13). Distributions of the total content of chlorophyll "a" and the integral primary production in Figure 12 show that the maximum of integral primary production and high chlorophyll "a" content occur in the frontal zone between the rings, i.e., between stations 7288 and 7289. The area of increased primary productivity correlates with the high nutrient content.

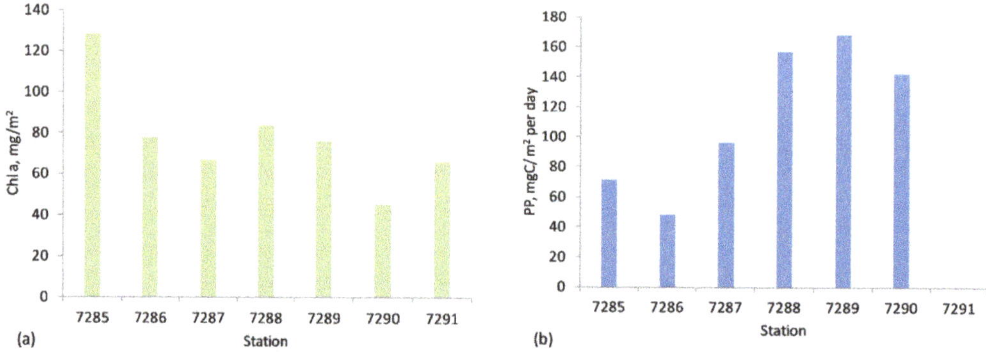

Figure 12. Distribution of the total content of chlorophyll "a" (determined by the extract method) (**a**) and the integral primary production (**b**) in the euphotic layer in the section.

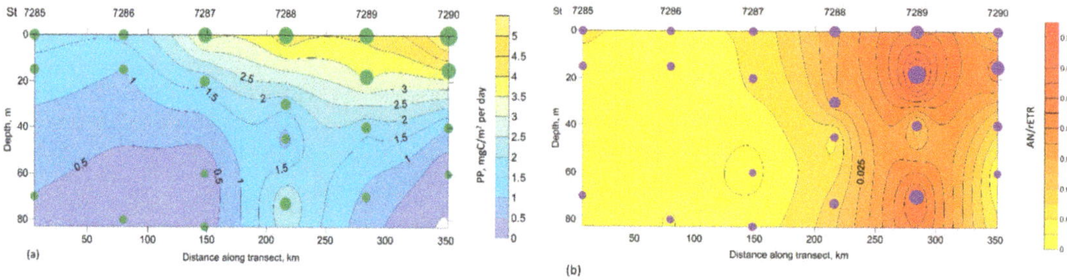

Figure 13. Spatial distribution of primary production (**a**) and photosynthetic efficiency (**b**) of phytoplankton in the euphotic layer. Circles indicate the location of sampling points, and their size is proportional to the values of the corresponding parameter.

The observed distribution of high values of primary production in this water mass is caused by high density stratification in the center of the eddy in the upper zone of the euphotic layer, which retains phytoplankton. In addition, the upward vertical velocity in the central part of the cyclonic (Subantarctic) eddy apparently promotes upwelling of nutrient-rich deep waters into the photosynthesis zone. In the interaction zone, where high velocities were measured, mixing occurs in the euphotic layer, and phytoplankton cells periodically appear in the zones of different illumination. Hence, active phytoplankton occupies the entire thickness of the euphotic layer.

3.6. Invertebrates

The catches of invertebrates at both stations contained elements of both Subantarctic *Primnoa micropa* (Amphipoda) and tropical *Vibia* spp., *Phronima* spp. (Amphipoda) fauna, as well as widespread species *Travisiopsis lanceolata* (Polychaeta). However, elements of tropical fauna prevailed at St. 7286; amphipods Platyscelidae gen. sp., pelagic polychaetes *Lopadorrhynchus appendiculatus*, *L.* cf. *nationalis*, *Rhynchonereella gracilis* and *Torrea candida* were encountered only at this station. St. 7290 was characterized by the predominance of the Subantarctic *Themisto gaudichaudi compressa* (Amphipoda) and the widespread *Nematoscelis megalops* (Euphausiidae) and *Parandania boecki* (Amphipoda). Polychaetes were represented by the large Antarctic species *Vanadis antarctica*, taxa with unclear distribution *Tomopteris* spp. and larvae of the families Amphinomodae, Polynoidae, Terebellidae.

It is hard to overestimate the significant ecological role of cold- and warm-core rings in large scale transport and small-scale transition, and finally biogeographical distribution of certain aquatic organisms supporting regional biodiversity and ecosystem functioning. One of the specific zooplankton species from the order Euphausiacea—*Nematoscelis megalops* is considered as a "sign" organism of transition regions between the tropical-subtropical and the subpolar waters in the Atlantic, Pacific, and Indian Oceans [55,56]. There are fragmental observations on *N. megalops* in the Malvinas and the Brazil Currents [57]. Thus, there is lack of information to discuss the comprehensive zoogeography in the SW Atlantic region. In our study, it was found that *N. megalops* was highly abundant in Subantarctic waters of cold-core water ring of the Malvinas Current. In contrast, this species was negligible in subtropical waters of warm-core water ring of the Brazil Current. Our record of *N. megalops* testifies to the selection of preferable water column and water properties by this species to maintain trophic and reproduction demands. It is known that *N. megalops* is able to occupy a depth from 800 m up to the surface layers but generally occurs at the depths from 300 to 600 m at favored temperatures ranging 8–10 °C [58,59]. Indeed, *N. megalops* was collected by us in a steam of cold-core water ring at depths from 640 m to the surface, within the preferable temperature range.

We noted that the occurrence and abundance of *N. megalops* were associated with the presence of top predators such as baleen whales (Mysticeti) and, in particular, the sei whale *Balaenoptera borealis*. According to our observations, eight sei whales were found in the cold-core water ring of the Malvinas Current. It is known that sei whales feed on zooplankton. Euphausiids are one of the major food items [60,61].

Moreover, our further studies will be focused on ecological and biochemical characteristics based on lipid and energetic status investigations of *N. megalops* in the studied area to determine general and specific patterns of euphausiids associated with environmental variables.

3.7. Ichthyoplankton

Species composition of ichthyoplankton samples at both stations considerably differed (Table 2). At St. 7286, warm water species distributed mostly in tropical and subtropical waters were caught, while the catch at St. 7290 was represented mainly by species occurring as a rule in cold waters of the notal zone, Subantarctic, and Antarctic. At the same time, *Cyclothone* spp. were found in both rings but more frequently within the warm waters of the Brazil Current.

Table 2. Species composition of ichthyoplankton at different stations.

Fish Taxa	Range Type (Source)	Number of Specimens(Fry/Larvae)	
		St. 7286	St. 7290
	Phosichthyidae		
Phosichthys argenteus	Subtropical-Notal [62]	19/1	0
	Sternoptychidae		
Argyropelecus hemigymnus	Temperate-subtropical [63]	3/17	0
Sternoptyx diaphana	Tropical-subtropical [64]	0/1	0
	Stomiidae		
Chauliodus schmidti	Tropical [65]	3/0	0
	Gonostomatidae		
Cyclothone braueri	Temperate-subtropical [63]	52/10	6/0
C. pseudopallida	Temperate-subtropical [63]	3/0	4/0
Cyclothone spp.	Not applicable	18/0	0

Table 2. *Cont.*

Fish Taxa	Range Type (Source)	Number of Specimens(Fry/Larvae)	
		St. 7286	St. 7290
	Myctophidae		
Diogenichthys atlanticus	Temperate-subtropical [63]	0/11	0
Krefftichthys anderssoni	Notal-Antarctic [43]	0	0/4
Protomyctophum normani	Notal [43]	0	0/14
P. bolini	Notal-Antarctic [43]	0	04
Hygophum bruni	Subantarctic [43]	0	0/2
H. hanseni	Notal [43]	0	0/2
Diaphus anderseni	Tropical-subtropical [43]	0	0/2
Diaphus sp.	Not applicable	0	0/1
Lampanyctus sp.	Not applicable	2/0	0
Fish larvae (inident.)	Not applicable	18	0

3.8. Sea Mammals and Birds

Two species of sea mammals and eight of the most numerous species of birds were recorded in the rings (Table 3).

Table 3. Species composition (number of individuals observed) of sea mammals and birds in the rings.

Species	Subtropical Water	Subantarctic Water	Frontal Zone
Sea Mammals			
Sei whale *Balaenoptera borealis*	0	8	0
Southern bottlenose whale *Hyperoodon planifrons*	3	0	0
Sea Birds			
Spectacled petrel *Procellaria conspicillata*	81	29	25
Soft-plumaged petrel *Pterodroma mollis*	78	17	13
Atlantic petrel *Pterodroma incerta*	16	7	5
Cory's shearwater *Calonectris diomedea*	5	9	7
White-chinned petrel *Procellaria aequinoctialis*	2	54	23
Great shearwater *Ardenna gravis*	1	24	12
Atlantic yellow-nosed albatross. *Thalassarche chlororhynchos*	7	5	5
Wandering albatross. *Diomedea exulans*	2	10	7
Albatross sp. *Diomedeidae* sp.	2	8	4

Marine mammals occurred in both rings. In the ring of subtropical water, three southern bottlenose whales (*Hyperoodon planifrons*) were recorded. At the same time, toothed whales were not found in Subantarctic waters, while eight sei whales were (*Balaenoptera borealis*) sighted in the ring. It is known that these two species have completely different trophic ecologies. The southern bottlenose whale has a circumpolar distribution in the southern hemisphere (south of 30° S). This species is most common in the region of 57–70° S probably migrating to Antarctica in summer. The southern bottlenose is characterized by feeding at great depths exceeding 1000 m, mainly on squid, but also on fish, such as Patagonian toothfish *Dissostichus eleginoides*. The sei whale is distributed from the tropical to the circumpolar zone in both hemispheres, mainly found in the open ocean far from the coast. The main food of the sei whale is zooplankton and small fish [60].

The majority of birds encountered in the areas of the two rings were represented by two families: albatrosses (Diomedeidae), and petrels (Procellariidae). It was found that the petrel family was the most numerous and diverse in terms of species composition (Table 3). However, a change in the occurrence of some species was noted during the transition from the anticyclonic ring to the cyclonic ring (Figure 14). For example, the number of spectacled petrel (*Procellaria conspicillata*), a mass bird species in the subtropical area, was 81 individuals, while their number decreased by almost three times in Subantarctic water (29 individuals). A decrease in the number of occurrences during the transition from subtropical to Subantarctic waters was also characteristic of the soft-fin typhoon (Pterodroma mollis) and Atlantic petrel (*P. incerta*): from 78 individuals to 17 individuals, and from 16 individuals to 7 individuals, respectively.

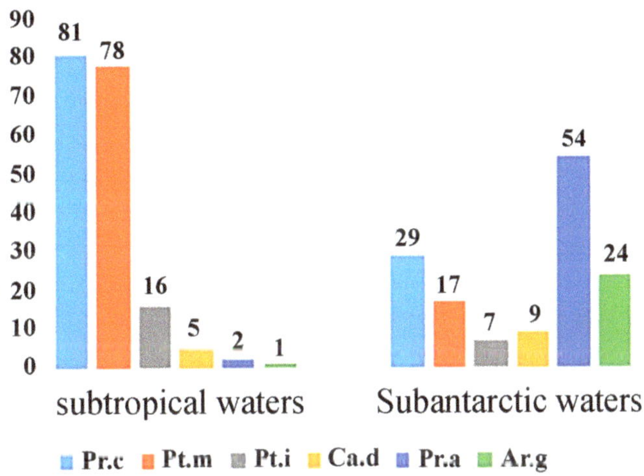

Figure 14. The Procellariidae family in the region of rings. Note: Pr.c is the spectacled petrel *Procellaria conspicillata*, Pt.m is the soft-plumaged petrel *Pterodroma mollis*, Pt.i is the Atlantic petrel *Pterodroma incerta*, Ca.d is Cory's shearwater *Calonectris diomedea*, Pr.a is the white-chinned petrel *Procellaria aequinoctialis*, Ar.g is the great shearwater *Ardenna gravis*.

On the contrary, some species of this family, when moving from one ring to another, began to occur much more frequently. For example, the white-chinned petrel (*Procellaria aequinoctialis*) was the most abundant bird species in the ring of Subantarctic water, 54 individuals, while this species was recorded only two times in the ring of subtropical water. An increase in occurrence in the ring of Subantarctic waters was noted for the gray petrel (*Ardenna gravis*) and Cory's shearwater (*Calonectris Diomedea*): from 1 to 24 individuals and from 5 to 9 individuals, respectively.

A change in the occurrence of some species of the albatross family during the transition from subtropical to Subantarctic waters was observed. This family was more numerous in the area of the Subantarctic ring. The wandering albatross (*Diomedea exulans*) was found five times more often in the Subantarctic ring than in subtropical waters.

Marine life at oceanic fronts is an interesting oceanographic object. Frontal zones are known as regions with favorable conditions for feeding of different species including pelagic fish, mammals and birds. A review of marine life at oceanic fronts is given in [66].

A frontal zone was found in the interval between stations 7287–7291. Marine mammals were not observed here. Most of the birds encountered in Subantarctic waters were found in this narrow frontal region (Table 3). Starting from the frontal zone, the number of some species of birds began to increase compared to in the subtropical water. The number of birds found in the frontal zone increased towards the cold ring. This increase was recorded for some petrels and the great shearwater (*Ardenna gravis*) (Table 3).

Based on the results of observations, we conclude that marine mammals and some species of birds were more common in the cyclonic ring; however, other species of birds were more common in the anticyclonic one. Some species of birds were common in the frontal zone and their numbers increased towards the cold ring. At the same time, the species composition and abundance of both marine mammals and birds in these two areas differed, indicating that these animals are confined to the conditions of their habitat.

4. Summary

We analyzed the properties of sea water in two intense rings of the South Atlantic in the region of the Brazil–Malvinas Confluence. One ring separated from the Brazil Current and the other from the Malvinas Current. Analysis is based on the CTD casts, ADCP measurements from onboard velocity profiler. The properties of subtropical and Subantarctic seawater were strongly different and a strong current was recorded at the interface between rings. Backward-in-time Lagrangian analysis revealed that waters of various origin from various surrounding regions were entrained in the rings. Analysis of different properties was supplemented by investigation of optical properties and chemical parameters such as dissolved oxygen, nitrates, silicates, pH and methane concentration. We also studied different biological properties in the rings such as primary production, plankton, fish, and invertebrates. The catches revealed different concentrations and species of plankton and other invertebrates. Analysis of the strong differences between the rings is supplemented by observations of whales and birds in the region.

Author Contributions: E.G.M., D.I.F.: Conceptualization, original draft preparation, writing, and editing; V.A.K., O.A.Z., O.S.M.: physical oceanography part; A.A.L., P.A.S., E.A.S.: optical part; A.A.K., M.V.B., V.I.P.: Lagrangian model; A.M.S., A.L.C., A.V.M. (Anna V. Masevich), N.I.T.: chemical part; A.M.O., S.A.M. (Svetlana A. Murzina), A.V.M. (Alexej V. Mishinand), N.Y.N., G.D.K., V.L.S.: biological parts; S.A.M. (Sergey A. Mosharovand), I.V.M.: primary production; A.O.K.: methane; P.V.C., A.V.T., L.G.T., A.D.C.; mammals and birds. All authors have read and agreed to the published version of the manuscript.

Funding: Ship measurements were supported by the targeted financial support from the Ministry of Science and Higher Education of the Russian Federation: State Tasks FMWE-2022-0001 (SIO), FNNN-20222-0001 (MHI), 0211-2019-0007 (POI), FFER-2019-0021 (IPEE), and Russian Science Foundation (ship operation, grant no. 21-77-2004).

Informed Consent Statement: No studies involving humans were performed. Birds and mammals were observed using binoculars and photo cameras. No experiments with animals have been carried out.

Data Availability Statement: Data are available upon request.

Conflicts of Interest: The authors declare no conflict of interest.

References

1. Kamenkovich, V.M.; Koshlyakov, M.N.; Monin, A.S. *Synoptic Eddies in the Ocean*; Springer: Berlin/Heidelberg, Germany, 1986; p. 444.
2. Stammer, D.; Wunsch, C. Temporal changes in eddy energy of the oceans. *Deep-Sea Res. I* **1999**, *46*, 77–108. [CrossRef]
3. Olson, D.B. Rings in the ocean. *Annu. Rev. Earth Planet. Sci.* **1991**, *19*, 283–311. [CrossRef]
4. Gordon, A.L.; Greengrove, C.L. Geostrophic circulation of the Brazil-Falkland confluence. *Deep-Sea Res. I* **1986**, *33*, 573–585. [CrossRef]
5. Orúe-Echevarría, D.; Pelegrí, J.L.; Alonso-González, I.J.; Benítez-Barrios, V.M.; Emelianov, M.; García-Olivares, A.; Rubinat, M.G.; De La Fuente, P.; Herrero, C.; Isern-Fontanet, J.; et al. A view of the Brazil-Malvinas confluence, March 2015. *Deep–Sea Res. I* **2021**, *172*, 103533. [CrossRef]

6. Valla, D.; Piola, A.R. Evidence of upwelling events at the northern Patagonian shelf break. *J. Geophys. Res. Oceans* **2015**, *120*, 7635–7656. [CrossRef]
7. Frey, D.I.; Piola, A.R.; Krechik, V.A.; Fofanov, D.V.; Morozov, E.G.; Silvestrova, K.P.; Tarakanov, R.Y.; Gladyshev, S.V. Direct measurements of the Malvinas Current velocity structure. *J. Geophys. Res. Oceans* **2021**, *126*, e2020JC016727. [CrossRef]
8. Piola, A.R.; Franco, B.C.; Palma, E.D.; Saraceno, M. Multiple jets in the Malvinas Current. *J. Geophys. Res. Oceans* **2013**, *118*, 2107–2117. [CrossRef]
9. Artana, C.; Provost, C.; Poli, L.; Ferrari, R.; Lellouche, J.-M. Revisiting the Malvinas Current upper circulation and water masses using a high-resolution ocean reanalysis. *J. Geophysical Res. Oceans* **2021**, *126*, e2021JC017271. [CrossRef]
10. Barré, N.; Provost, C.; Saraceno, M. Spatial and temporal scales of the Brazil–Malvinas Current confluence documented by simultaneous MODIS Aqua 1.1-km resolution SST and color images. *Adv. Space Res.* **2006**, *37*, 770–786. [CrossRef]
11. Gordon, A.L. Brazil-Malvinas Confluence–1984. *Deep Sea Res. Part. A* **1989**, *36*, 359–384. [CrossRef]
12. Garcia, C.A.; Sarma, Y.; Mata, M.M.; Garcia, V.M. Chlorophyll variability and eddies in the Brazil–Malvinas Confluence region. *Deep Sea Res. II* **2004**, *51*, 159–172. [CrossRef]
13. Jullion, L.; Heywood, K.J.; Naveira Garabato, A.C.; Stevens, D.P. Circulation and water mass modification in the Brazil–Malvinas Confluence. *J. Phys. Oceanogr.* **2010**, *40*, 845–864. [CrossRef]
14. Manta, G.; Speich, S.; Karstensen, J.; Hummels, R.; Kersalé, M.; Laxenaire, R.; Meinen, C.S. The South Atlantic Meridional Overturning Circulation and Mesoscale Eddies in the First GO-SHIP Section at 34.5° S. *J. Geophys. Res. Oceans* **2021**, *126*, e2020JC016962. [CrossRef]
15. Valla, D.; Piola, A.R.; Meinen, C.S.; Campos, E. Strong mixing and recirculation in the northwestern Argentine Basin. *J. Geophys. Res. Oceans* **2018**, *123*, 4624–4648. [CrossRef]
16. Manta, G.; Speich, S.; Barreiro, M.; Trinchin, R.; de Mello, C.; Laxenaire, R.; Piola, A.R. Shelf water export at the Brazil-Malvinas Confluence evidenced from combined in-situ and satellite observations. *Front. Mar. Sci.* **2022**, *9*, 857594. [CrossRef]
17. Meinen, C.S.; Speich, S.; Perez, R.C.; Dong, S.; Piola, A.R.; Garzoli, S.L.; Baringer, M.O.; Gladyshev, S.; Campos, E.J.D. Temporal variability of the meridional overturning circulation at 34. 5oS: *Results from two pilot boundary arrays in the South Atlantic. J. Geophys. Res. Oceans* **2013**, *118*, 6461–6478. [CrossRef]
18. Lumpkin, R.; Garzoli, S. Interannual to decadal changes in the western South Atlantic's surface circulation. *J. Geophys. Res. Oceans* **2011**, *116*, C01014. [CrossRef]
19. Artana, C.; Provost, C.; Lellouche, J.-M.; Rio, M.-H.; Ferrari, R.; Sennéchael, N. The Malvinas Current at the confluence with the Brazil Current: Inferences from 25 years of Mercator Ocean reanalysis. *J. Geophys. Res. Oceans* **2019**, *124*, 7178–7200. [CrossRef]
20. Matano, R.P.; Combes, V.; Piola, A.R.; Guerrero, R.; Palma, E.D.; Strub, P.T.; James, C.; Fenco, H.; Chao, Y.; Saraceno, M. The salinity signature of the cross-shelf exchanges in the Southwestern Atlantic Ocean: Numerical simulations. *J. Geophys. Res. Oceans* **2014**, *119*, 7949–7968. [CrossRef]
21. Gaube, P.; Barceló, C.; McGillicuddy, D.J., Jr.; Domingo, A.; Miller, P.; Giffoni, B.; Marcovaldi, N.; Swimmer, Y. The use of mesoscale eddies by juvenile loggerhead sea turtles (Caretta caretta) in the southwestern Atlantic. *PLoS ONE* **2017**, *12*, e0172839. [CrossRef]
22. Lentini, C.A.D.; Olson, D.B.; Podestá, G.P. Statistics of Brazil Current rings observed from AVHRR: 1993 to 1998. *Geophys. Res. Lett.* **2002**, *29*, 1811. [CrossRef]
23. Lentini, C.A.D.; Goni, G.J.; Olson, D.B. Investigation of Brazil Current rings in the confluence region. *J. Geophys. Res. Oceans* **2006**, *111*, C06013. [CrossRef]
24. Mason, E.; Pascual, A.; Gaube, P.; Ruiz, S.; Pelegrí, J.L.; Delepoulle, A. Subregional characterization of mesoscale eddies across the Brazil-Malvinas Confluence. *J. Geophys. Res. Oceans* **2017**, *122*, 3329–3357. [CrossRef]
25. Frenger, I.; Münnich, M.; Gruber, N.; Knutti, R. Southern Ocean eddy phenomenology. *J. Geophys. Res. Oceans* **2015**, *120*, 7413–7449. [CrossRef]
26. Chelton, D.B.; Schlax, M.G.; Samelson, R.M.; de Szoeke, R.A. Global observations of large oceanic eddies. *Geophys. Res. Lett.* **2007**, *34*, L15606. [CrossRef]
27. Saraceno, M.; Provost, C. On eddy polarity distribution in the southwestern Atlantic. *Deep Sea Res. I* **2012**, *69*, 62–69. [CrossRef]
28. Willson, H.R.; Rees, N.W. Classification of mesoscale features in the Brazil-Falkland Current confluence zone. *Prog. Oceanogr.* **2000**, *45*, 415–426. [CrossRef]
29. Wessel, P.; Smith, W.H.F. A global, self-consistent, hierarchical, high-resolution shoreline database. *J. Geophys. Res.* **1996**, *101*, 8741–8743. [CrossRef]
30. Pujol, M.-I.; Faugère, Y.; Taburet, G.; Dupuy, S.; Pelloquin, C.; Ablain, M.; Picot, N. DUACS DT2014: The new multi-mission altimeter data set reprocessed over 20 years. *Ocean. Sci.* **2016**, *12*, 1067–1090. [CrossRef]
31. Prants, S.V. Backward-in-time methods to simulate large-scale transport and mixing in the ocean. *Phys. Scr.* **2015**, *90*, 074054. [CrossRef]
32. Chereskin, T.K.; Harris, C.L. *Shipboard Acoustic Doppler Current Profiling during the WOCE Indian Ocean. Expedition*; Scripps Institution of Oceanography, University of California: San Diego, CA, USA, 1997; Volume 10, p. 137.
33. Egbert, G.D.; Erofeeva, S.Y. Efficient inverse modeling of barotropic ocean tides. *J. Atmosph. Oceanic Tech.* **2002**, *19*, 183–204. [CrossRef]
34. Vereshchagina, O.F.; Korovitskaya, E.V.; Mishukova, G.I. Methane in water columns and sediments of the north western Sea of Japan. *Deep-Sea Res. II* **2013**, *86*, 25–33. [CrossRef]

35. Yamamoto, S.; Alcauskas, J.B.; Crozier, T.E. Solubility of methane in distilled water and sea water. *J. Chem. Eng. Data* **1976**, *21*, 78–80. [CrossRef]
36. Wiesenburg, D.A.; Guinasso, N.L. Equilibrium solubility of methane, carbon monooxide, and hydrogen in water and sea water. *J. Chem. Eng. Data* **1979**, *24*, 356–360. [CrossRef]
37. Jerlov, N.G. *Marine Optics*, 2nd ed.; Elsevier Oceanography Series: Amsterdam, The Netherlands, 1976; Volume 14, p. 371.
38. JGOFS (Joint Global Ocean Flux Study Protocols). *Protocols for the Joint Global Ocean Flux Study Protocols (JGOFS) Core Measurements*; UNESCO: Paris, France, 1994; pp. 119–122.
39. Steemann-Nielsen, E. The use of radio-active carbon (C14) for measuring organic production in the sea. *J. Cons. /Cons. Perm. Int. L'exploration Mer.* **1952**, *18*, 117–140. [CrossRef]
40. Genty, B.; Briantais, J.-M.; Baker, N.R. The relationship between the quantum yield of photosynthetic electron transport and quenching of chlorophyll fluorescence. *Biochim. Biophys. Acta* **1989**, *990*, 87–92. [CrossRef]
41. Schreiber, U. Pulse amplitude (PAM) fluorometry and saturation pulse method: An overview. In *Chlorophyll a Fluorescence: A Signature of Photosynthesis*; Advances in Photosynthesis and Respiration Series; Papageorgiou, G., Govindjee, G., Eds.; Kluwer Academic Publishers: Dordrecht, Germany, 2004; pp. 270–319.
42. Bouchard, C.; Mollard, S.; Suzuki, K.; Robert, D.; Fortier, L. Contrasting the early life histories of sympatric Arctic gadids Boreogadus saida and Arctogadus glacialis in the Canadian Beaufort Sea. *Pol. Biol.* **2016**, *39*, 1005–1022. [CrossRef]
43. Becker, V.E. *Miktofovye Ryby Mirovogo Okeana (Myctophid Fish of the World Ocean)*; Nauka: Moscow, Russia, 1983; p. 246.
44. Moser, H.G.; Ahlstrom, E.H. The role of larval stages in systematic investigations of marine teleosts: The Myctophidae, a case study. In *The Early Life History of Fish*; Springer: Berlin/Heidelberg, Germany, 1974; pp. 605–607.
45. Olivar, M.P.; Fortuño Alós, J.M. Guide to ichthyoplankton of the Southeast Atlantic (Benguela Current region). *Sci. Mar.* **1991**, *55*, 1–383.
46. Pertseva-Ostroumova, T.A. Some morphological characteristics of myctophid larvae (Myctophidae, Pisces). Fishes of the Pacific and Indian Oceans, biology and distribution. *Tr. Inst. Okeanol. Akad. Nauk SSSR* **1964**, *73*, 79–97.
47. Richards, W.J. *Early Stages of Atlantic Fishes: An Identification Guide for the Western Central North Atlantic*; CRC Press: Boca Raton, FL, USA, 2006; p. 1824.
48. Sutton, T.T.; Hulley, P.A.; Wienerroither, R.; Zaera-Perez, D.; Paxton, J.R. *Identification Guide to the Mesopelagic Fishes of the Central and South East Atlantic Ocean*; FAO: Rome, Italy, 2020; p. 346. [CrossRef]
49. Ponomarev, V.I.; Fayman, P.A.; Prants, S.V.; Budyansky, M.V.; Uleysky, M.Y. Simulation of mesoscale circulation in the Tatar Strait of the Japan Sea. *Ocean Model.* **2018**, *126*, 43–55. [CrossRef]
50. Prants, S.V.; Budyansky, M.V.; Uleysky, M.Y. Lagrangian simulation and tracking of the mesoscale eddies contaminated by Fukushima-derived radionuclides. *Ocean Sci.* **2017**, *13*, 453–463. [CrossRef]
51. Davis, K.; Anderson, M.A.; Yates, M.V. Distribution of indicator bacteria in Canyon Lake, California. *Water Res.* **2005**, *39*, 1277–1288. [CrossRef] [PubMed]
52. Elci, S. Effects of thermal stratification and mixing on reservoir water quality. *Limnology* **2008**, *9*, 135–142. [CrossRef]
53. Latushkin, A.A.; Artamonov, Y.V.; Lee, R.I.; Sysoev, A.A.; Sysoeva, I.V.; Fedirko, A.V.; Martynov, O.V. The Northern Part of the Black Sea Waters' Bio-optical and Hydrology Structure Features in the Autumn Period 2016. *Fundam. I Prikl. Gidrofiz.* **2020**, *13*, 78–82. [CrossRef]
54. Lipinskaia, N.A.; Salyuk, P.A. Analysis of Hydro-Biological and Hydro-Optical Characteristics in Submesoscale Eddies in the Gulf of Peter the Great with the Use of Simultaneous in-situ and Remote Measurements. *Fundam. Appl. Hydrophys.* **2021**, *14*, 111–121. (In Russian) [CrossRef]
55. Barange, M.; Gibbons, M.J.; Carola, M. Diet and feeding of Euphausia hanseni and Nematoscelis megalops (Euphausiacea) in the northern Benguela Current: Ecological significance of vertical space partitioning. *Mar. Ecol. Prog. Ser.* **1991**, *73*, 173–181. [CrossRef]
56. Wiebe, P.H.; Boyd, S.H. Limits of Nematoscelis megalops in the Northwestern Atlantic in relation to Gulf Stream cold core rings. I. Horizontal and vertical distributions. *J. Mar. Res.* **1978**, *36*, 119–142.
57. Curtulo, L.M.; Dadon, J.R.; Mazzoni, H.E. Distribution and abundance of euphausiids off Argentina in spring 1978. *Neritica* **1990**, *5*, 1–14.
58. Werner, T.; Huenerlage, K.; Verheye, H.; Buchholz, F. Thermal constraints on the respiration and excretion rates of krill, Eupausia hanseni and Nematoscelis megalops, in the northern Benguela upwelling system off Namibia. *Afr. J. Mar. Sci.* **2012**, *34*, 391–399. [CrossRef]
59. Yoon, W.D.; Nival, P.; Choe, S.A.; Picheral, M.; Gorsky, G. Vertical distribution and nutritional behavior of Cyclothone braueri, Nematoscelis megalops, Meganyctiphanes norvegica and Salpa fusiformis in the NW Mediterranean mesopelagic zone. *ICES CM F* **2007**, *3*, 1–28.
60. Jefferson, T.A.; Webber, M.A.; Pitman, R.L. Marine Mammals of the World. In *A Comprehensive Guide to Their Identification*; Elsevier: Amsterdam, The Netherlands, 2015; 608p.
61. Kawamura, A. A review of food of Balaenopterid whales. *Sci. Rep. Whales Res. Inst.* **1980**, *32*, 155–197.
62. Balushkin, A.V. Findings of the silver lightfish Phosichthys argenteus (Phosichthyidae) at the Pulkovo Ridge (South-Central Pacific). *J. Ichthyol.* **2017**, *57*, 821–826. [CrossRef]

63. Porteiro, F.M.; Sutton, T.T.; Byrkjedal, I.; Orlov, A.M.; Heino, M.; Menezes, G.; Bergstad, O.A. Fishes of the Northern Mid-Atlantic Ridge collected during the MAR-ECO cruise in June–July 2004: An annotated checklist. Arquipel. *Life Mar. Sci.* **2017**, *10*, 126.
64. Weitzman, S.H. Family No. 75: Sternoptychidae. In *Smiths' Sea Fishes*; Smith, M.M., Heemstra, P.C., Eds.; J.L.B. Smith Institute of Ichthyology: Grahamstown, South Africa, 1986; pp. 253–259.
65. Ege, V. Chauliodus Schn., bathypelagic genus of fishes. A systematic, phylogenetic and geographical study. *Dana Rep.* **1948**, *31*, 1–146.
66. Prants, S.V. Marine life at Lagrangian fronts. *Prog. Oceanogr.* **2022**, *204*, 102790. [CrossRef]

water

MDPI

Article

Recent Changes in Composition and Distribution Patterns of Summer Mesozooplankton off the Western Antarctic Peninsula

Valentina V. Kasyan

Laboratory of Systematics and Morphology, A.V. Zhirmunsky National Scientific Center of Marine Biology, Far Eastern Branch, Russian Academy of Sciences (NSCMB FEB RAS), 690041 Vladivostok, Russia; valentina-k@yandex.ru

Abstract: The Southern Ocean has undergone significant climate-related changes in recent decades. As a result, pelagic communities inhabiting these waters, particularly mesozooplankton, have adapted to new conditions. The present study considers the patterns of horizontal and vertical (up to 1000 m) distribution, the composition, abundance, and biomass of mesozooplankton, and the relationships of these parameters to the extreme environmental conditions off the western Antarctic Peninsula throughout the record-warm austral summer season of 2022. Sampling was conducted using the opening/closing Multinet system (0.25 m^2 aperture) equipped with five 150-μm mesh nets and a WP-2 net. The mesozooplankton was represented by the three most abundant groups: eggs and larvae of euphausiids such as *Euphausia superba*, small copepods such as *Oithona similis*, and large calanoid copepods such as *Calanoides acutus*, *Calanus propinquus*, *Metridia gerlachei*, and *Rhincalanus gigas*. The composition and quantitative distribution of the mesozooplankton significantly varied: the copepods were abundant in the west, off the Antarctic Peninsula, while eggs and larvae of euphausiids were abundant in the east, off the South Orkney Islands. Most mesozooplankton occurred in the upper 200 m layer, and each taxon showed characteristic depth preference: small copepods, euphausiids larvae, and *cirripeds* cypris larvae were abundant in the epipelagic layer, while large calanoid copepods, euphausiids eggs, amphipods, pelagic polychaetes, and ostracods were found mostly in the mesopelagic layer. The composition and quantitative distribution of mesozooplankton had clear relationships with environmental factors, particularly with a combination of variables such as water salinity, temperature, and chlorophyll *a* concentration.

Keywords: mesozooplankton; copepods; euphausiid larvae; distribution; abundance; biomass; Bransfield Strait; Antarctic Sound; Weddell Sea; South Orkney Islands; Southern Ocean

Citation: Kasyan, V.V. Recent Changes in Composition and Distribution Patterns of Summer Mesozooplankton off the Western Antarctic Peninsula. *Water* **2023**, *15*, 1948. https://doi.org/10.3390/w15101948

Academic Editor: Epaminondas D. Christou

Received: 19 April 2023
Revised: 13 May 2023
Accepted: 19 May 2023
Published: 21 May 2023

1. Introduction

The condition of Antarctic marine ecosystems has recently attracted major attention from scientists, the public, nature conservation activists, and authorities alarmed by the dramatic ecosystem shifts that took place in the region under the effect of global climate change associated with warming and increasing anthropogenic pressure [1–4]. In the West Antarctic sector of the Southern Ocean, climate change has caused the following consequences over the past few decades: an increase in sea surface temperature (SST) of up to 0.2 °C per decade, with projections indicating further extensive increase at a rate of 0.27–1.08 °C per decade that may occur by the late 21st century [5–7], a reduction in sea-ice cover near the Antarctic Peninsula [8,9], and significant interannual variations in chlorophyll *a* (Chl *a*) concentrations [10]. Some significant changes in the plankton communities were reported previously, including replacement of dominant phytoplankton species by others [11–13], variations in the dominance and abundance of taxonomic groups of zooplankton, in particular copepods [14–20], variations in the abundance, biomass, and spatial distribution of the pelagic tunicate *Salpa thompsoni* [21,22], krill *Euphausia superba* [23–26], and krill larvae [18,27–29]. Such changes might be either part of a natural ecosystem mechanism [30,31] or a negative trend that caused communities to adapt to new habitat conditions [15,32,33].

Recent studies in the Southern Ocean have focused on the most important zooplankton taxa such as krill, salps, and copepods that constitute the largest portion of the total abundance and biomass [15,17,21,34]. These taxa form the major food supply in the food web between primary producers and predators and are used as a commercially valuable source of marine-derived animal protein [14,35–37]. To date, the distribution patterns, life cycles, and interannual variability of the most common zooplanktonic species in the Antarctic marine ecosystems have been sufficiently well studied [33,38]. Nevertheless, Antarctica requires continuous scientific research in order to identify emerging trends and provide predictions. Annual surveys of zooplanktonic communities will help timely identify the responses of Antarctic marine ecosystems to the ongoing climate changes.

Previously, data on the mesozooplankton together with the other plankton categories (phytoplankton, macroplankton, and ichthyoplankton) were obtained as preliminary results, e.g., [20,29]; in the present study, we provide more details and carried out additional analyses.

We hypothesize that the mesozooplankton communities in the Southern Ocean have changed habitat conditions in these ecosystems. The main objectives of this study were as follows: update existing information on the summer mesozooplankton composition, abundance, biomass, and distribution patterns during the record-warm SSTs [9]; study the vertical distribution patterns of the dominant taxa to a 1000 m depth; and identify the environmental factors having no/having effects on the composition and distribution of mesozooplankton in the study region.

2. Materials and Methods

Plankton samples were collected in the research expedition on board the R/V Akademik Mstislav Keldysh during a period from 19 January to 10 February (austral summer), 2022. The study region included the Bransfield Strait, the Antarctic Sound, the Powell Basin of the Weddell Sea, and off the South Orkney Islands (Figure 1). Sampling was performed using an opening/closing Multinet system (150-μm mesh, 0.25 m^2 aperture) equipped with five nets [39]. Vertical tows of the nets from 1000 (800) m to 500, 500–200, 200–100, 100–50, and 50–0 m were made at night. In addition, at near-shore stations (depths < 200 m) sampling was carried out with a WP-2 net (150 μm mesh, 0.25 m^2 aperture) [40]. A total of 117 plankton samples were collected and processed. The samples delivered to the laboratory were viewed in a Bogorov's chamber under a SZX7 binocular microscope (Olympus, Tokyo, Japan), and then fixed in a 4% formaldehyde solution. The mesozooplankton taxa were identified using the relevant identification keys [41,42]. Copepods and euphausiids larvae (such as the krill *E. superba*) at different development stages, as the dominant groups of mesozooplankton, were subjected to more detailed taxonomic analysis using the Marine Planktonic Copepods database [43] and special identification keys [44,45], respectively. Copepods were identified into species; morphologically indistinguishable small copepodites from some genera were grouped together (such as *Oithona* spp. and *Oncaea* spp.). The taxonomic groups (Appendicularia, Ctenophora, Echinodermata, Hydromedusae, Polychaeta, Pteropoda, Siphonophorae, eggs, and larvae of fish) that were represented at sampling stations by less than 5% of the total mesozooplankton abundance were categorized as "Other". Abundance (expressed in terms of ind. m^{-3}) and biomass (expressed in terms of mg wet weight (WW) m^{-3}) of mesozooplankton were calculated by the standard hydrobiological methods according to [46,47]. The environmental parameters such as water temperature (T, °C), salinity (S, psu), dissolved oxygen (O_2, mg/L), and chlorophyll *a* (Chl *a*, μg/L) concentrations at the stations were measured using a CTD multiparameter probe (Idronaut, Italy) with a calibrated fluorometer (Seapoint Sensors Inc., Exeter, NH, USA) [20]. The data of water temperature, salinity, oxygen, and Chl *a* concentrations between the 500 and 1000 m layers were not significantly different and, therefore, we here provide values of the environmental variables for the 500 m layer.

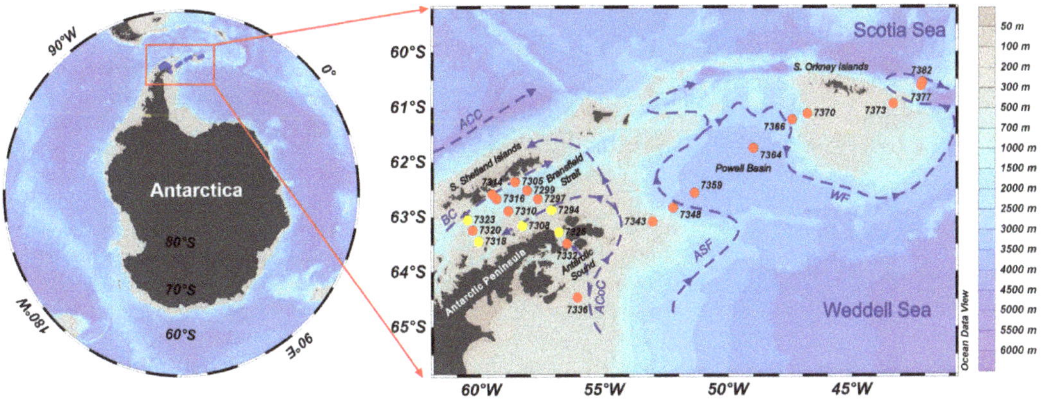

Figure 1. Map of the sampling stations and the main ocean currents designated according to [20,48–50]. Numerals are codes of the stations. Circles indicate plankton nets used at the stations: red is the Multinet, and yellow is WP-2. Dashed lines indicate currents in the study region: ACC is the Antarctic Circumpolar Current; BC, the Bransfield Current; ACoC, the Antarctic Coastal Current; ASF, the Antarctic Slope Front; WF, the Weddell Front.

Statistical Analysis

We hypothesized that the environmental variables such as water temperature, salinity, oxygen, and chlorophyll *a* concentrations at the surface (upper 50 m depth) and in the 50–200 and 200–500 m layers may be key factors that determine distribution patterns of mesozooplankton communities in the study region. To assess the similarity between the mesozooplankton communities in the study region, cluster analysis and non-metric multidimensional scaling (nMDS) were used [51]. The Bray–Curtis similarity was calculated as a log10 (x + 1) transformation of the species/taxa abundance [52]. Cluster analysis was applied through average group sorting. To assess the reliability of clustering, the SIMPROF (Similarity Profile Routine) permutation test (number of repeats 999, $p = 0.05$) was performed [51]. The SIMPER (Similarity Percentage) procedure was used to assess the average percent contribution of each mesozooplankton species to the overall dissimilarities, the contribution of each species to the average intra-group similarity (SIM) with the standard deviation (SD), and the contribution of each species to the average between-group dissimilarity (DISS/SD) [53].

The relationships between the species/taxa abundances and the environmental variables at different layers of the water column were assessed using Spearman's rank order correlation [53]. The BEST analysis (BIO-ENV algorithm) was used to test relationships between the environmental variables and the biological data (abundance) which best explain the observed patterns of mesozooplankton distribution. Environmental variables were log-transformed, normalized (divided by SD), and calculated for Euclidean distances [53]. To determine the significance of relationships between the similarity matrices (biological data and environmental variables), the RELATE test was used [53]. Additionally, the Canonical Correspondence analysis (CCA) was carried out to reveal the environmental variables that determine the distribution of mesozooplankton communities [53]. Multivariate analyses were performed in the PRIMER ver. 6 [54] and PAST ver. 4 [55] software. The map of the sampling stations was composed using Ocean Data View ver. 5 [56].

3. Results

3.1. Dynamics of Environmental Conditions

The detailed dynamics of some environmental variables and the pattern of main currents in the study region were described earlier elsewhere [20,26,29,57–60]. During the summer of 2022, the average values of the environmental variables at the sampling stations were as follows: water temperature, −0.02 ± 0.70 °C; water salinity, 34.56 ± 0.15 psu; oxygen concentration, 8.84 ± 1.46 mg/L; and chlorophyll *a* concentration, 0.30 ± 0.62 μg/L. The highest values of temperature were recorded from off the South Orkney Islands and from the Bransfield Strait off the South Shetland Islands; the lowest values, from the Antarctic Sound and off the Antarctic Peninsula, where one extreme temperature minimum (Stn. 7336) was recorded (Figure 2a). The highest salinity values were recorded from the Bransfield Strait off the South Shetland Islands and the Powell Basin; the extreme lowest values, were off the Antarctic Peninsula (Figure 2b). The average values of the water temperature and salinity variables increased from off the Antarctic Peninsula to off the South Orkney Islands, except for the decreasing sea surface salinity and temperature in the 200 m layer (Table 1). The highest values of oxygen concentrations (>10 mg/L) were recorded at most stations; they decreased by half at the greatest depth (Figure 2c). The highest chlorophyll *a* concentrations were observed in the Antarctic Sound, Powell Basin, and off the South Orkney Islands, with extreme values at Stn. 7336 off the Antarctic Peninsula (Figure 2d). The average values of oxygen and chlorophyll *a* concentrations decreased from off the Antarctic Peninsula to off the South Orkney Islands, except for the increasing surface Chl *a* concentration (Table 1).

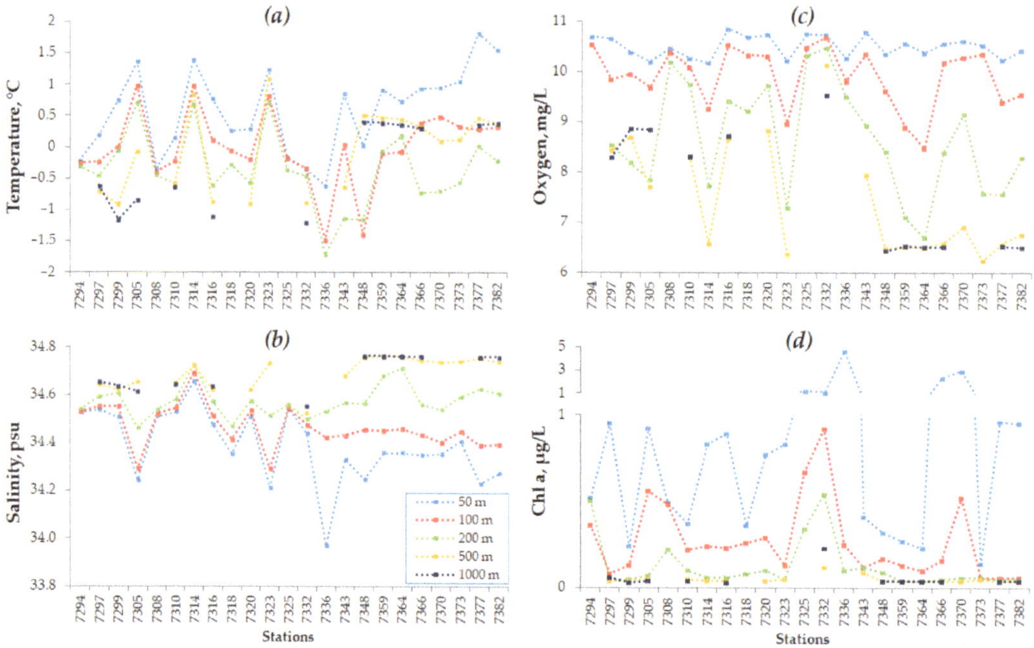

Figure 2. Values of water temperature (**a**), salinity (**b**), oxygen concentration (**c**), and chlorophyll *a* concentration (**d**) in the 50-, 100-, 200-, 500-, and 1000 m layers during the austral summer of 2022.

Table 1. The environmental variables (mean ± SD) in the Bransfield Strait (BS), Antarctic Sound (AS), Weddell Sea (WS), and off the South Orkney Islands (SOI) during the austral summer of 2022.

Variables	Study Region			
	BS	AS	WS	SOI
50 m S	34.46 ± 0.14	34.49 ± 0.07	34.27 ± 0.15	34.32 ± 0.08
200 m S	34.56 ± 0.07	34.53 ± 0.04	34.60 ± 0.07	34.59 ± 0.04

Table 1. *Cont.*

Variables	Study Region			
	BS	AS	WS	SOI
500 m S	34.62 ± 0.08	34.54 ± 0.03	34.70 ± 0.09	34.74 ± 0.01
50 m T	0.53 ± 0.61	−0.25 ± 0.14	0.47 ± 0.63	1.35 ± 0.41
200 m T	−0.11 ± 0.55	−0.41 ± 0.07	−0.77 ± 0.72	−0.37 ± 0.33
500 m T	−0.28 ± 0.68	−0.63 ± 0.37	−0.10 ± 0.91	0.27 ± 0.19
50 m O_2	10.49 ± 0.25	10.76 ± 0.01	10.50 ± 0.19	10.46 ± 0.16
200 m O_2	8.75 ± 0.95	10.41 ± 0.12	8.17 ± 1.08	8.15 ± 0.75
500 m O_2	8.12 ± 0.90	9.74 ± 0.59	7.09 ± 0.9	6.63 ± 0.29
50 m Chl *a*	0.65 ± 0.26	1.17 ± 0.09	1.38 ± 1.82	1.15 ± 1.1
200 m Chl *a*	0.12 ± 0.14	0.44 ± 0.14	0.07 ± 0.03	0.06 ± 0.01
500 m Chl *a*	0.07 ± 0.08	0.11 ± 0.01	0.06 ± 0.03	0.04 ± 0.01

Notes: Abbreviations of environmental variables are as follows: salinity (S), temperature (T), chlorophyll *a* concentration (Chl *a*), and oxygen (O_2) concentration in the surface (to 50 m), 200-, and 500 m layers.

3.2. Composition and Distribution Patterns of Mesozooplankton Abundance and Biomass

The total mesozooplankton abundance varied between 47.3 and 2797.8 ind. m^{-3} with a mean value of 970.3 ± 376.8 ind. m^{-3}. The maximum abundances were recorded from off the South Orkney Islands (Figure 3a). At the stations of this area, euphausiids larvae numerically dominated (Figure 3b). At the rest of the stations, mesozooplankton abundances ranged from 300 to 1200 ind. m^{-3} with a trend to increase from west to east. Euphausiids larvae, being the most abundant group of mesozooplankton, constituted 39.9% of total abundance. It was noted that the greatest euphausiids larvae abundances mostly matched the stations characterized by high phytoplankton concentrations; as, e.g., at station (Stn.) 7332 eggs and nauplii of euphausiids accounted for >80% of total abundance. Phytoplankton cells were not counted, and phytoplankton concentration was estimated visually in the samples. The second abundant group of mesozooplankton, comprised of small-sized copepods, such as *Oithona* spp., *Oncaea* spp., and copepod nauplii (hereafter referred to as small copepods), constituted 27.2% of total abundance. Contributions of small copepods at the stations sampled in the Bransfield Strait were higher compared to other study areas (Figure 3b). The third abundant group of mesozooplankton was comprised of large-sized calanoid copepods, e.g., *Calanoides acutus, Calanus propinquus, Metridia gerlachei,* and *Rhincalanus gigas* (hereafter referred to as large copepods), that made up 24.7% of total abundance. At the stations, their distribution did not have any clear pattern. Other mesozooplankton taxa rarely accounted for more than 5% of total abundance (Supplementary Material).

The total mesozooplankton biomass varied between 20.1 and 12,287.8 mg WW m^{-3} with a mean of 1757.5 ± 869.1 mg WW m^{-3}. The highest biomasses were recorded from two groups of stations off the South Orkney Islands (1st group, Stns. 7377 and 7382; 2nd group, Stns. 7370 and 7373) (Figure 4a). In the 1st group, euphausiids larvae accounted for >90% of total mesozooplankton biomass; the stations of the 2nd group were characterized by generally high concentrations of large calanoid copepods and euphausiids larvae. At the other stations, the mesozooplankton biomass ranged from 100 to 1100 mg WW m^{-3} with no clear distribution pattern. Generally, at stations with concentrations less than 500 mg WW m^{-3}, large calanoid copepods accounted for more than 50% of total biomass (Figure 4b). At some stations, other taxa of mesozooplankton contributed significantly to total biomass: e.g., small copepods constituted up to 48% (Stn. 7318); pelagic tunicates, mainly Salpa thompsoni, up to 12% (Stn. 7359); chaetognaths, up to 6% (Stn. 7314); amphipods, mainly Themisto gaudichaudii, up to 5% (Stn. 7299). We observed a tendency of total mesozooplankton biomass to increase, similarly to abundance, from the Antarctic Peninsula towards the South Orkney Islands.

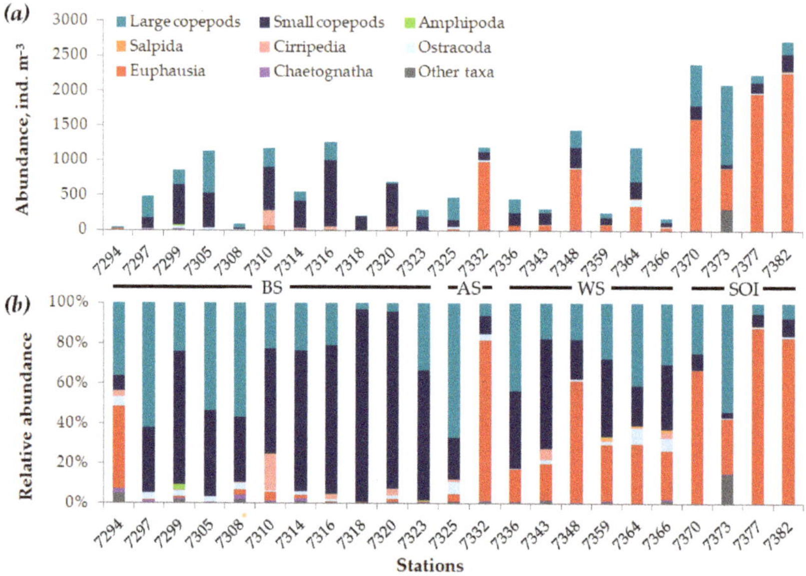

Figure 3. The total abundance (**a**) and contribution of the major taxa (**b**) of mesozooplankton in the Bransfield Strait (BS), Antarctic Sound (AS), Weddell Sea (WS), and off the South Orkney Islands (SOI) during the austral summer 2022.

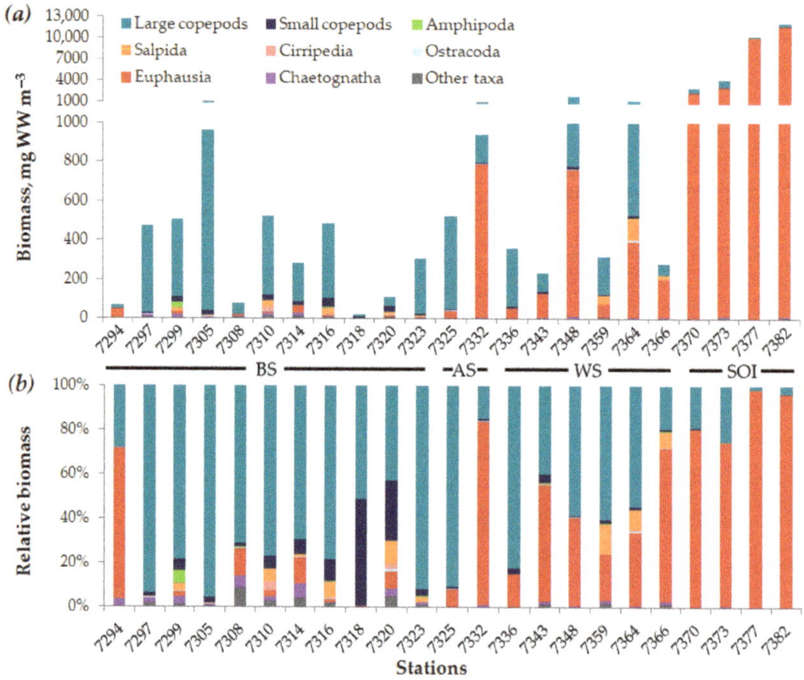

Figure 4. The total biomass (**a**) and contribution of the major taxa (**b**) of mesozooplankton in the study region. For an explanation of acronyms, see Figure 3.

3.3. Dynamics of Copepod Community

Copepods were the most abundant taxa and prominent components of the mesozooplankton communities in the study region (Figure 3). Usually, the copepod abundance ranged between 200 and 700 ind. m^{-3} in the Powell Basin (Weddell Sea), Antarctic Sound, and off the Antarctic Peninsula. The highest copepod abundance was recorded at the stations in the Bransfield Strait and off the South Orkney Islands (Figure 5a). At most stations, small copepods, e.g., *Oithona* spp. dominated and showed substantial values in the Bransfield Strait and Powell Basin, while large calanoid copepods, e.g., *C. acutus*, *C. propinquus*, *M. gerlachei*, and *R. gigas* numerically dominated off the South Orkney Islands. We also noted that the lowest copepod abundances mostly matched the stations with high concentrations of pelagic tunicates, *S. thompsoni* (up to 13 ind. m^{-3}).

The highest copepod biomasses were recorded from the Bransfield Strait, Powell Basin, and off the South Orkney Islands. The larger calanoid copepods significantly contributed to the total biomass of copepods in the study region. Metridia gerlachei dominated most stations; with several exceptions, the samples from the Powell Basin and off the South Orkney Islands were dominated by *C. acutus*, *C. propinquus*, and *R. gigas* (Figure 5b).

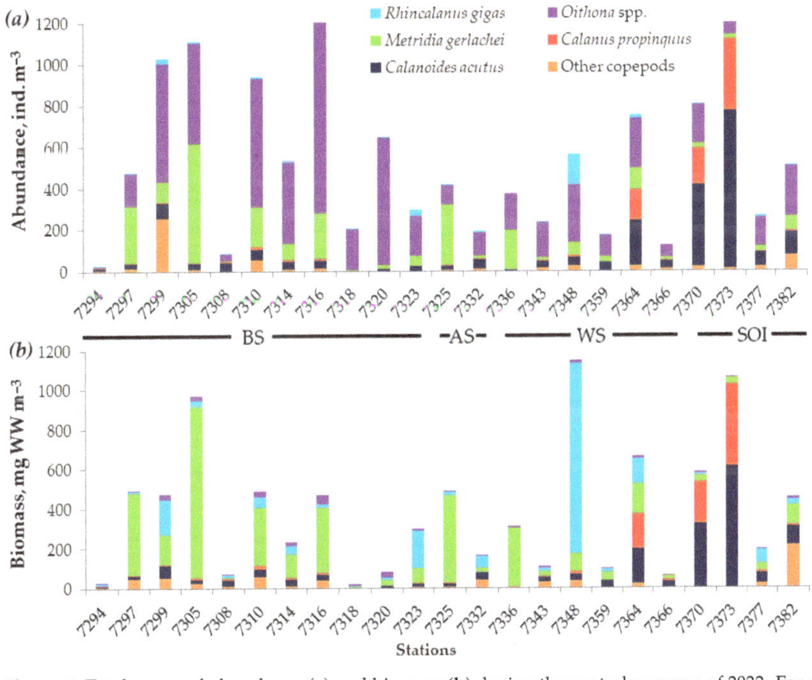

Figure 5. Total copepod abundance (**a**) and biomass (**b**) during the austral summer of 2022. For an explanation of acronyms, see Figure 3.

3.4. Vertical Distribution Patterns of Mesozooplankton Abundance

The average mesozooplankton abundance in the waters off the western Antarctic Peninsula reduced with increasing depth (Figure 6a). A major part of mesozooplankton occurred in the upper 200 m layer. Each mesozooplankton taxon showed characteristic depth preference: small copepods had high average abundance values in the depth layer 100–200 m, while large calanoid copepods were abundant in the layers deeper than 200 m (Figure 6b). Usually, euphausiid eggs were concentrated in the layers deeper than 500 m, with some exceptions off the South Orkney Islands (Stn. 7370) where eggs occurred in the surface layer. Euphausiids larvae were abundant in the upper 200 m layer, with a maximum of 100–200 m. The copepods *C. acutus* and *C. propinquus* were abundant in the

200–500 m layer, with several exceptions off the South Orkney Islands where these copepods dominated the warm surface layer. *Metridia gerlachei* was abundant in the 100–500 m layer; *R. gigas*, deeper than the 200 m layer; and *Calanus simillimus*, *Candacia* spp., *Paraeuchaeta antarctica*, and *Euchirella rostomagna*, in the 500–1000 m layer. Amphipods, ostracods, and pelagic polychaetes concentrated deeper than the 200 m layer; *cirripeds* cypris larvae were abundant within the 100–200 m layer.

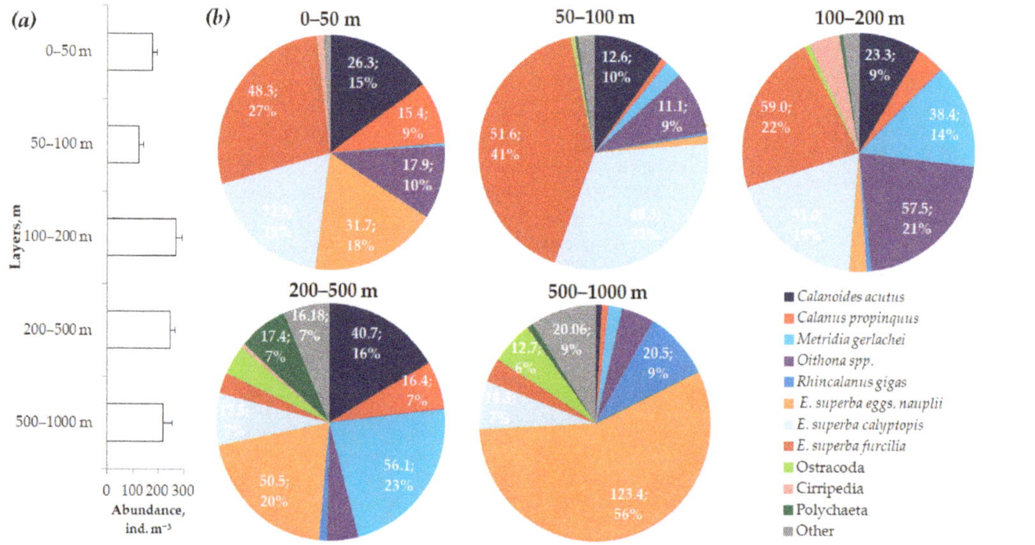

Figure 6. Vertical distribution (up to 1000 m) of the average abundance of mesozooplankton (**a**) and contribution of the major species/taxa (abundance, ind. m^{-3} and proportions, %) (**b**) off the western Antarctic Peninsula during the austral summer of 2022.

The pattern of vertical distribution of mesozooplankton major species/taxa for the four study areas (BS—AS—WS—SOI) is shown in Figure 7. The maximum values of the average abundance of mesozooplankton were recorded from the warm 0–100 m layer in the waters off the South Orkney Islands (Figure 7f,g) influenced by the warm and saline-modified Antarctic Surface Water (AASW). Relatively high concentrations of mesozooplankton were recorded from the deep-sea (200–500 and 500–1000 m) layers in the cold and saline Transitional Zonal Water with Weddell Sea influence (TWW) of the Antarctic Coastal Current (ACoC) and modified circumpolar deep water (mCDW) (Figure 7a), freezing and freshening Shelf Water (SW) and TWW (Figure 7c,d), warm and saline Warm Deep Water (WDW) (Figure 7e), except waters off the South Shetland Islands in the Bransfield Strait influenced by the Transitional Bellingshausen Water (TBW) of the warm Bransfield Current (BC) and mCDW (Figure 7b).

3.5. Effects of Environmental Variables on Distribution of Summer Mesozooplankton

The BEST analysis between the environmental variables and the biological data (abundance) showed that significant variations in quantitative distribution could be explained by three environmental variables such as salinity, temperature, and Chl *a* concentration (correlation coefficient from 0.353 to 0.400). In contrast, the other variables did not increase the correlation coefficient (Table 2). These relationships were statistically significant for each of the four models (RELATE; $p < 0.01$). The best combination of variables explaining the variations in mesozooplankton abundance was the salinity and Chl *a* concentration at 500 m (correlation coefficient, 0.400). As our study showed, the abundances of mesozooplankton in particular copepods negatively correlated with the Chl *a* concentration at

500 m, while the mesozooplankton biomass positively correlated with the salinity at 500 m ($p < 0.05$) (Table 3). The abundance of euphausiids larvae showed positive relationships with the salinity at 500 m and negative ones with the sea surface salinity and also with the temperature at 200 m. The surface Chl *a* concentration and the salinity at 500 m positively correlated with the biomass of euphausiids larvae. The correlations between copepod biomass and all variables considered in the analysis were non-significant.

Table 2. Combination of environmental variables leading to the best results (BEST analysis).

Number of Variables	Correlation Coefficient	Selections
2	0.400	500 m S, 500 m Chl *a*
3	0.385	500 m S, 50 m Chl *a*, 500 m Chl *a*
3	0.361	500 m S, 200 m T, 500 m Chl *a*
3	0.353	500 m S, 50 m S, 500 m Chl *a*

Table 3. Spearman's rank order correlations between the quantitative value of mesozooplankton and the environmental variables. Significant correlations ($p < 0.05$) are highlighted in bold.

	50 m T	50 m S	50 m O_2	50 m Chl *a*	200 m T	200 m S	200 m O_2	200 m Chl *a*	500 m T	500 m S	500 m O_2	500 m Chl *a*
Abn Meso	0.38	−0.16	−0.20	0.05	−0.06	0.35	−0.13	−0.23	0.11	0.40	−0.24	**−0.46**
Bm Meso	0.37	−0.28	−0.24	0.12	0.01	0.31	−0.27	−0.34	0.21	**0.50**	−0.31	−0.41
Abn Cop	0.29	0.08	−0.21	−0.19	0.00	0.26	−0.13	−0.23	−0.10	0.21	−0.12	**−0.43**
Bm Cop	0.14	−0.07	−0.32	−0.10	0.02	0.18	−0.21	−0.25	0.11	0.38	−0.29	−0.29
Abn Euph	0.08	**−0.36**	−0.03	0.12	**−0.39**	0.34	−0.09	−0.02	0.21	**0.44**	−0.28	−0.14
Bm Euph	0.17	−0.26	0.02	**0.40**	−0.26	0.36	−0.17	−0.07	0.27	**0.49**	−0.33	−0.15

Notes: Abn, abundance (ind. m^{-3}); Bm, biomass (mg WW m^{-3}); Meso, mesozooplankton; Cop, copepods; and Euph, euphausiids. For explanations for abbreviations of environmental variables, see Table 1.

3.6. Mesozooplankton Community Structure

Using the cluster and nMDS (stress value of 0.14) analyses, two significant groups of stations were distinguished that geographically matched the inner sector off the Antarctic Peninsula (ISAP) and the outer sector off the Antarctic Peninsula (OSAP) (Figure 8). The ISAP was located in the Bransfield Strait, in the northern Antarctic Sound, and the Powell Basin of the Weddell Sea, while the OSAP was located in the deep-sea area of the southern Antarctic Sound, in the Powell Basin, and off the South Orkney Islands. A non-recognized group (Stns. 7294, 7318, and 7336) stood out against the other stations. The ISAP and OSAP groups off the western Antarctic Peninsula overlapped somewhat but were well separated from the non-recognized group of stations.

By comparing the locations of the distinguished groups of stations with the key currents in the study region (Figure 1), we have found that the OSAP was influenced by the modified warm waters from the Powell Basin: Antarctic Slope Front (ASF) and the Weddell Front (WF) with the warm and saline modified AASW. On the contrary, the ISAP was influenced by the cold and freshening AASW and by the TBW of the warm Bransfield Current and the cool and saline TWW of the cold Antarctic Coastal Current. Stations of the non-recognized group were influenced by the freezing and freshening water.

Typical species (SIM/SD ratio of >3) of the ISAP community were *Oithona* spp., *M. gerlachei*, and *C. acutus* that together accounted for 53.1% of the average intra-group similarity (SIMPER; average similarity, 64.81%). The OSAP community was characterized by euphausiids larvae, *Oithona* spp., *M. gerlachei*, *C. acutus*, and ostracods, which together accounted for 67.66% of the average intra-group similarity (SIMPER; average similarity, 68.78%). The dissimilarity between the tested sectors (OSAP vs. ISAP) was driven by 15 species that represented such taxa as euphausiids, copepods, ostracods, pelagic polychaetes, cirripeds cypris larvae, and pelagic tunicates, which collectively accounted for 41.8% of the dissimilarity, as shown by a SIMPER analysis (Table 4). The samples (non-recognized group) collected from the very cold and freshening water were characterized by a minimum abundance and a low species richness represented by Antarctic species [61,62].

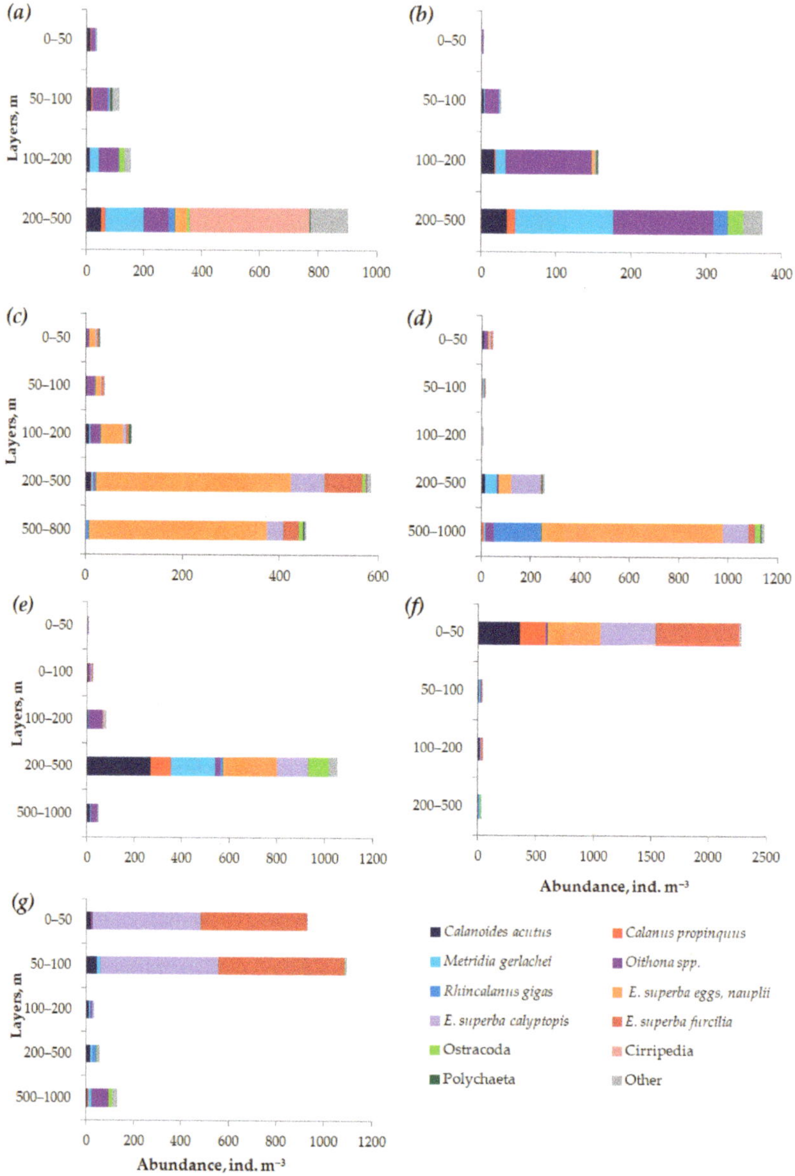

Figure 7. Vertical distribution of average abundances of mesozooplankton (ind. m^{-3}): (**a**) in the middle of the Bransfield Strait (BS); (**b**) in the Bransfield Strait off the South Shetland Islands; (**c**) in the deep-sea waters of the Antarctic Sound (AS); (**d**) in the southwestern Powell Basin (WS); (**e**) in the northeastern Powell Basin (WS); (**f**) in the coastal waters off the South Orkney Islands (SOI); (**g**) in the deep-sea waters off the South Orkney Islands (SOI).

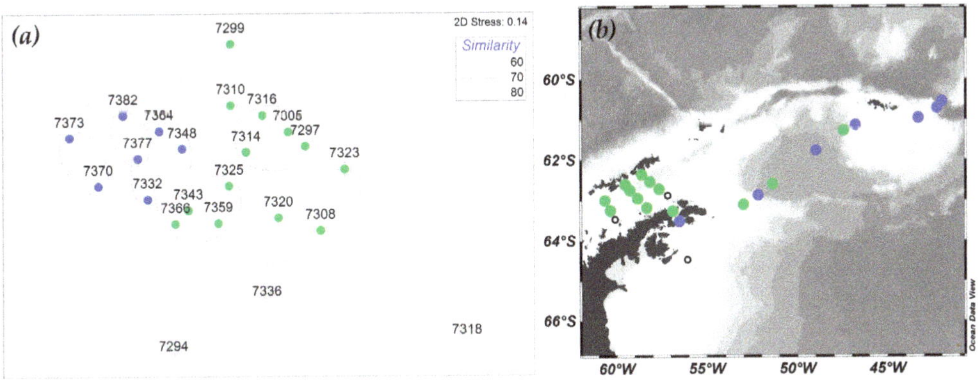

Figure 8. nMDS ordination plot (**a**) based on abundance of mesozooplankton by the Bray–Curtis similarity matrix, and location of the groups of stations in the inner (ISAP, green circles) and outer (OSAP, blue circles) sectors off the western Antarctic Peninsula (**b**). Blank circles mean non-recognized group.

Table 4. Result of a SIMPER analysis of average abundance (N, ind. m^{-3}) of mesozooplankton species/taxa with contributions (Contr, %) to and cumulative total (Cum, %) of intra-group similarity (SIM/SD) and between-group dissimilarity (DISS/SD) within the station groups of the inner (ISAP) and outer (OSAP) sectors off the western Antarctic Peninsula. The listed taxa contributed to at least 80% of the dissimilarity between the sectors.

Taxa	OSAP			ISAP			OSAP vs. ISAP		
	N	SIM/SD	Contr, %	N	SIM/SD	Contr, %	DISS/SD	Contr, %	Cum, %
	Similarity: 68.78			Similarity: 64.81			Dissimilarity: 41.82		
E. superba calyptopis	5.58	9.42	14.81	1.33			2.84	12.48	12.48
E. superba furcilia	4.73	1.73	9.91	0.77			1.82	11.89	24.37
E. superba eggs	4.08	0.92	6.82	1.33			1.43	10.30	34.67
Calanus propinquus	3.42	2.04	6.89	1.67	3.24	6.45	1.13	5.57	40.24
Calanoides acutus	4.93	6.74	12.45	3.39	4.56	14.67	1.31	4.62	44.86
Rhincalanus gigas	2.19			1.64	1.81	5.18	1.28	4.32	49.18
Cirripedia	0.79			1.45			1.20	4.21	53.39
Polychaeta	2.08			1.05			0.85	4.02	57.42
Metridia gerlachei	3.63	8.77	9.41	4.22	3.30	15.77	1.45	3.95	61.37
Oncea spp.	1.00			0.53			0.92	3.54	64.91
Ostracoda	3.00	3.78	7.13	2.38	1.98	8.46	1.10	3.26	68.17
Calanus simillimus	1.03			1.25			1.28	2.99	71.15
Oithona spp.	5.08	11.43	13.95	5.42	5.48	22.66	1.51	2.90	74.06
Metridia spp.	0.88			0.22			0.69	2.81	76.87
Salpa thompsoni	0.96			0.74			1.27	2.63	82.01

A canonical correspondence analysis (CCA) performed to assess relationships between the mesozooplankton composition and the environmental variables in the two sectors (ISAP and OSAP) together showed that the revealed models explained 97.1% of the total variation in the dataset (64.4 and 32.7% for the CCA1 and CCA2 axes, respectively) (Figure 9). The increase in the abundance of euphausiid eggs was associated with the high oxygen and Chl *a* concentrations and the low temperature in the 200 m (T 200 m) layer. In general, the higher concentrations of euphausiids larvae were associated with the high temperature in the surface (T 50 m) and 500 m (T 500 m) layers, the high salinity in the 200 m (S 200 m) and 500 m (S 500 m) layers, and a low oxygen concentration. The abundance of large copepods was poorly represented on the first two CCA axes (it was located close to the origin). The high salinity in the surface layer (S 50 m) was associated with the abundance of small copepods and other mesozooplankton aggregations, ostracods, amphipods, and chaetognaths. The abundance of pelagic tunicates was associated with the high temperature in the 200 m layer (T 200 m) and low Chl *a* conditions. The abundance of cirripeds cypris larvae was associated with the high salinity in the surface layer (S 50 m) and the high oxygen in the 500 m layer (O_2 500 m), but the low temperature and the low Chl *a* concentration which is the predominant environmental conditions in the ISAP (green circles in Figure 9).

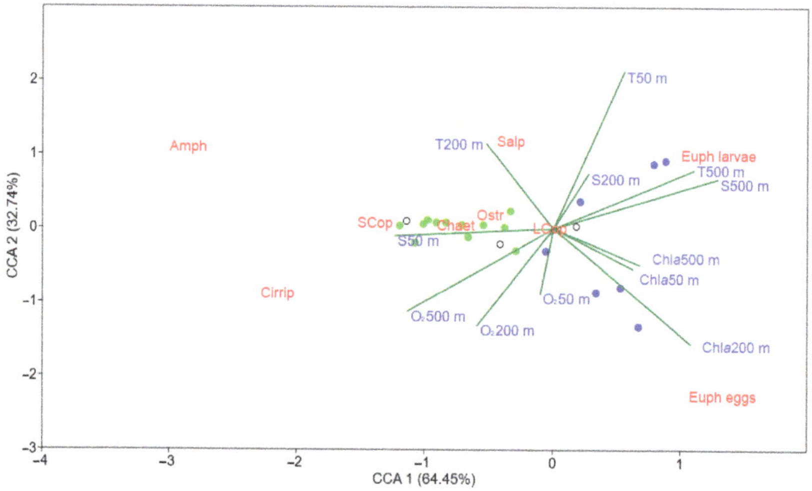

Figure 9. Canonical correspondence analysis (CCA) of mesozooplankton composition (in red) in relationships with the environmental variables (in blue). The symbols of the mesozooplankton communities (circles), as they were disclosed by nMDS analysis (see Figure 8), were superimposed on the station labels. Abbreviations of the taxa are as follows: large copepods (LCop), small copepods (SCop), ostracods (Ostr), *cirripeds* cypris larvae (Cirrip), pelagic tunicates (Salp), amphipods (Amph), chaetognaths (Chaet), and euphausiid (Euph) eggs and larvae. For explanations for abbreviations of environmental variables, see Table 1.

4. Discussion

Climate warming [5] and freshening of surface waters due to the extremely rapid melting of glaciers over the past few decades [2,63,64] have induced marked changes in the pelagic zone of the Southern Ocean such as a decrease in the photosynthesis efficiency [65] and in the total phytoplankton biomass, and also a change in the phytoplankton structure from the predominance of large diatoms to cryptophytes and small flagellates [66–68]. This has exerted a significant impact on the abundance of krill and salps [33,69]. Overall, the composition and distribution of mesozooplankton communities in the Austral summer of 2022 changed slightly compared to data for other years [16,18,19,61,70]. It should, however, be pointed out that while average abundance and biomass levels were within the documented range, they showed the highest estimated values. This can be explained by the 150 µm mesh used in this study. For comparison, most of the other estimates were derived from net samples where mesh size was often 200 µm or greater, and a 200 µm mesh preserves by an average of ~10 and ~20% less biomass and density, respectively, compared to the similarly designed 150 µm mesh net [71]. It is possible that much of small-size mesozooplankton is underrepresented in many present and historical data sets [72].

According to NOAA [9], the Austral summer of 2022 in the region off the western Antarctic Peninsula was the second warmest period after 2016 (observations since 1910). The SST recorded in this period proved to be the highest for the past 113 years, with a positive anomaly of 1.35 °C [9]. Furthermore, the record lows in sea-ice cover in the Antarctic for the past 44 years, including the retreat of the sea-ice edge up to 64°30′ S in the Weddell Sea, and the increased concentration of chlorophyll *a* were also observed in the Austral summer of 2022 [73]. Copepods, eggs, and larvae of euphausiids were the most abundant taxa and prominent components of the mesozooplankton communities throughout the record-warm summer season of 2022. The spatial distributions of copepods and euphausiids larvae together drove distribution patterns for the total mesozooplankton abundance. Copepods showed maximum abundances immediately off the Antarctic Peninsula, whereas abundances of euphausiid larvae were highest east of the Antarctic Peninsula, at the Weddell-Scotia Confluence. The trends of variations in total mesozooplankton abundance over time were quite consistent, i.e., a steady increase from 1998 to 2009, a leveling-off, and then a decrease between 2010 and 2014, followed by a second period of increase that occurred in 2015–2018 [17]. During the warmest summer of 2022, the distribution of mesozooplankton varied considerably, and abundances were generally high at the easternmost stations. There was no surprise that small copepods accounted for a major part of the mesozooplankton abundance off the Antarctic Peninsula [15,16,61,74,75]. A positive trend of increasing small copepod contributions to the total mesozooplankton abundance off the Antarctic Peninsula has been observed over the last 20 years [17,76,77]. Small copepods and nauplii of large species constitute a very important group of mesozooplankton, providing a food supply for fish larvae and macrozooplankton [36,78]. The average abundance of euphausiid eggs and larvae proved to be 3–5-fold lower than previously reported [27,28,79]. It is likely that the survey in 2022 could coincide with a period between years of high abundance of the krill population, as was noted earlier [22,28]. Previous studies showed that the euphausiid abundance did not increase after 2010 [17,33]. Currently, a trend of krill population declines due to climate changes observed since the late 20th century is very probable [15,17,22,62].

The stability of mesozooplankton communities in the Southern Ocean is related to changes in the phytoplankton structure, in particular, variations in food availability, which may be a major factor responsible for differences in the composition and quantitative distribution of mesozooplankton [17,68,80,81]. Larger-sized phytoplankton organisms (e.g., diatoms) are typically grazed by calanoid copepods and krill [82,83], whereas smaller-sized ones (e.g., flagellates) are grazed by small copepods and pelagic tunicates [84,85]. In our study, a maximum abundance of small copepods was located in areas with high concentrations of Haptophyta and Dinophyta microalgae, e.g., *Phaeocystis antarctica*, *Polarella glacialis*, and *Protoperidinium* spp., while a maximum abundance of large copepods and euphausiid larvae was in waters with high concentrations of diatoms, e.g., *Chaetoceros criophilus*, *C. concavicornis*, *Fragilariopsis kerguelensis*, *Odontella weissflogii*, and *Rhizosolenia* spp., which is consistent with a previous study [20]. If the change in the phytoplankton structure trend continues, we may observe the increasing concentration of flagellates microalgae, which may be a trigger for an increase in abundance and expansive distribution of small copepods and pelagic tunicates in the Southern Ocean.

Differences in the vertical distribution of mesozooplankton may be caused by seasonal migration of plankton organisms, food availability, and seasonal pycnocline [86–88]. The calanoid copepods *C. acutus*, *C. propinquus*, and *M. gerlachei* commonly occur in the epipelagic layer of 0–200 m off the Antarctic Peninsula in summer [61,89]. In the warmest summer of 2022, these calanoid copepods were ubiquitously distributed in the warm and saline water of the mesopelagic layer (200–500 m), except for a few stations off the South Orkney Islands where these copepods dominated the warm surface layer of 0–50 m. The presence of a sharp seasonal pycnocline in the study region, where gradients of thermohaline characteristics reached 0.5 °C and 0.06 psu per 10 m [20], may have prevented calanoid copepods from migrating to the surface for feeding and breeding. Nevertheless, stratification of the water column can contribute to the active vegetation of phytoplankton as essential food for the development of macro- and mesozooplankton [12,35], as, e.g., off the South Orkney Islands, where the highest density of euphausiid larvae can be associated with bloom events of diatoms of the genera *Chaetoceros* and *Rhizosolenia* in the warm near-surface layer. On the contrary, in the Antarctic Sound, characterized by high concentrations of microalgae (mainly diatoms), with an increase from the surface to the bottom (800 m depth), the abundance of eggs and nauplii of euphausiids was recorded deeper than 500 m.

Yet, there is a limited understanding of how environmental conditions can structure the dynamics of mesozooplankton communities in the Antarctic waters. As was previously reported, mesozooplankton communities are thermally resilient to the present levels of sea surface warming, whereas other selective pressures, in particular food availability and the properties of underlying water masses, imposed greater constraints on the distribution of mesozooplankton in the Antarctic region [16,90]. Nevertheless, we found that water salinity, temperature, and chlorophyll *a* concentration are important factors structuring the composition and the quantitative distribution of mesozooplankton in the waters off the Antarctic Peninsula during the warmest austral summer. Each mesozooplankton taxon showed characteristic environmental preference. Copepods, ostracods, and chaetognaths were associated with salinity; pelagic tunicates, with temperature; *cirripeds* cypris larvae and amphipods, with salinity and oxygen concentration; euphausiids larvae, with temperature and salinity; euphausiids eggs, with oxygen and chlorophyll *a* concentrations. The highest euphausiid egg abundances were observed at the deep-sea stations which were associated with the highest oxygen concentrations (>10 mg/L), very cold water (−1 °C), and high, bloom-like phytoplankton concentrations. Mackintosh [91] found that euphausiid eggs and larvae at early stages tend to certain zones as the thermocline disappears; in addition, an abundance of krill larvae at early developmental stages may be associated with temperature and Chl *a* concentration [27,29,92,93]. Further studies are expected to be carried out to elucidate the relative contributions of the factors driving the abundance of copepods that dominate the mesozooplankton of the Southern Ocean and form the major food supply in the food chain between primary producers and predators [94,95].

5. Conclusions

Eggs and larvae of euphausiids such as Euphausia superba, small copepods such as *O. similis*, and large calanoid copepods such as *C. acutus*, *C. propinquus*, *M. gerlachei*, and *R. gigas* proved to be the most abundant taxa and prominent components of the mesozooplankton communities off the western Antarctic Peninsula during the record-warm summer season of 2022. Small copepods dominated substantially in the Bransfield Strait and off the Antarctic Peninsula, while large calanoid copepods and euphausiid eggs and larvae numerically dominated in the Powell Basin and off the South Orkney Islands. The other taxa of mesozooplankton, e.g., Appendicularia, Ctenophora, Echinodermata, Hydromedusae, Polychaeta, Pteropoda, Siphonophorae, and eggs and larvae of fish did not play any substantial role in the total abundance. With a few exceptions, the contributions of amphipods, chaetognaths, ostracods, salps, pelagic polychaetes, and cirripeds cypris larvae were significant. The total abundance and biomass of the mesozooplankton increased in an easterly direction, from the Antarctic Peninsula to the South Orkney Islands. Most mesozooplankton occurred in the upper 200 m layer. Each taxon showed characteristic depth preference: small copepods, euphausiids, and cirripeds cypris larvae were abundant in the epipelagic layer (up to 200 m), while large calanoid copepods, euphausiid eggs, amphipods, ostracods, and pelagic polychaetes were concentrated in the mesopelagic layer (up to 1000 m). Two significant groups of sampling stations were identified off the western Antarctic Peninsula. The first group, located in the inner sector off of the Antarctic Peninsula (ISAP), was represented by copepods *Oithona* spp., *M. gerlachei*, and *C. acutus*; the second group, located in the outer sector off of the Antarctic Peninsula (OSAP), by euphausiids larvae, copepods *Oithona* spp., *M. gerlachei*, *C. acutus*, and ostracods. The composition and quantitative distribution of mesozooplankton showed clear relationships with environmental factors, particularly with a combination of variables such as water salinity, temperature, and Chl *a* concentration. The increase in the abundance of euphausiid eggs was associated with the high oxygen and Chl *a* concentrations and the low temperature in the 200 m layer; euphausiid larvae, with the high temperature (>1 °C) in the surface and 500–m layers, the high salinity in the 200 m (S 200 m) and 500 m layers, and a low oxygen concentration; small copepods, together with ostracods and chaetognaths, with the high salinity in the surface layer; pelagic tunicates, with the high temperature in the 200 m layer and low Chl *a* conditions; cirripeds cypris larvae, with the high salinity in the surface layer and the high oxygen concentration in the 500 m layer, but a low temperature and low Chl *a* concentration.

The results of the present and future studies do and will provide invaluable data as regards the life of planktonic organisms in the extreme conditions of the changing Southern Ocean. These are also expected to have important practical implications for predicting long-term changes in pelagic communities and designing future programs for the conservation of Antarctic marine ecosystems and sustainable management of commercial fisheries.

Supplementary Materials: The following supporting information can be downloaded at: https://www.mdpi.com/article/10.3390/w15101948/s1, Table S1: Composition and mean abundance, ind. m^{-3} (±SD) of mesozooplankton in the Bransfield Strait (BS), Antarctic Sound (AS), Weddell Sea (WS), and off the South Orkney Islands (SOI) during the austral summer of 2022.

Funding: This study was supported by the Russian State Assignment no. FMWE-2022-0001 no. 122072000067-9 (for the A.V. Zhirmunsky National Scientific Center of Marine Biology, Far Eastern Branch, Russian Academy of Sciences) and FWFE-2021-0003 no. 1021062912502-3.

Data Availability Statement: Not applicable.

Acknowledgments: We are grateful to the administration of the Shirshov Institute of Oceanology, Russian Academy of Sciences, for the organization of the expedition and field operations, to the plankton and benthos research teams for collecting and providing plankton samples, to the hydrology research team for providing environmental parameters, to the primary production research team, especially to O. Kalinina, for advice on phytoplankton and also to the captain and the crew of the R/V Akademik Mstislav Keldysh for their comprehensive assistance. Special thanks are due to E. Shvetsov for his help in improving the English. The author thanks anonymous reviewers whose valuable comments and suggestions substantially improved the manuscript.

Conflicts of Interest: The author declares no conflict of interest.

References

1. Vaughan, D.; Marshall, G.; Connolley, W.; Parkinson, C.; Mulvaney, R.; Hodgson, D.; King, J.C.; Pudsey, C.J.; Turner, J. Recent rapid regional climate warming on the Antarctic Peninsula. *Clim. Chang.* **2003**, *60*, 243–274. [CrossRef]
2. Cook, A.J.; Fox, A.J.; Vaughan, D.G.; Ferrigno, J.G. Retreating Glacier fronts on the Antarctic Peninsula over the past half-century. *Science* **2005**, *308*, 541–544. [CrossRef] [PubMed]
3. Xiao, C. Changes in Antarctic climate system: Past, present and future. *Adv. Clim. Chang. Res.* **2008**, *4*, 1–7.
4. Constable, A.J.; Melbourne-Thomas, J.; Corney, S.P.; Arrigo, K.R.; Barbraud, C.; Barnes, D.K.A.; Bindoff, N.L.; Boyd, P.W.; Brandt, A.; Costa, D.P.; et al. Climate change and Southern Ocean ecosystems I: How changes in physical habitats directly affect marine biota. *Glob. Chang. Biol.* **2014**, *20*, 3004–3025. [CrossRef] [PubMed]
5. Meredith, M.P.; King, J.C. Rapid climate change in the ocean west of the Antarctic Peninsula during the second half of the 20th century. *Geophys. Res. Lett.* **2005**, *32*, L19604. [CrossRef]
6. Turner, J.; Colwell, S.; Marshall, G.; Lachlan-Cope, T.; Carleton, A.; Jones, P.; Lagun, V.; Reid, P.A.; Iagovkina, S. Antarctic climate change during the last 50 years. *Int. J. Climatol.* **2005**, *25*, 279–294. [CrossRef]
7. Hill, S.L.; Phillips, T.; Atkinson, A. Potential climate change effects on the habitat of Antarctic krill in the Weddell Quadrant of the Southern Ocean. *PLoS ONE* **2013**, *8*, e72246. [CrossRef] [PubMed]
8. Stammerjohn, S.; Martinson, D.; Smith, R.; Iannuzzi, R. Sea ice in the western Antarctic Peninsula region: Spatio-temporal variability from ecological and climate change perspectives. *Deep-Sea Res. II* **2008**, *55*, 2041–2058. [CrossRef]
9. NOAA National Centers for Environmental Information, Monthly Global Climate Report for January 2022, Published Online February 2022. Available online: https://www.ncei.noaa.gov/access/monitoring/monthly-report/global/202201 (accessed on 13 March 2023).
10. Constable, A.J.; Nicol, S.; Strutton, P.G. Southern Ocean productivity in relation to spatial and temporal variation in the physical environment. *J. Geophys. Res.* **2003**, *108*, 8079.
11. Montes-Hugo, M.; Vernet, M.; Martinson, D.; Smith, R.; Iannuzzi, R. Variability on phytoplankton size structure in the western Antarctic Peninsula (1997–2006). *Deep Sea Res. II* **2008**, *55*, 2106–2117. [CrossRef]
12. Deppeler, S.L.; Davidson, A.T. Southern Ocean phytoplankton in a changing climate. *Front. Mar. Sci.* **2017**, *4*, 40. [CrossRef]
13. Sapozhnikov, P.V.; Kalinina, O.Y.; Morozova, T.V. Phytopelagic ommunities of the Powell Basin in the summer of 2020. In *Antarctic Peninsula Region of the Southern Ocean*; Morozov, E.G., Flint, M.V., Spiridonov, V.A., Eds.; Springer: Cham, Switzerland, 2021; Volume 6, pp. 279–306. [CrossRef]

14. Rombola, E.F.; Franzosi, C.A.; Tosonotto, G.V.; Alder, V.A.; Marschoff, E.R. Variability of euphausiid larvae densities during the 2011, 2012, and 2014 summer seasons in the Atlantic sector of the Antarctic. *Polar. Sci.* **2019**, *19*, 86–93. [CrossRef]
15. Garcia, M.D.; Sofia Dutto, M.; Chazarreta, C.J.; Berasategui, A.A.; Schloss, I.R.; Hoffmeyer, M.S. Micro- and mesozooplankton successions in the Antarctic coastal environment during a warm year. *PLoS ONE* **2020**, *15*, e0232614. [CrossRef]
16. Pakhomov, E.; Pshenichnov, L.; Krot, A.; Paramonov, V.; Slypko, I.; Zabroda, P. Zooplankton distribution and community structure in the Pacific and Atlantic Sectors of the Southern Ocean during austral summer 2017–2018: A Pilot Study Conducted from Ukrainian Long-Liners. *J. Mar. Sci. Eng.* **2020**, *8*, 488. [CrossRef]
17. Pinkerton, M.H.; Decima, M.; Kitchener, J.A.; Takahashi, K.T.; Robinson, K.V.; Stewart, R.; Hosie, G.W. Zooplankton in the Southern Ocean from the continuous plankton recorder: Distributions and long-term change. *Deep Sea Res.* **2020**, *162*, 103303. [CrossRef]
18. Yakovenko, V.A.; Spiridonov, V.A.; Gorbatenko, K.M.; Shadrin, N.V.; Samyshev, E.Z.; Minkina, N.I. Macro- and mesozooplankton in the Powell Basin (Antarctica): Species composition and distribution of abundance and biomass in February 2020. In *Antarctic Peninsula Region of the Southern Ocean*; Morozov, E.G., Flint, M.V., Spiridonov, V.A., Eds.; Springer: Cham, Switzerland, 2021; Volume 6, pp. 131–141. [CrossRef]
19. Criales–Hernandez, M.I.; Jerez–Guerrero, M.; Latandret–Solana, S.A.; Gomez–Sanchez, M.D. Spatial distribution of meso- and macrozooplankton in the Bransfield Strait and around Elephant Island, Antarctic Peninsula, during the 2019–2020 austral summer. *Polar. Sci.* **2022**, *32*, 100821. [CrossRef]
20. Kasyan, V.V.; Bitiutskii, D.G.; Mishin, A.V.; Zuev, O.A.; Murzina, S.A.; Sapozhnikov, P.V.; Kalinina, O.Y.; Syomin, V.L.; Kolbasova, G.D.; Voronin, V.P.; et al. Composition and Distribution of Plankton Communities in the Atlantic Sector of the Southern Ocean. *Diversity* **2022**, *14*, 923. [CrossRef]
21. Loeb, V.J.; Santora, J.A. Population dynamics of *Salpa thompsoni* near the Antarctic Peninsula: Growth rates and interannual variations in reproductive activity (1993–2009). *Prog. Oceanogr.* **2012**, *96*, 93–107. [CrossRef]
22. Siegel, V.; Watkins, J.L. Distribution, Biomass and Demography of Antarctic Krill, *Euphausia superba*. In *Biology and Ecology of Antarctic Krill*; Advances in Polar Ecology; Siegel, V., Ed.; Springer: Cham, Switzerland, 2016; pp. 21–100. [CrossRef]
23. Pakhomov, E.A.; Froneman, P.W.; Perissinotto, R. Salp/krill interactions in the Southern Ocean: Spatial segregation and implications for the carbon flux. *Deep Sea Res. II* **2002**, *49*, 1881–1907. [CrossRef]
24. Atkinson, A.; Hill, S.L.; Pakhomov, E.A.; Siegel, V.; Reiss, C.S.; Loeb, V.J.; Steinberg, D.K.; Schmidt, K.; Tarling, G.A.; Gerrish, L.; et al. Krill (*Euphausia superba*) distribution contracts southward during rapid regional warming. *Nat. Clim. Chang.* **2019**, *9*, 142–147. [CrossRef]
25. McBride, M.M.; Schram Stokke, O.; Renner, A.H.H.; Krafft, B.A.; Bergstad, O.; Biuw, M.; Lowther, A.; Stiansen, J. Antarctic krill *Euphausia superba*: Spatial distribution, abundance, and management of fisheries in a changing climate. *Mar. Ecol. Prog. Ser.* **2021**, *668*, 185–214. [CrossRef]
26. Bitiutskii, D.G.; Samyshev, E.Z.; Minkina, N.I.; Melnikov, V.V.; Chudinovskih, E.S.; Usachev, S.I.; Salyuk, P.A.; Serebrennikov, A.N.; Zuev, O.A.; Orlov, A.M. Distribution and demography of Antarctic krill and salps in the Atlantic sector of the Southern Ocean during austral summer 2021–2022. *Water* **2022**, *14*, 3812. [CrossRef]
27. Gao, Q.; Xu, Z.; Huang, H.; Chen, X.; Feng, C.; Li, L. Geographical distribution and age composition of *Euphausia superba* larvae (Crustacea: Euphausiacea) in the South Shetland Islands region and southern Scotia Sea in relation to environmental conditions. *Acta Oceanol. Sin.* **2013**, *32*, 59–67. [CrossRef]
28. Sologub, D.O. Distribution and age structure of euphausiids larvae at the Western Antarctic Peninsula (Subarea 48.1). *Trudy VNIRO* **2015**, *154*, 3–15.
29. Kasyan, V.V. Age Structure and Spatial Distribution of *Euphausia superba* Larvae off the Antarctic Peninsula, Southern Ocean. *Water* **2022**, *14*, 3196. [CrossRef]
30. Voronina, N.M. *Pelagic Ecosystems of the Southern Ocean*; Nauka: Moscow, Russia, 1984; pp. 5–206.
31. McBride, M.; Dalpadado, P.; Drinkwater, K.; Godo, O.; Hobday, A.; Hollowed, A.; Kristiansen, T.; Murphy, E.; Ressler, P.; Subbey, S.; et al. Krill, climate, and contrasting future scenarios for Arctic and Antarctic fisheries. *ICES J. Mar. Sci.* **2014**, *71*, 1934–1955. [CrossRef]
32. Takahashi, K.; Takamura, T.; Odate, T. Zooplankton communities along a Southern Ocean monitoring transect at 110 E from three CPR surveys (Dec 2014, Jan 2015, Mar 2015). *Polar. Biol.* **2021**, *44*, 1069–1081. [CrossRef]
33. Johnston, N.M.; Murphy, E.J.; Atkinson, A.; Constable, A.J.; Cotte, C.; Cox, M.; Daly, K.L.; Driscoll, R.; Flores, H.; Halfter, S.; et al. Status, change, and futures of zooplankton in the Southern Ocean. *Front. Ecol. Evol.* **2022**, *9*, 624692. [CrossRef]
34. Quetin, L.B.; Ross, R.M.; Frazer, T.K.; Haberman, K.L. Factors affecting distribution and abundance of zooplankton, with an emphasis on Antarctic krill, *Euphausia superba*. In *Antarctic Research Series*; Hofmann, E.E., Ross, R.M., Quetin, L.B., Eds.; American Geophysical Union: Washington, DC, USA, 1996; pp. 357–371. [CrossRef]
35. Atkinson, A. Life cycle strategies of epipelagic copepods in the Southern Ocean. *J. Mar. Syst.* **1998**, *15*, 289–311. [CrossRef]
36. Kawaguchi, S.; Nicol, S. Learning about Antarctic krill from the fishery. *Antarct. Sci.* **2007**, *19*, 219–230. [CrossRef]
37. Siegel, V.; Reiss, C.S.; Dietrich, K.S.; Haraldsson, M.; Rohardt, G. Distribution and abundance of Antarctic krill (*Euphausia superba*) along the Antarctic Peninsula. *Deep Res. Part I* **2013**, *77*, 63–74. [CrossRef]

38. Steinberg, D.K.; Ruck, K.E.; Gleiber, M.R.; Garzio, L.M.; Cope, J.S.; Bernard, K.S.; Stammerjohn, S.E.; Schofield, O.M.E.; Quetin, L.B.; Ross, R.M. Long-term (1993–2013) changes in macrozooplankton off the Western Antarctic Peninsula. *Deep Sea Res. I* **2015**, *101*, 54–70. [CrossRef]
39. Weikert, H.; John, H.C. Experiences with a modified be multiple opening–closing plankton net. *J. Plankton Res.* **1981**, *3*, 167–176. [CrossRef]
40. Tranter, D.J. (Ed.) Reviews on zooplankton sampling methods. In *Zooplankton Sampling*; Part I (2); UNESCO: Paris, France, 1968; pp. 11–144.
41. Boltovskoy, D. *South Atlantic Zooplankton*; Backhuys Publishers: Leiden, The Netherlands, 1999; p. 868.
42. WoRMS Editorial Board. World Register of Marine Species. 2015. Available online: http://www.marinespecies.org (accessed on 29 August 2022).
43. Razouls, C.; De Bovee, F.; Kouwenberg, J.; Desreumaux, N. *Diversity and Geographic Distribution of Marine Planktonic Copepods*; CNRS—Sorbonne University: Paris, France, 2005–2022. Available online: http://copepodes.obs--banyuls.fr/en/ (accessed on 3 February 2022).
44. Makarov, R.R. Larvae of *Euphausia superba* Dana in plankton from the Sea of Scotia. *Tr. Vses. Nauchno-Issled. Ryb. Okeanogr.* **1974**, *99*, 84–103.
45. Pertsova, K.N. Larvae of euphausiids of the Antarctic. In *Ecology and Biogeography of Plankton: Proceedings of the Institute of Oceanology RAS*; Nauka: Moscow, Russia, 1976; Volume 105, pp. 147–170.
46. Suntov, A.P. *Manual on Quantitative Processing of Marine Net Plankton*; TINRO: Vladivostok, Russia, 1982; pp. 1–29. (In Russian)
47. Mizdalski, E. Weight and length data of zooplankton in the Weddell Sea in austral spring 1986 (ANT V/3). *Ber. Polarforsch.* **1988**, *55*, 1–72.
48. Heywood, K.J.; Garabato, A.C.N.; Stevens, D.P.; Muench, R.D. On the fate of the Antarctic Slope Front and the origin of the Weddell Front. *J. Geophys. Res.* **2004**, *109*, C06021. [CrossRef]
49. Thompson, A.F.; Heywood, K.J.; Thorpe, S.E.; Renner, A.H.H.; Trasvica, A. Surface circulation at the tip of the Antarctic Peninsula from drifters. *J. Phys. Oceanog.* **2009**, *39*, 3–26. [CrossRef]
50. Morozov, E.G.; Flint, M.V.; Spiridonov, V.A. *Antarctic Peninsula region of the Southern Ocean*; Advances in Polar Ecology; Springer: Cham, Switzerland, 2021; Volume 6, p. 433. [CrossRef]
51. Clark, K.R.; Warwick, R.M. *Change in Marine Communities: An Approach to Statistical Analysis and Interpretation*; PRIMER–E Ltd.: Plymouth, UK, 2001; p. 144.
52. Bray, J.R.; Curtis, J.T. An ordination of the upland forest communities of Southern Wisconsin. *Ecol. Monogr.* **1957**, *27*, 325–349. [CrossRef]
53. Clarke, K.R.; Ainsworth, M. A method of linking multivariate community structure to environmental variables. *Mar. Ecol. Prog. Ser.* **1993**, *92*, 205–219. [CrossRef]
54. Clarke, K.R.; Gorley, R.N. *PRIMER v6: User Manual/Tutorial*; PRIMER–E Ltd.: Plymouth, UK, 2006; p. 192.
55. Hammer, O.; Harper, D.A.T.; Ryan, P.D. PAST: Paleontological Statistics Software Package for Education and Data Analysis. *Palaeontol. Electron.* **2001**, *4*, 262. Available online: http://palaeo--electronica.org/2001_1/past/issue1_01.htm (accessed on 1 March 2023).
56. Schlitzer, R. *Ocean Data View*; Alfred Wegener Institute: Bremerhaven, Germany, 2012; p. 151. Available online: http://odv.awi.de (accessed on 13 February 2023).
57. Zhou, X.Q.; Zhu, G.P.; Hu, S. Influence of tides on mass transport in the Bransfield Strait and the adjacent areas. *Antarct. Polar. Sci.* **2020**, *23*, 100506. [CrossRef]
58. Fedotova, A.A.; Stepanova, S.V. Water mass transformation in the Powell Basin. In *Antarctic Peninsula Region of the Southern Ocean*; Morozov, E.G., Flint, M.V., Spiridonov, V.A., Eds.; Springer: Cham, Switzerland, 2021; Volume 6, pp. 115–129. [CrossRef]
59. Krechik, V.A.; Frey, D.I.; Morozov, E.G. Peculiarities of water circulation in the central part of the Bransfield Strait in January 2020. *Dokl. Earth Sci.* **2021**, *496*, 92–95. [CrossRef]
60. Frey, D.I.; Krechik, V.A.; Morozov, E.G.; Drozd, I.D.; Gordey, A.S.; Latushkin, A.A.; Mekhova, O.S.; Mukhametianov, R.Z.; Murzina, S.A.; Ostroumova, S.A.; et al. Water Exchange between Deep Basins of the Bransfield Strait. *Water* **2022**, *14*, 3193. [CrossRef]
61. Lee, W.; Kim, S.; Kang, S.; Bang, H.; Lee, K.; Kwak, I. Distribution and abundance of zooplankton in the Bransfield Strait and the western Weddell Sea during austral summer. *Ocean Polar. Res.* **2004**, *26*, 607–618. [CrossRef]
62. Atkinson, A.; Ward, P.; Hunt, B.P.V.; Pakhomov, E.A.; Hosie, G.W. An overview of Southern Ocean zooplankton data: Abundance, biomass, feeding and functional relationships. *CCAMLR Sci.* **2012**, *19*, 171–218.
63. Edwards, M.; Bresnan, E.; Cook, K.; Heath, M.; Helaouet, P.; Lynam, C.; Raine, R.; Widdicombe, C. Impacts of climate change on plankton. *MCCIP Sci. Rev.* **2013**, *4*, 98–112. [CrossRef]
64. Sallee, J.B. Southern Ocean warming. *Oceanography* **2018**, *31*, 52–62. [CrossRef]
65. Schofield, O.; Brown, M.; Kohut, J.; Nardelli, S.; Saba, G.; Waite, N.; Ducklow, H. Changing upper ocean mixed layer depth and phytoplankton productivity along the West Antarctic Peninsula. *Philos. Trans. R. Soc. A* **2018**, *376*, 20170174. [CrossRef]
66. Moline, M.A.; Claustre, H.; Frazer, T.K.; Schofield, O.; Vernet, M. Alteration of the food web along the Antarctic Peninsula in response to a regional warming trend. *Glob. Chang. Biol.* **2004**, *10*, 1973–1980. [CrossRef]

67. Montes-Hugo, M.; Doney, S.C.; Ducklow, H.W.; Fraser, W.; Martinson, D.; Stammerjohn, S.E.; Schofield, O. Recent changes in phytoplankton communities associated with rapid regional climate change along the Western Antarctic Peninsula. *Science* **2009**, *323*, 1470–1473. [CrossRef]
68. Biggs, T.E.G.; Alvarez-Fernandez, S.; Evans, C.; Mojica, K.D.A.; Rozema, P.D.; Venables, H.J.; Pond, D.W.; Brussaard, C.P.D. Antarctic phytoplankton community composition and size structure: Importance of ice type and temperature as regulatory factors. *Polar. Biol.* **2019**, *42*, 1997–2015. [CrossRef]
69. Atkinson, A.; Hill, S.L.; Pakhomov, E.A.; Siegel, V.; Anadon, R.; Chiba, S.; Daly, K.L.; Downie, R.; Fielding, S.; Fretwell, P.; et al. KRILLBASE: A circumpolar database of Antarctic krill and salp numerical densities, 1926–2016. *Earth Syst. Sci. Data* **2017**, *9*, 193–210. [CrossRef]
70. Morozov, E.G.; Krechik, V.A.; Frey, D.I.; Polukhin, A.A.; Artemiev, V.A.; Kasyan, V.V.; Sapozhnikov, P.V.; Mukhametianov, R.Z. Frontal Zone between Relatively Warm and Cold Waters in the Northern Weddell Sea. In *Antarctic Peninsula Region of the Southern Ocean*; Morozov, E.G., Flint, M.V., Spiridonov, V.A., Eds.; Springer: Cham, Switzerland, 2021; Volume 6, pp. 31–54. [CrossRef]
71. Gallienne, C.P.; Robins, D.B. Is Oithona the most important copepod in the world's oceans? *J. Plankton Res.* **2001**, *23*, 1421–1432. [CrossRef]
72. Skjoldal, H.R.; Wiebe, P.H.; Postel, L.; Knutsen, T.; Kaartvedt, S.; Sameoto, D.D. Intercomparison of zooplankton (net) sampling systems: Results the ICES/GLOBEC sea-going workshop. *Prog. Oceanogr.* **2013**, *108*, 1–42. [CrossRef]
73. Morozov, E.G.; Flint, M.V.; Orlov, A.M.; Frey, D.I.; Molodtsova, T.N.; Krechik, V.A.; Latushkin, A.A.; Salyuk, P.A.; Murzina, S.A.; Minin, K.V.; et al. Oceanographic and Ecosystem Studies in the Atlantic Sector of Antarctica (Cruise 87 of the Research Vessel Akademik Mstislav Keldysh). *Oceanology* **2022**, *62*, 825–827. [CrossRef]
74. Hopkins, T.L.; Torres, J.J. The zooplankton community in the vicinity of the ice edge, western Weddell Sea, March 1986. *Polar. Biol.* **1988**, *9*, 79–87. [CrossRef]
75. Voronina, N.M.; Kosobokova, K.N.; Pakhomov, E.A. Composition and biomass of summer metazoan plankton in the 0–200 m layer of the Atlantic sector of the Antarctic. *Polar. Biol.* **1994**, *14*, 91–95. [CrossRef]
76. Fransz, H.G.; Gonzalez, S.R. The production of *Oithona similis* (Copepoda: Cyclopoida) in the Southern Ocean. *ICES (Int. Counc. Explor. Sea) J. Mar. Sci.* **1995**, *52*, 549–555.
77. Pinkerton, M.H.; Smith, A.N.H.; Raymond, B.; Hosie, G.W.; Sharp, B.; Leathwick, J.R.; Bradford-Grieve, J.M. Spatial and seasonal distribution of adult *Oithona similis* in the Southern Ocean: Predictions using boosted regression trees. *Deep Sea Res. I* **2010**, *57*, 469–485. [CrossRef]
78. Takahashi, M.; Nemoto, T. The food of some Antarctic fish in the western Ross Sea in summer 1979. *Polar. Biol.* **1984**, *3*, 237–239. [CrossRef]
79. Spiridonov, V.A.; Zalota, A.K.; Yakovenko, V.A.; Gorbatenko, K.M. Composition of population and transport of juveniles of Antarctic krill in Powell Basin region (northwestern Weddell Sea) in January 2020. *Tr. VNIRO* **2020**, *181*, 33–51. [CrossRef]
80. Tarling, G.A.; Peck, V.; Ward, P.; Ensor, N.; Achterberg, E.; Tynan, E.; Poulton, A.J.; Mitchell, E.; Zubkov, M.V. Effects of acute ocean acidification on spatially–diverse polar pelagic foodwebs: Insights from on–deck microcosms. *Deep-Sea Res. II* **2016**, *127*, 75–92. [CrossRef]
81. Trifoglio, N.L.; Olguin Salinas, H.F.; Franzosi, C.A.; Alder, V.A. Annual cycle of phytoplankton, protozoa and diatom species from Scotia Bay (South Orkney Islands, Antarctica): Community structure prior to, during and after an anomalously low sea ice year. *Prog. Oceanogr.* **2022**, *204*, 102807. [CrossRef]
82. Quetin, L.B.; Ross, R.M. Feeding by Antarctic krill, *Euphausia superba*: Does size matter? In *Antarctic Nutrient Cycles and Food Webs*; Siegfried, W.R., Condy, P.R., Laws, R.M., Eds.; Springer: Berlin, Germany, 1985; pp. 372–377.
83. Granli, E.; Graneli, W.; Rabbani, M.M.; Daugbjerg, N.; Fransz, G.; Roudy, J.C.; Alder, V.A. The influence of copepod and krill grazing on the species composition of phytoplankton communities from the Scotia Weddell Sea. *Polar. Biol.* **1993**, *13*, 201–213. [CrossRef]
84. Froneman, P.W.; Perissinotto, R. Microzooplankton grazing and protozooplankton community structure in the South Atlantic and in the Atlantic sector of the Southern Ocean. *Deep Sea Res.* **1996**, *43*, 703–721. [CrossRef]
85. Perissinotto, R.A.; Pakhomov, E.A. The trophic role of the tunicate *Salpa thompsoni* in the Antarctic marine ecosystem. *J. Mar. Syst.* **1998**, *17*, 361–374. [CrossRef]
86. Flores, H.; Hunt, B.P.V.; Kruse, S.; Pakhomov, E.A.; Siegel, V.; van Franeker, J.A.; Strass, V.; Van de Putte, A.P.; Meesters, E.H.W.G.; Bathmann, U. Seasonal changes in the vertical distribution and community structure of Antarctic macrozooplankton and micronekton. *Deep Sea Res.* **2014**, *84*, 127–141. [CrossRef]
87. Takahashi, K.T.; Ojima, M.; Tanimura, A.; Odate, T.; Fukuchi, M. The vertical distribution and abundance of copepod nauplii and other micro- and mesozooplankton in the seasonal ice zone of Lützow-Holm Bay during austral summer 2009. *Polar. Biol.* **2017**, *40*, 79–93. [CrossRef]
88. Conroy, J.A.; Steinberg, D.K.; Thibodeau, P.S.; Schofield, O. Zooplankton diel vertical migration during Antarctic summer. *Deep Sea Res. I* **2020**, *162*, 103324. [CrossRef]
89. Delgado, L.E.; Jana, R.; Mari, V.H. Testing hypotheses on life-cycle models for Antarctic calanoid copepods, using qualitative, winter, zooplankton samples. *Polar. Biol.* **1998**, *20*, 74–76. [CrossRef]
90. Tarling, G.A.; Ward, P.; Thorpe, S.E. Spatial distributions of Southern Ocean mesozooplankton communities have been resilient to long-term surface warming. *Glob. Chang. Biol.* **2018**, *24*, 132–142. [CrossRef]

91. Mackintosh, N.A. Life cycle of Antarctic krill in relation to ice and water conditions. *Discov. Rep.* **1972**, *36*, 1–94.
92. Ross, R.M.; Quetin, L.B.; Kirsch, E. Effect of temperature on developmental times and survival of early larval stages of *Euphausia superba* Dana. *J. Exp. Mar. Biol. Ecol.* **1988**, *121*, 55–71. [CrossRef]
93. Ross, R.M.; Quetin, L.B.; Baker, K.S.; Vernet, M.; Smith, R.C. Growth limitation in young *Euphausia superba* under field conditions. *Limnol. Oceanogr.* **2000**, *45*, 31–43. [CrossRef]
94. Kouwenberg, J.H.M.; Razouls, C.; Desreumaux, N. Southern Ocean pelagic copepods. In *The Biogeographic Atlas of the Southern Ocean*; De Broyer, C., Koubbi, P., Griffith, H.J., Raymond, B., d'Udekem d'Acoz, C., Van de Putte, A.D., Danis, B., David, B., Grant, S., Gutt, J., et al., Eds.; SCAR: Cambridge, UK, 2014; pp. 209–296.
95. Cornils, A.; Sieger, R.; Mizdalski, E.; Schumacher, S.; Grobe, H.; Schnack-Schiel, S.B. Copepod species abundance from the Southern Ocean and other regions (1980–2005)—A legacy. *Earth Syst. Sci. Data* **2018**, *10*, 1457–1471. [CrossRef]

water

MDPI

Article

Distribution and Phylogenetic Position of the Antarctic Ribbon Worm *Heteronemertes longifissa* (Nemertea, Pilidiophora)

Alexei V. Chernyshev * and Neonila E. Polyakova

A.V. Zhirmunsky National Scientific Center of Marine Biology, Far Eastern Branch, Russian Academy of Sciences, ul. Palchevskogo 17, Vladivostok 690041, Russia
* Correspondence: nemertea1969@gmail.com

Abstract: To date, a total of 23 valid species of heteronemerteans belonging to 15 genera have been recorded in Antarctic and Subantarctic waters. The ribbon worm *Heteronemertes longifissa* (Hubrecht, 1887) is the only heteronemertean species reported to have bipolar distribution, but this statement is doubtful. The phylogenetic relationships of *H. longifissa* to other heteronemerteans remain uncertain. A genetic analysis of specimens from Antarctica has shown that the name *H. longifissa* refers to two sibling species with an uncorrected *p*-distance of 5.3% in COI. These species differ in body color: one is whitish, and the other is grayish-pink. The species with the whitish body has been reliably identified from off the Norway coast (as *Cerebratulus* sp. NemBar1383 (BOLD: ACM5920)), i.e., it has a bipolar distribution. A molecular phylogenetic analysis of Lineidae based on five gene markers (COI, 16S, 18S, 28S, and histone H3) has shown the genus *Heteronemertes* to belong to Lineage D of Clade 2 sensu Kajihara et al., 2022 (crown Lineidae). The phylogenetic positions of four more species of unidentified lineids are currently under discussion.

Keywords: benthos; nemerteans; larvae; bipolar distribution

Citation: Chernyshev, A.V.; Polyakova, N.E. Distribution and Phylogenetic Position of the Antarctic Ribbon Worm *Heteronemertes longifissa* (Nemertea, Pilidiophora). *Water* **2023**, *15*, 809. https://doi.org/10.3390/w15040809

Academic Editor: Morozov Eugene

Received: 23 January 2023
Revised: 14 February 2023
Accepted: 16 February 2023
Published: 19 February 2023

1. Introduction

The phylum Nemertea currently comprises approximately 1340 species [1] inhabiting mainly the littoral and sublittoral zones of the world's oceans. Studies on Antarctic and Subantarctic nemerteans have been conducted since the late 19th century [2,3], but the most important works were published in the period from 1905 to 1985 [4–14]. Nemerteans of the class Heteronemertea have always attracted special attention as these animals reach large sizes and are most frequently found in benthic samples. Thus, *Parborlasia corrugata* (McIntosh, 1876), encountered mostly in Antarctic and Subantarctic waters, grows to 1–2 m in length. A total of 23 valid species of heteronemerteans belonging to 15 genera have been recorded from Antarctic and Subantarctic waters south of latitude 50° S [14,15]. Among them, the ribbon worm *Heteronemertes longifissa* (Hubrecht, 1887) (=*Cerebratulus longifissus*, *Lineus longifissus*) is the only heteronemertean species with bipolar distribution reported. Known from Antarctic and Subantarctic waters, this species has also been found in the North and Barents seas [16,17] and even in the coastal waters of Japan [18,19]. The specimens from Japan belong to another species [14] which has recently been described as *Corsoua takakurai* Natsumi and Kajihara, 2000 [20]. Gibson [14] assumed that the nemerteans from the Arctic seas were also misidentified as *C. longifissus*. Genetic studies could address this issue, but, of all the described Antarctic and Subantarctic heteronemerteans, sequences are currently available only for *Parborlasia corrugata* s.l. [21–23]. A substantial number of sequences were obtained from adult individuals and larvae of unidentified heteronemerteans from the coastal waters of Antarctica [22,24,25]. In the present study, we discuss the results of a molecular phylogenetic analysis of Lineidae conducted in order to clarify whether *H. longifissa* actually has a bipolar distribution, and also to identify its relationships with other heteronemerteans. The systematic positions of several unidentified heteronemertean species from Antarctica are also considered.

2. Materials and Methods

2.1. Specimen Collection

Samples were collected with a Sigsbee trawl from five localities in Antarctic waters during research cruises #79 (January and February 2020) and #87 (January–April 2022) aboard the R/V *Akademik Mstislav Keldysh* (Table 1). On board, the sediment was carefully sieved through a 1000 μm mesh screen, sorted out in seawater, and nemerteans were photographed and fixed in 96% ethanol.

Table 1. Specimens and sampling points of some Lineidae collected during the research cruises aboard the R/V *Akademik Mstislav Keldysh*.

Specimen	Station	Coordinates	Depth, m	Date
Heteronemertes longifissa 11	6615	60.8879 S. 45.5342 W	370	30 January 2020
Heteronemertes longifissa 20	6615	60.8879 S. 45.5342 W	370	30 January 2020
Heteronemertes longifissa 25	6615	60.8879 S. 45.5342 W	370	30 January 2020
Heteronemertes longifissa 103	7371	61.1977 S. 47.1032 W	1459	8 February 2022
Heteronemertes longifissa 107	7371	61.1977 S. 47.1032 W	1459	8 February 2022
Heteronemertes longifissa 75	6615	60.8879 S. 45.5342 W	370	30 January 2020
Heteronemertes longifissa 78	6615	60.8879 S. 45.5342 W	370	30 January 2020
Heteronemertes longifissa 81	6615	60.8879 S. 45.5342 W	370	30 January 2020
Parborlasia corrugata 111	7371	61.1977 S. 47.1032 W	1459	8 February 2022
Cerebratulus sp. Antarctica 28	6614	60.8862 S. 45.5282 W	367	29 January2020
Lineidae sp. Antarctica 1	6615	60.8879 S. 45.5342 W	370	30 January 2020
Lineidae sp. Antarctica 5	6658	61.0387 S. 50.6875 W	740	18 February 2020
Lineidae sp. Antarctica 14	6652.	63.2897 S. 53.6021 W	364	15 February 2020
Lineidae sp. Antarctica 16	6615	60.8879 S. 45.5342 W	370	30 January 2020
Lineidae sp. Antarctica 18	6615	60.8879 S. 45.5342 W	370	30 January2020

2.2. DNA Extraction, PCR Amplification, and Sequencing

Total genomic DNA was extracted from ethanol-fixed specimens using a DNA-sorb-B nucleic acid extraction kit (AmpliSens, Moscow, Russia) and DNeasy Blood according to the manufacturer's protocol. Five markers of partial nuclear 18S rRNA (18S), 28S rRNA (28S), histone H3 (H3), and mitochondrial 16S rRNA (16S) and cytochrome *c* oxidase subunit I (COI) gene sequences were amplified from the genomic DNA. Amplification of polymerase chain reaction (PCR) was carried out using the primers listed in Table S1. PCR cycling profiles were as follows: for COI, 2 min at 94 °C, 40 cycles (40 s at 94 °C, 40 s at 50 °C and 1 min at 72 °C) and 7 min at 72 °C; for 16S, 2 min at 94 °C, 40 cycles (40 s at 94 °C, 40 s at 48 °C and 1 min at 72 °C) and 7 min at 72 °C; for 18S, 2 min at 94 °C, 40 cycles (1 min at 94 °C, 1 min at 52 °C for primer pairs Tim A/1100R and 3F/18Sbi, [1 min at 49 °C for the primer pair 18Sa2.0/9R], and 1 min at 72 °C) and 7 min at 72 °C; for 28S, 2 min at 94 °C, 40 cycles (40 s at 94 °C, 40 s at 52 °C, and 1 min at 72 °C) and 7 min at 72 °C; for H3, 2 min at 94 °C, 35 cycles (40 s at 94 °C, 40 s at 55 °C and 1 min at 72 °C) and 7 min at 72 °C.

The amplified products were purified using ExoSAP (Thermo Fisher Scientific, Waltham, MA, USA). Sequencing in forward and reverse directions was carried out on an ABI Prism 3500 Genetic Analyzers (Applied Biosystems, Waltham, MA, USA) under conditions recommended by the manufacturer, using a BigDye Terminator ver. 3.1 Cycle Sequencing Kit (Applied Biosystems) and the same primers as for PCR. BLAST searches [26], as implemented in the NCBI website (http://www.ncbi.nlm.nih.gov accessed on 17 February 2023), were conducted to check for putative contamination.

2.3. Phylogenetic Analysis

The sequences for the five gene markers (16S, 18S, 28S, COI, and H3) were aligned separately using MAFFT ver. 7 [27] with default parameters. We put more weight on unedited alignment including all positions, as suggested by [28]. A supermatrix with a total length of 6656 bp was formed by concatenating the five markers using SequenceMatrix [29], wherein external gaps were coded as 'missing data'. Simultaneous selection of partition schemes and the search for optimal nucleotide substitution models for the supermatrix obtained were carried out using PartitionFinder [30,31] with implementation of the 'greedy' search scheme. According to the best-suggested scheme, the final supermatrix was divided into seven partitions (Table S2). A combined analysis based on the five concatenated gene markers was conducted using Bayesian inference (BI) and maximum likelihood analyses (ML). BI was carried out in MrBayes 3.2 [32] by launching two parallel runs with four Markov chains in each run (three cold and one hot) during 10,000,000 generations.

The values of run convergence indicated that a sufficient number of trees and parameters were sampled. Based on the convergence of likelihood scores, 25% of sampled trees were discarded as burn-in. The rest was used to build the consensus tree, while the nodes with posterior probabilities of less than 50% collapsed. The maximum likelihood phylogenetic tree was inferred using the edge-147 linked partition model on the IQ-TREE web server [33]; branch supports with the 1000 ultrafast bootstrap replicates were obtained in the IQ-TREE software [34]. The BI topology was chosen as the main phylogenetic scheme for the present study. A total of 89 taxa of Lineidae from different genera were included in the molecular phylogenetic analyses, with *Sonnenemertes cantelli* and *Baseodiscus mexicanus* used as outgroups (Table 2).

We reconstructed the haplotype network based on the COI gene sequences including both all original *H. longifissa* sequences and one accessed from GenBank trimmed to the length of the shortest sequence, 600 bp, using TCS v1.21 software under 95% connection limit. The haplotype network was visualized as a pie chart with geographic information taken into account using the tcsBU web-based program.

The uncorrected pairwise *p*-distances were calculated in MEGA ver. 6.0.

Table 2. List of species included in the phylogenetic analysis of Lineidae s.l., with GenBank accession numbers for sequences (the sequences new to this study are highlighted in bold).

Species	16S	18S	28S	COI	H3	Source
Apatronemertes albimaculosa	JF277587	JF293030	HQ856860	HQ848584	JF277733	[21]
Cerebratulus lacteus	JF277575	JF293044	HQ856857	HQ848576	JF277728	[21]
Cerebratulus marginatus	JF277576	JF293042	HQ856858	HQ848575	JF277729	[21]
Cerebratulus orochi	LC538101	LC538103	LC538104	LC538102	LC538105	[35]
Cerebratulus mordukhovichi	OM422971	OM423090	OM423029	OM456681	OM468125	[36]
Cerebratulus sp. NemBar1383	–	–	–	KP697728	–	Strand unpubl.
***Cerebratulus* sp. Antarctica28**	**OQ449306**	**OQ449292**	**OQ449324**	**OQ450482**	**OQ446609**	**Present study**
Cerebratulus sp. DH-2009 isolate A4pilidia04	GU227009			GU227125		[24]
Cf. Heteronemertea sp. DH-2009_isolate_D4pilidia04	GU227013	–	–	GU227120	–	[24]
Cf. Heteronemertea sp. DH-2009_isolate_A4larva04	GU227014	–	–	GU227126	–	[24]
Corsoua takakurai	LC520112	–	LC520126	LC520106	LC520128	[20]
Dushia wijnhoffae	EF124878	–	EF178494	EF124967	–	[37]
Dushia cf. *nigra*	LC389832	LC389840	LC389844	LC389867	LC389851	[37]
Euborlasia maycoli	LC520114	LC520121	LC520125	LC520108	–	[38]
Gorgonorhynchus albocinctus	–	LC010650	LC010651	LC010649	–	[39]
Gorgonorhynchus cf. *bermudensis*	KF935467	KF935300	KF935356	KF935517	KF935412	[40]
Gorgonorhynchus cf. *repens*	LC520115	LC520122	LC520123	LC520105	LC520131	[41]
Heteronemertea gen. sp. 4 TCH-2015_isolate_119	KU197548	–	KU365690	KU197835	–	[42]
Heteronemertea sp. 17	LC625672	LC625688	LC625699	LC625640	LC625729	[43]
***Heteronemertes longifissa* 11**	–	**OQ449293**	**OQ449325**	**OQ450483**	–	**Present study**
***Heteronemertes longifissa* 20**	**OQ449307**	**OQ449294**	–	**OQ450484**	**OQ446610**	**Present study**
***Heteronemertes longifissa* 25**	**OQ449308**	**OQ449295**	**OQ449326**	**OQ450485**	**OQ446611**	**Present study**
***Heteronemertes longifissa* 75**	–	–	–	**OQ450486**	–	**Present study**
***Heteronemertes longifissa* 78**	**OQ449309**	**OQ449296**	**OQ449327**	**OQ450487**	**OQ446612**	**Present study**
***Heteronemertes longifissa* 81**	–	–	–	**OQ450488**	–	**Present study**
***Heteronemertes longifissa* 103**	–	–	–	**OQ450489**	–	**Present study**
***Heteronemertes longifissa* 107**	–	–	–	**OQ450490**	–	**Present study**
***Hinumanemertes kikuchii* ***	**OQ449310**	**OQ449297**	**OQ449328**	**OQ450491**	**OQ446613**	**Present study**
Kulikovia alborostrata	KU821503	-	KU856679	KU821529	KU821552	[44]
Kulikovia manchenkoi	KU821497	KY468934	KU856671	KU821523	KU821546	[44]

Table 2. *Cont.*

Species	16S	18S	28S	COI	H3	Source
Kulikovia cf. *montgomeryi*	OM422978	OM423098	OM423037	OM456685	OM456685	[36]
Kulikovia torguata LtUr1	KU821486	KY468935	KU856673	KU821511	KU821534	[44]
Lineidae KuramBio1 12-5	MN211473	MN211375	MN211427	MN205497	MN205448	[45]
Lineidae KuramBio2 85	MN211475	MN211377	MN211429	MN205498	–	[45]
Lineidae KuramBio2 77	MN211481	MN211383	MN211434	MN205504	MN205454	[45]
Lineidae sp. 41DS	MF512050	MF512076	MF512102	–	MF512144	[45]
Lineidae sp. Antarctica 1	OM422984	OM423104	OM423043	OM456687	OM468138	[36]
Lineidae_sp. Antarctica 5	OM422988	OM423108	–	OM456691	OM468142	[36]
Lineidae sp. Antarctica 14	OM422986	OM423106	OM423045	OM456689	OM468140	[36]
Lineidae sp. Antarctica 16	OM422987	OM423107	OM423046	OM456690	OM468141	[36]
Lineidae sp. Antarctica 18	OM422985	OM423105	OM423044	OM456688	OM468139	[36]
Lineidae sp. G06 Bering	OM422979	OM423099	OM423038	–	OM468133	[36]
Lineidae sp. H02 IceAGE	OM422982	OM423102	OM423041	–	OM468136	[36]
Lineidae sp. KGK-2	LC625651	–	–	LC625624	LC625709	[43]
Lineidae sp. KGK-4	LC625653	LC625683	–	LC625626	LC625711	[43]
Lineidae sp. KGK-6	LC625656	–	LC625690	LC625627	LC625714	[43]
Lineidae sp. KGK-7	LC625657	–	LC625691	–	–	[43]
Lineidae sp. KGK-8	LC625658	–	–	LC625628	LC625715	[43]
Lineidae sp. Kuril O11	OM422989	OM423109	OM423047	OM456692	OM468143	[36]
Lineus acutifrons	JF277573	JF304778	HQ856855	GU590937	JF277727	[21]
Lineus clandestinus	**MK064103**	**OQ449298**	**OQ449329**	**MK078739**	**OQ446614**	[46] **present study**
Lineus flavescens	KP682165	–	EF178497	KP682050	–	[47,48]
Lineus lacteus	JF277584	JF293065	HQ856850	HQ848583	JF277725	[21]
Lineus longissimus	MW073006	KY468932	MW077245	KY561813	KY606234	[44,49]
Lineus sanquineus	KF935468	KF935301	KF935301	KF935518	KF935413	[40]
Lineus ruber	MK064093	KY468933 *	KY468929 *	MK078684	KY606235 *	[44,46]
Lineus sp. Guam	KU821507	–	KY468928	–	KY561818	[44]
Lineus viridis	MK064101	**OQ449299**	**OQ449330**	MK078733	**OQ446615**	[46] **present study**
Maculaura aquilonia **	**OQ449311**	**OQ449300**	**OQ449331**	**OQ450492**	**OQ446616**	**Present study**
***Maculaura* sp.** *	–	**OQ449301**	**OQ449332**	**OQ450493**	**OQ446617**	**Present study**
Micrura bathyalis	MN211479	MN211381	MN211432	MN205502	–	[45]
Micrura bella ***	**OQ449312**	**OQ449302**	**OQ449333**	**OQ450494**	**OQ446618**	**Present study**
Micrura callima	MN211472	MN211374	MN211426	MN205496	MN205447	[45]
Micrura chlorapardalis	KF935459	KF935292	KF935348	KF935512	KF935404	[40]
Micrura dellechiajei	KF935461	KF935294	KF935350	KF935514	KF935406	[40]
Micrura fasciolata	JF277585	JF293038	HQ856846	HQ848578	JF277721	[21]

Table 2. *Cont.*

Species	16S	18S	28S	COI	H3	Source
Micrura ignea ****	**OQ449313**	**OQ449303**	**OQ449334**	**OQ450495**	**OQ446619**	**Present study**
Micrura purpurea	JF277577	JF293036	HQ856845	HQ848586	JF277726	[21]
Micrura rubramaculosa	KF935460	KF935293	KF935349	KF935513	KF935405	[40]
Micrura sp. albocephala	KU197574	–	KU365712	KU197849	–	[42]
Micrura sp. dark	KU197586	–	KU365713	KU197858	–	[42]
Micrura sp. 3	KU197563	–	KU365710	KU197841	–	[42]
Micrura sp. 4	KU197581	–	KU365711	KU197857	–	[42]
Micrura sp. IceAGE	OM422994	OM423114	OM423052	OM456696	OM468146	[36]
Micrura sp. IZ 132532	KF935457	KF935290	KF935346	KF935510	KF935402	[42]
Micrura sp. IZ 132529	KF935458	KF935291	KF935347	KF935511	KF935403	[40]
Micrura verrilli	KF935455	KF935288	KF935344	KF935508	KF935400	[40]
Micrura wilsoni	KU197535	–	KU365716	KU197827	–	[42]
Notospermus geniculatus	KF935462	KF935295	KF935351	–	KF935407	[40]
Notospermus mitellatus	LC625660	LC625685	LC625693	LC625629	LC625717	[43]
Nipponomicrura* sp.	**OQ449314**	**OQ449304**	**OQ449335**	**OQ450496**	**OQ446620**	**Present study**
Nipponomicrura uchidai	KU821509	–	KY468930	KY561815	KY561819	[44]
Parborlasia corrugata	JF277578	JF293037	HQ856851	EU194826 *	JF277732	[21,22]
***Parborlasia corrugata* 111**	**OQ449315**	**OQ449305**	**OQ449336**	**OQ450497**	**OQ446621**	**Present study**
Parvicirrus dubius	AJ436830	–	AJ436885	AJ436940	–	[50]
Polydendrorhynchus zhanjiangensis	MT659662	MT648831	MT648832	MT648511	MT655749	[41]
Pseudomicrura afzelii	GU445914	GU445924	GU445919	GU392013	–	[51]
Riseriellus occultus	JF277581	JF293031	HQ856848	HQ848581	JF277724	[21]
Siphonenteron nakanoi	LC625678	–	LC625706	LC625646	LC625737	[43]
Tenuilineus bicolor	AJ436823	–	EF124960	AJ436933	AJ436980	[48,50]
Zygeupolia rubens	JF277574	JF293045	HQ856861	HQ848585	JF277735	[21]
Yininemertes pratensis	KY274025	KY274047	KY274069	KY274003	KY274091	[40]
Outgroups						
Baseodiscus mexicanus	KF935449	KF935281	KF935337	KF935503	KF935393	[40]
Sonnenemertes cantelli	MF512048	MF512073	MF512099	MF512118	MF512141	[45]

* Sea of Japan, Vostok Bay; ** Sea of Okhotsk, Magadan; *** Sea of Japan, Spokoynaya Bay; **** Panama.

3. Results

3.1. Phylogenetic Analysis of Lineidae

The combined aligned sequences comprise 636 bp for 16S rDNA, 1935 bp for 18S rDNA, 3096 bp for 28S rDNA, 658 bp for COI, and 331 bp for histone H3. They contain 2449 variable sites, of which 1640 are informative; the frequencies of variable and informative sites are 36.8% and 24.6%, respectively. The frequencies of variable sites in the aligned sequences of mitochondrial 16S rDNA (63.2%) and COI (50.9%) are greater than those in sequences of nuclear 18S rDNA (26.8%), 28S rDNA (34.4%) and histone 3 (38.4%).

The general topology of the lineid phylogenetic tree is as follows: *Micrura ingnae* + (Lineage A + Lineage C + (Clade 2 + Lineage B)) (Figure 1). Clade "2 + B" is strongly supported (PP = 1, BS = 99%). Relationships of Lineage D with other Lineages of Clade 2 sensu Kajihara et al., 2022 remain unresolved; Clade 3 has low support. Three strongly supported clades, each including two or more Lineages, can be distinguished within Clade 2 (crown Lineidae). Clade N + P (PP = 1, BS = 90%) comprises representatives of Lineages N and P, and also *Hinumanemertes kikuchii*. The monophyly of Lineage N sensu Kajihara et al., 2022 is not confirmed, since *Lineus acutifrons* and *H. kikuchii* form a strongly supported subclade, whose relationships inside Clade N+P remain unclear. Clade 4 (PP = 1, BS = 93%) comprises Lineages E, H, G, I, J, K, and Q, and also Lineidae KuramBio2 77. Clade J + K (PP = 1, BS = 93%) comprises representatives of sister Lineages J and K. *Micrira dellechiajei* forms a strongly supported subclade (PP = 1, BS = 100%) with *Notospermus* species (Lineage E). Some of the lineids are not included in any of the previously identified Lineages: Lineidae sp. H02 IceAGE, Lineidae KuramBio2 77, Heteronemertea sp. 17 HA-2021, Heteronemertea gen. sp. 4 TCH-2015, *Pseudomicrura afzelii*, and *Zygeupolia rubens*.

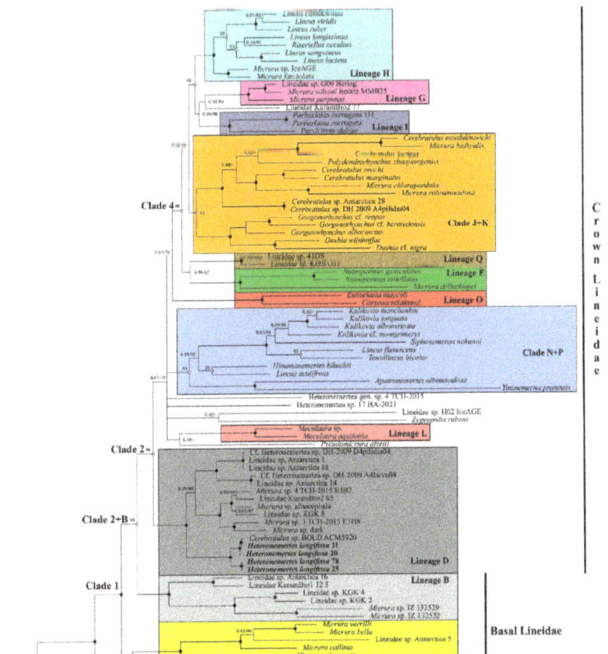

Figure 1. Bayesian inference (BI) phylogenetic tree for the supermatrix Lineidae of the five-marker dataset (16S, COI, 18S, 28S, H3). Numerals near the branches are nodal support values (Bayesian posterior probability/ML bootstrap value). Black circles indicate nodal support of 0.1/100; white circles indicate nodal support of 1.0/. Antarctic *Heteronemertes longifissa* specimens are highlighted in bold.

3.2. Antarctic Lineids

All eight individuals identified on the basis of external characters as *Heteronemertes longifissa* belong to Lineage D and are sisters of the rest of the representatives of this clade. Their distinctive feature is the very long cephalic horizontal slits. Three individuals (*H. longifissa* 25, 78, and 81) have a grayish-pink body color (Figure 2a–d), while five individuals (*H. longifissa* 11, 20, 75, 103, and 107) have a whitish color; reddish brain and beige gut are visible through translucent body wall (Figure 2e–g). Haplotypes for the COI gene are grouped into two networks (A and B) (Figure 3). The uncorrected *p*-distances between networks A and B are 5.2–5.7% in COI; the *p*-distances between the samples within each of the networks amount to 0.2–0.7% in COI (Table S3). The uncorrected *p*-distances between *H. longifissa* A and B are 3.0% and 1.3% in 16S and histone H3, respectively.

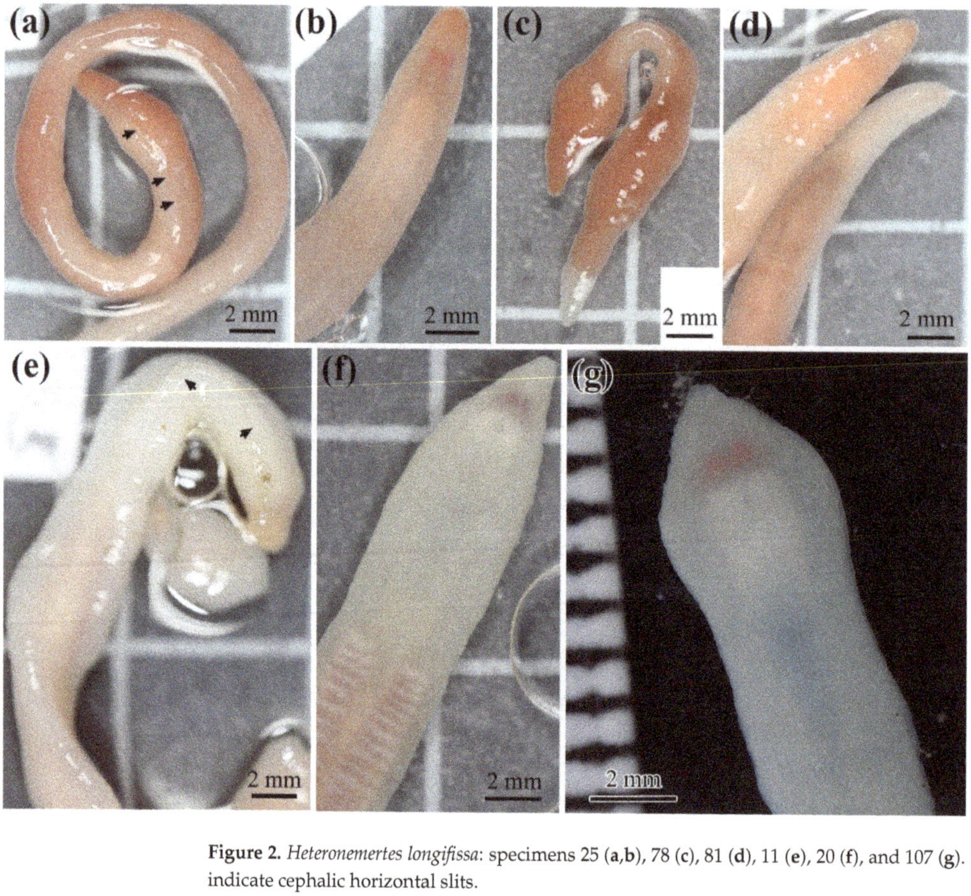

Figure 2. *Heteronemertes longifissa*: specimens 25 (**a**,**b**), 78 (**c**), 81 (**d**), 11 (**e**), 20 (**f**), and 107 (**g**). Arrows indicate cephalic horizontal slits.

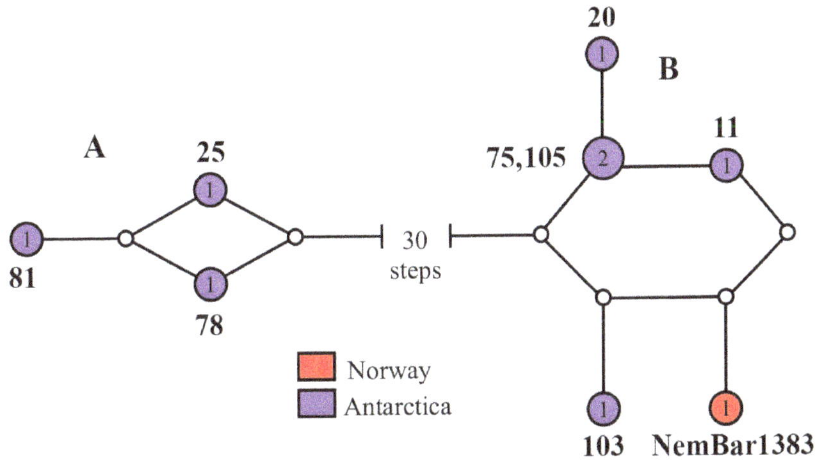

Figure 3. Statistical parsimony haplotype network based on the mitochondrial DNA cytochrome c oxidase subunit I gene of the examined *Heteronemertes longifissa* specimens, colored in accordance with their geographic distribution. The connecting limit is set at 95%. The line connecting the haplotype pie charts represents a single mutational change; each white dot on the line represents one additional mutational change. The numerals within the pie charts represent the number of specimens within each haplotype. The numerals near haplotypes correspond to the specimens' numbers (Table 2).

In GenBank (NCBI) and BOLD (The Barcode of Life Data System), three sequences are available that match the sequences of *H. longifissa*: two specimens of unidentified heteronemertean Nemertea sp. Antarctic ARM-2008 (accession nos. EU718394.1 and EU718388.1) from Antarctica [22] (with the uncorrected *p*-distance between this sample and *H. longifissa* B being 0.2% in 16S) and a specimen *Cerebratulus* sp. NemBar1383 BOLD: ACM5920 (accession no. KP697728) from Norway, belonging to network B (Figure 3) (with the uncorrected *p*-distances between this sample and *H. longifissa* B being 0.5–0.8% in COI—see Table S3).

The three unidentified heteronemerteans (Lineidae spp. Antarctica 1, 14, and 18) belong to Lineage D (Figure 1). Lineidae sp. Antarctica 1 and 18 have a pale olive body color with a reddish brain visible through a translucent body wall (Figure 4a,b); the body of Lineidae spp. Antarctica 14 is pale beige with a reddish brain (Figure 4c,d). A BLAST analysis has shown that the sequences of COI of pilidia from the Ross Sea under the names DH-2009 GU227120 isolate D4pilidia04 and DH-2009 GU227126 isolate A4larva04 [24] are identical to the sequences of Lineidae sp. Antarctica 1 (=Lineidae sp. Antarctica 18) and Lineidae sp. 14, respectively (Figure 1). Lineidae sp. Antarctica 16 with the deep, rose body (Figure 4e,f) is closely related to Lineidae KuramBio1 12 5 from the abyssal plain adjacent to the Kuril–Kamchatka Trench (Lineage B) (Figure 1). Lineidae sp. Antarctica 5 with a yellowish body and orange-yellowish head (Figure 4g,h) belongs to Lineage A and the subclade of the "*Micrura*" (*Evelineus*) species with a fragile soft body and red or orange anterior head end (Figure 1). *Cerebratulus* sp. Antarctica 28 with a grayish-pink body (Figure 4i–k) belongs to Lineage J (Figure 2). The pilidium from the Ross Sea under the name *Cerebratulus* sp. DH-2009 isolate A4pilidia04 [24] is conspecific to *Cerebratulus* sp. Antarctica 28 (Figure 1). One specimen with a grayish-red body and a light-colored transverse band on the head (Figure 4l) is conspecific to *Parborlasia corrugata*.

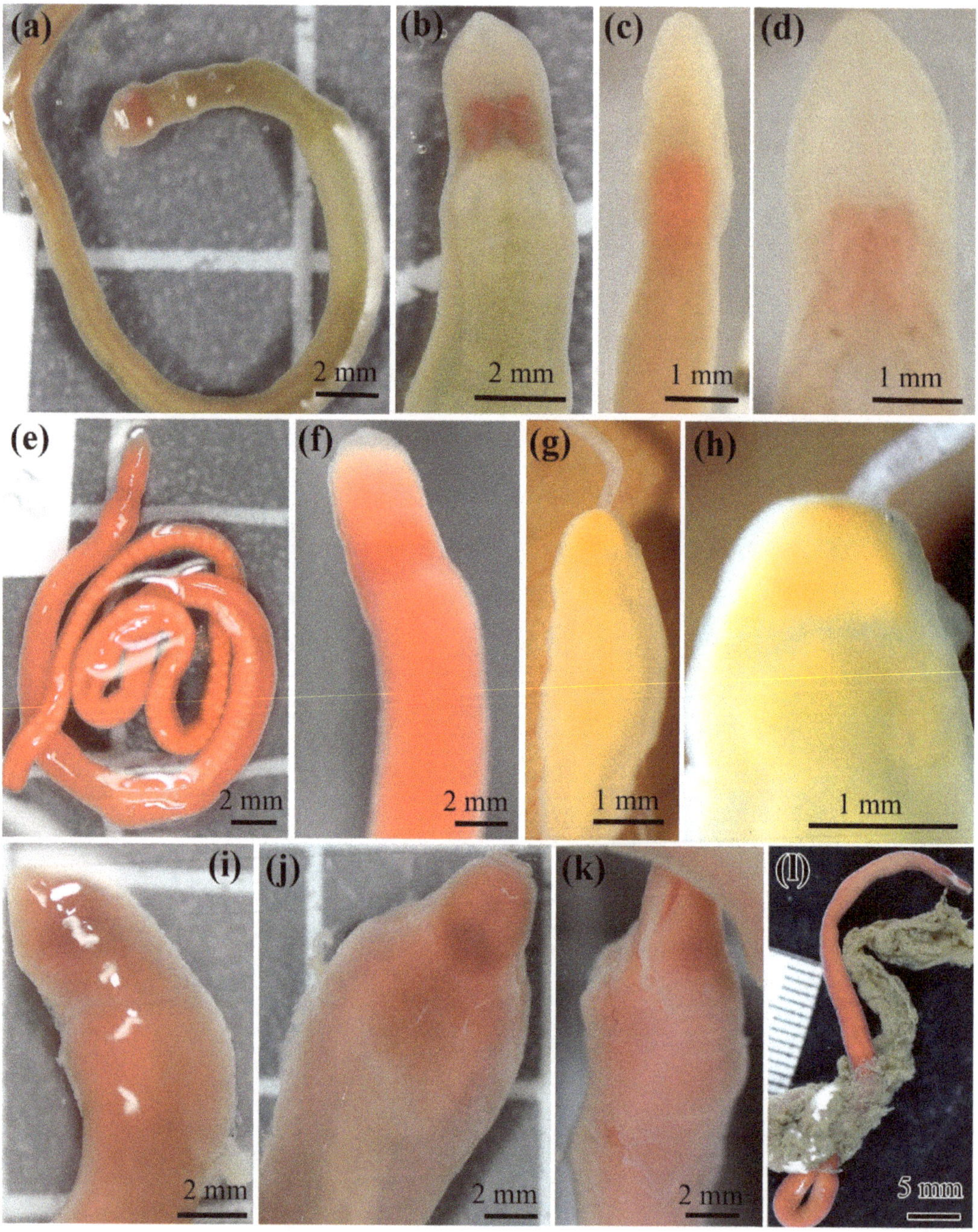

Figure 4. Antarctic lineids: Lineidae sp. Antarctica 1 (**a**,**b**), Lineidae sp. Antarctica 14 (**c**,**d**), Lineidae sp. Antarctica 16 (**e**,**f**), Lineidae sp. Antarctica 5 (**g**,**h**), *Cerebratulus* sp. Antarctica 28 (**i**–**k**), and *Parborlasia corrugata* 111 (**l**).

4. Discussion

4.1. Phylogenetic Analysis of Lineids

The molecular phylogenetic analyses of heteronemerteans carried out in 2019–2022 revealed the major clades within this order [36,43,45,52]. Chernyshev and Polyakova [36] proposed the names *Lineus*, *Cerebratulus*, and *Siphonenteron* for the three clades without providing definitions for them. An extensive phylogenetic analysis by Kajihara et al. [43] made it possible to identify 16 phylogenetic lineages (high-supported clades) designated with letters (A, B, C, D, etc.) within the family Lineidae. In addition, three large clades were identified: Clades 1, 2, and 3 [43]. Another Lineage (Q) was identified later [36]. The results obtained by Kajihara et al. [43] are largely confirmed by our studies with some changes and supplements. Inside Clade 2 (we suggest referring to it as "crown Lineidae"), our analysis does not support Clade 3 which should be a sister of Lineage D. In the analysis by Kajihara et al. [43], *Lineus acutifrons* belongs to Lineage N, holding a basal position in it. In our analysis, *L. acutifrons* and *Hinumanemertes kikuchii* form an independent subclade within Clade N + P. We have established the phylogenetic position of *H. kikuchii* for the first time, and this brackish-water species has not got in Lineage P that includes the freshwater species *Apatronemertes albimaculosa* and the brackish-water *Yininemertes pratensis*. We assume that these differences result from the fact that Kajihara et al. [43] included many species for which 1–2 sequences are known in their phylogenetic analysis, while for our analysis we selected a smaller number of species with 3–5 sequences known.

4.2. Systematic Position of the Genus Heteronemertes

Identification of heteronemerteans based on fixed specimens is challenging since in formalin and alcohol they usually lose the external traits characteristic of the species (body color pattern, head shape, etc.). In this regard, *Heteronemertes longifissa* has a characteristic feature that is clearly visible in both live and fixed individuals: very long lateral head slits reaching far behind the mouth [9,14] (Figure 2a,e). Our specimens have been identified on the basis of this feature. We have described the body color of live individuals for the first time.

The systematic position of *Heteronemertes longifissa* remains unclear. This species was described as a member of the genus *Cerebratulus* [53] and then transferred to the genus *Lineus* [9]. Gibson [14] described in detail the internal structure of *H. longifissa* and confirmed its affiliation to the genus *Lineus*, although the proboscis organization in this species is not typical of the species of *Lineus* s. str. Chernyshev [54] transferred *H. longifissa* into a new genus, *Heteronemertes*, and added such traits as the presence of very long head slits to its diagnosis. Our phylogenetic analysis has shown that *H. longifissa* belongs to Lineage D and is not closely related to either *Lineus* s. str. (Lineage H) or *Cerebratulus* s. str. (Lineage J). It is currently the only described species for which affiliation to Lineage D has been proven. Below is the revised and supplemented diagnosis of the genus *Heteronemertes*.

Genus *Heteronemertes* Chernyshev, 1995.

Type species: *Heteronemertes longifissa* (Hubrecht, 1887).

Diagnosis. Body flattened, caudal cirrus absent, and horizontal cephalic slits reach far behind the mouth. The cutis glands are not separated from the body wall musculature by a connective tissue layer. Longitudinal muscle plate between rhynchocoel and foregut absent. Proboscis typical heterotype (see [55]) with outer longitudinal musculature well developed, muscle crosses present, and neural sheath not developed into separate nerves. Foregut with subepithelial glands. Cerebral ganglia have outer and inner neurilemma; neurochords and neurochord cells are absent.

4.3. Is Heteronemertes longifissa a Bipolar Species?

Heteronemertes longifissa was first described from materials collected off Marion Island [39] and then found in different parts of the Antarctic and Subantarctic regions [9,10,13,14]. However, this species under the name *Cerebratulus longifissus* was reported for the North and Barents seas [16,17], and also as *Lineus longifissus* for the coastal waters of Japan [18,19].

The reason why the Japanese specimens were assigned to *Lineus longifissus* remains unclear, since the horizontal cephalic slits do not reach behind the mouth in them. It has been found that *Lineus longifissus* sensu Takakura, 1898 belongs to the recently described species *Corsoua takakurai*, which is attributed to Lineage O [20].

Friedrich [56] suggested that *Cerebratulus longifissus* from off Norway and the Barents Sea "... may be *Lineus longifissus*, or perhaps *Cerebratulus fissuralis*" (p. 18). *Cerebratulus fissuralis* Friedrich, 1958 from off Iceland, was described extremely briefly and without illustrations; this species differs from *Heteronemertes longifissa* by the presence of a neurochord. *Cerebratulus longifissus* sensu Punnett, 1903 from off Norway differs from *Heteronemertes longifissa* by having a thick layer of connective tissue between the dermal glands and the body wall outer longitudinal muscles, a short rhynchocoel, and a distinct cephalic vascular loop [16]. Therefore, Gibson [14] assumed *Cerebratulus longifissus* sensu Punnett, 1903 and *C. longifissus* sensu Uschakov, 1928 to belong to an independent species.

The molecular genetics analysis has shown significant genetic differences between the eight individuals of *H. longifissa* that we studied: uncorrected *p*-distances between the two *H. longifissa* networks of 5.2–5.7% in COI vs. 0.2–0.7% between the samples in each of the networks. These individuals also differ in body color: *H. longifissa* 11, 20, 75, 103, and 107 have a whitish color vs. a grayish-pinkish color in *H. longifissa* 25, 78, and 81. The low values of the COI uncorrected *p*-distances between networks A and B are a boundary between interspecific and intraspecific distances ("barcoding gap") recorded for heteronemerteans (4–9%) [57]. All three individuals were collected at the same station (Table 1) and, therefore, there is every reason to assume that *H. longifissa* includes two sibling species that differ in body color. Because of the lack of data on the body color pattern of the type specimens, the question as to which of the specimens should be assigned the name *H. longifissa* still remains open.

According to genetic data, *Cerebratulus* sp. NemBar1383 (BOLD: ACM5920), collected off Norway at a depth of 40 m north of Tromsø, is conspecific with *H. longifissa* 11, 20, 75, 103, and 107 from Antarctica. All these individuals have a whitish body color with a reddish brain visible through a translucent body wall. The snow-white body color was described also for *Cerebratulus longifissus* from off Norway [16]. The color pattern of the specimen from the Barents Sea is unknown [17]. Thus, one of the two sibling species of *H. longifissa* s.l. has a bipolar distribution. This is the first case where the bipolar distribution of nemerteans has been proven genetically. In addition to *H. longifissa*, bipolar distribution was reported for the hoplonemertean *Nipponnemertes pulchra* (Johnston, 1837) [58], but this conclusion needs genetic confirmation. A recent phylogenetic analysis has shown that *Nipponnemertes* sp. 17 from Antarctica, which looks very similar to *N. pulchra*, is an independent species [36].

4.4. Problems of Identification of Antarctic Heteronemerteans

One of the major challenges in the study of Antarctic and Subantarctic heteronemerteans is their species identification. First, this is explained by the fact that a substantial portion of the species was described only from fixed specimens. Thus, Gibson [14] described seven new heteronemertean species from Antarctica and Subantarctica, but none of the descriptions had information about the color and head shape of live individuals. Recent studies using gene markers show the importance of these traits in discriminating sibling species of heteronemerteans [44,46,47,52,59]. Using only internal morphology traits does not guarantee accurate identification of the sibling species. Even such a unique for heteronemerteans trait as very long horizontal cephalic slits is present in two sibling species, which is evidenced by our studies. Another issue is the great number of cryptic species among nemerteans. As has been shown recently, *Parborlasia corrugata*, widely distributed in Antarctica and the Subantarctic region, includes two cryptic species that differ genetically [22,23]. Thus, it is unclear which of these two cryptic species should be assigned the name *Parborlasia corrugata*. The same applies to *Heteronemertes longifissa*.

The problems in the identification of Antarctic heteronemerteans have resulted in a situation where sequences were obtained for only one described species (*Parborlasia corrugata*) [21–23] and a large number of species unidentified to date [22,25,36], including the four Lineidae species from our study. Most of these unidentified species are probably new to science, but their description is unlikely to be possible in the following decade. In addition to the sequences of adult heteronemerteans, sequences of heteronemertean larvae have also been obtained. However, researchers note great differences in the species diversity of adult and larval samples [25]. Even a common Antarctic species such as *Parborlasia corrugata* was found to comprise only 4.3% of larvae sampled [25]. A similar phenomenon of significant differences in the species composition of larvae and adult heteronemerteans has been reported for the Pacific coast of North America [42]. In this regard, the identity of the sequences of three heteronemerteans' larvae from Antarctica with the sequences of Lineidae sp. Antarctica 1, Lineidae sp. Antarctica 18, Lineidae sp. 14, and *Cerebratulus* sp. Antarctica 28 is of certain interest. These three adult lineids were collected from a depth of 370 m. The significant differences between pilidia and adult heteronemerteans can apparently be explained by the fact that a large number of larvae of the deep-sea species, which have been studied much more poorly than shallow-water nemerteans, are found in plankton.

Supplementary Materials: The following supporting information can be downloaded at https://www.mdpi.com/article/10.3390/w15040809/s1, Table S1: List of primers used in the present study. Forward primer sequences are highlighted in bold; Table S2: The main parameters of model-based inferred phylogenies; Table S3: Uncorrected COI *p*-distances (%) within *Heteronemertes longifissa*.

Author Contributions: Conceptualization, A.V.C.; methodology, A.V.C. and N.E.P.; software, N.E.P.; validation, A.V.C. and N.E.P.; formal analysis, A.V.C. and N.E.P.; investigation, A.V.C. and N.E.P.; resources, A.V.C. and N.E.P.; data curation, A.V.C.; writing—original draft preparation, A.V.C. and N.E.P.; writing—review and editing, A.V.C.; visualization, A.V.C. and N.E.P.; supervision, A.V.C.; project administration, A.V.C.; funding acquisition, A.V.C. All authors have read and agreed to the published version of the manuscript.

Funding: This research was funded by the Russian Science Foundation (grant no. 22-24-00184).

Data Availability Statement: The datasets studied during the present study are available from the corresponding author upon reasonable request.

Acknowledgments: The authors wish to thank Anna E. Vlasenko and Grigorii V. Malykin for providing the material from Antarctica. We are also grateful to Evgeniy P. Shvetsov for proofreading the English of this manuscript.

Conflicts of Interest: The authors declare no conflict of interest.

References

1. Chernyshev, A.V. An updated classification of the phylum Nemertea. *Invertebr. Zool.* **2021**, *18*, 188–196. [CrossRef]
2. McIntosh, W.C. Descriptions of some new species of Annelida from Kerguelen's Island. *Ann. Mag. Nat. Hist.* **1876**, *17*, 318–323. [CrossRef]
3. Bürger, O. Südgeorgische und andere exotische Nemertinen. *Zool. Jahrb. Abt. Syst. Geog. Biol. Tiere* **1893**, *7*, 207–240. [CrossRef]
4. Joubin, L. Note préliminaire sur les Némertiens recueillis par l'expédition antarctique française du Dr Charcot. *Bull. Mus. Natl. Hist. Nat.* **1905**, *11*, 431–437.
5. Joubin, L. Némertiens. In *Expédition Antarctique Francaise 1903–1905; Vers et Brachiopodes*; Masson: Paris, France, 1908; pp. 1–16.
6. Joubin, L. Nemertinea. In *National Antarctic Expedition 1901–1904*; Royal Society: London, UK, 1910; Volume 5, pp. 1–15.
7. Joubin, L. Némertiens. In *Deuxième expédition antarctique francaise (1908–1910)*; Masson: Paris, France, 1914; Volume 4, pp. 1–33.
8. Baylis, H.A. Nemertinea. In *British Antarctic ("Terra Nova") Expedition, 1910. Natural History Reports. Zoology*; British Museum: London, UK, 1915; Volume 2, pp. 113–134.
9. Wheeler, J.F.G. Nemerteans from the South Atlantic and southern oceans. *Discov. Rep.* **1934**, *9*, 215–294.
10. Wheeler, J.F.G. Nemerteans of Kerguelen and the Southern Ocean. In *Report, B.A.N.Z. Antarctic Research Expedition 1929–1931*; Series B; B.A.N.Z.A.R. Expedition Committee: Adelaide, Australia, 1940; Volume 4, pp. 233–256.
11. Korotkevitsch, V.S. Pelagic nemerteans of Antarctic and temperate waters of the Southern Hemisphere. *Issled. Fauny Morey* **1964**, *2*, 132–167. (In Russian)
12. De Esteban, S.; De La, C.J.; Moretto, H.J.A. Heteronemertea en la bahia de Ushuaia. I. *Parborlasia fueguina* sp. nov. *Huilkia ushuaiensis* gen. et sp. nov. *Physis. Buenos Aires* **1968**, *28*, 171–181.

13. Friedrich, H. Nemertinen aus Chile. *Sarsia* **1970**, *40*, 1–80. [CrossRef]
14. Gibson, R. Antarctic nemerteans: Heteronemertea—descriptions of new taxa, reappraisals of the systematic status of existing species and a key to the heteronemerteans recorded south of latitude 50° S. *Zool. J. Linn. Soc.* **1985**, *83*, 95–227. [CrossRef]
15. Fernandez-Álvarez, F.A.; Anadón, N. *Oligodendrorhynchus hesperides* gen. et sp. n. (Heteronemertea) from the Bellingshausen Sea. *Pol. Polar Res.* **2012**, *33*, 81–98. [CrossRef]
16. Punnett, R.C. On the nemerteans of Norway. *Bergens Mus. Årbog.* **1903**, *2*, 1–35.
17. Ushakov, P.V. To the fauna of nemerteans of the Barentz Sea. *Tr. Inst. Po Izucheniju Krainego Sev.* **1928**, *37*, 56–66. (In Russian)
18. Takakura, U. A classification of nemerteans from the vicinity of Misaki. *Zool. Mag.* **1898**, *10*, 38–44, 116–120, 184–187, 331–337, 424–429. (In Japanese)
19. Iwata, F. Nemertini from the coasts of Kyusyu. *J. Fac. Sci. Hokkaido Univ. Ser. 6 Zool.* **1952**, *11*, 126–148.
20. Hookabe, N.; Kajihara, H. Taxonomic reappraisal of *Lineus longifissus* auct. (Nemertea: Pilidiophora) from Japan for the first time in 122 years. *Zool. Sci.* **2020**, *37*, 467–475. [CrossRef]
21. Andrade, S.C.S.; Strand, M.; Schwartz, M.; Chen, H.-X.; Kajihara, H.; von Dohren, J.; Sun, S.; Junoy, J.; Thiel, M.; Norenburg, J.L.; et al. Disentangling ribbon worm relationships: Multi-locus analysis supports traditional classification of the phylum Nemertea. *Cladistics* **2012**, *28*, 141–159. [CrossRef]
22. Thornhill, D.J.; Mahon, A.R.; Norenburg, J.L.; Halanych, K.M. Open-ocean barriers to dispersal: A test case with the Antarctic Polar Front and the ribbon worm *Parborlasia corrugatus* (Nemertea: Lineidae). *Mar. Ecol.* **2008**, *17*, 5194–5197. [CrossRef]
23. Hookabe, N.; Watanabe, K.; Tsujimoto, M.; Kajihara, H. Molecular identity of the antarctic heteronemertean *Parborlasia corrugata* (Nemertea: Pilidiophora) from Lützow-Holm Bay. *Polar Sci.* **2020**, *25*, 100535. [CrossRef]
24. Heimeier, D.; Lavery, S.; Sewell, M.A. Using DNA barcoding and phylogenetics to identify Antarctic invertebrate larvae: Lessons from a large scale study. *Mar. Genom.* **2010**, *3*, 165–177. [CrossRef]
25. Mahon, A.R.; Thornhill, D.J.; Norenburg, J.L.; Halanych, K.M. DNA uncovers Antarctic nemertean biodiversity and exposes a decades-old cold case of asymmetric inventory. *Polar Biol.* **2010**, *33*, 193–202. [CrossRef]
26. Altschul, S.F.; Madden, T.L.; Schaffer, A.A.; Zhang, J.; Zhang, Z.; Miller, W.; Lipman, D.J. Gapped BLAST and PSI-BLAST: A new generation of protein database search programs. *Nucl. Acids Res.* **1997**, *25*, 3389–3402. [CrossRef]
27. Katoh, K.; Standley, D.M. MAFFT multiple sequence alignment software version 7: Improvements in performance and usability. *Mol. Biol. Evol.* **2013**, *30*, 772–780. [CrossRef]
28. Lindgren, A.R.; Daily, M. The impact of length-variable data and alignment criterion on the phylogeny of Decapodiformes (Mollusca: Cephalopoda). *Cladistics* **2007**, *23*, 464–476. [CrossRef]
29. Vaidya, G.; Lohman, D.J.; Meier, R. SequenceMatrix: Concatenation software for the fast assembly of multi-gene datasets with character set and codon information. *Cladistics* **2011**, *27*, 171–180. [CrossRef] [PubMed]
30. Lanfear, R.; Calcott, B.; Ho, S.Y.W.; Guindon, S. PartitionFinder: Combined selection of partitioning schemes and substitution models for phylogenetic analyses. *Mol. Biol. Evol.* **2012**, *29*, 1695–1701. [CrossRef] [PubMed]
31. Lanfear, R.; Calcott, B.; Kainer, D.; Mayer, C.; Stamatakis, A. Selecting optimal partitioning schemes for phylogenomic datasets. *BMC Evol. Biol.* **2014**, *14*, 1–14. [CrossRef]
32. Ronquist, F.; Teslenko, M.; van der Mark, P.; Ayres, D.L.; Darling, A.; Höhna, S.; Larget, B.; Liu, L.; Suchard, M.A.; Huelsenbeck, J.P. MrBayes 3.2: Efficient Bayesian phylogenetic inference and model choice across a large model space. *Syst. Biol.* **2012**, *61*, 539–542. [CrossRef]
33. Nguyen, L.T.; Schmidt, H.A.; von Haeseler, A.; Minh, B.Q. IQ-TREE: A fast and effective stochastic algorithm for estimating maximum likelihood phylogenies. *Mol. Biol. Evol.* **2015**, *32*, 268–274. [CrossRef]
34. Minh, B.Q.; Nguyen, M.A.T.; von Haeseler, A. Ultrafast approximation for phylogenetic bootstrap. *Mol. Biol. Evol.* **2013**, *30*, 1188–1195. [CrossRef]
35. Kajihara, H. Redescription of *Cerebratulus marginatus* auct. (Nemertea: Pilidiophora) from Hokkaido, Japan, as a new species. *Zootaxa* **2020**, *4819*, 295–315. [CrossRef]
36. Chernyshev, A.V.; Polyakova, N.E. Nemerteans collected in the Bering Sea during the research cruises aboard the R/V Akademik, M.A. Lavrentyev in 2016, 2018, and 2021 with an analysis of deep-sea heteronemertean and hoplonemertean species. *Deep-Sea Res. II* **2022**, *199*, 105081. [CrossRef]
37. Hookabe, N.; Schwartz, M.L.; Kajihara, H.; Norenburg, J.L. Molecular systematics of the heteronemertean genus *Dushia* (Nemertea, Pilidiophora), with descriptions of *D. wijnhoffae* sp. nov. and *D. nigra* species complex comb. nov. *Zootaxa* **2019**, *4691*, 333–358. [CrossRef]
38. Hookabe, N.; Kajihara, H. *Euborlasia* Vaillant, 1890 (Nemertea: Pilidiophora) from Bocas del Toro: Description of a new species, with comments on the systematics of the genus. *Mar. Biodivers.* **2020**, *50*, 1–13. [CrossRef]
39. Kajihara, H. A histology-free description of the branched proboscis ribbon worm *Gorgonorhynchus albocinctus* sp. nov. (Nemertea: Heteronemertea). *Publ. Seto Mar. Biol. Lab.* **2015**, *43*, 92–102. [CrossRef] [PubMed]
40. Kvist, S.; Laumer, C.E.; Junoy, J.; Giribet, G. New insights into the phylogeny, systematics and DNA barcoding of Nemertea. *Invertebr. Syst.* **2014**, *28*, 287–308. [CrossRef]
41. Hookabe, N.; Xu, C.-M.; Tsuyuki, A.; Jimi, N.; Sun, S.-C.; Kajihara, H. A new nemertean with a branched proboscis, *Gorgonorhynchus citrinus* sp. nov. (Nemertea: Pilidiophora), with molecular systematics of the genus. *Invertebr. Syst.* **2021**, *35*, 350–359. [CrossRef]

42. Hiebert, T.C. New nemertean diversity discovered in the Northeast Pacific, using surveys of both planktonic larvae and benthic adults. Ph.D. Thesis, University of Oregon, Eugene, OR, USA, 2016.
43. Kajihara, H.; Ganaha, I.; Kohtsuka, H. Lineid heteronemerteans (Nemertea: Pilidiophora) from Sagami Bay, Japan, with some proposals for the family-level classification system. *Zool. Sci.* **2022**, *39*, 62–80. [CrossRef]
44. Chernyshev, A.V.; Polyakova, N.E.; Turanov, S.V.; Kajihara, H. Taxonomy and phylogeny of *Lineus torquatus* and allies (Nemertea, Lineidae) with descriptions of a new genus and a new cryptic species. *System. Biodivers.* **2018**, *16*, 55–68. [CrossRef]
45. Chernyshev, A.V.; Polyakova, N.E. Nemerteans from the deep-sea expedition KuramBio II with descriptions of three new hoplonemerteans from the Kuril-Kamchatka Trench. *Prog. Oceanogr.* **2019**, *178*, 102148. [CrossRef]
46. Cherneva, I.A.; Chernyshev, A.V.; Ekimova, I.A.; Polyakova, N.E.; Schepetov, D.M.; Turanov, S.V.; Neretina, T.V.; Chaban, E.M.; Malakhov, V.V. Species identity and genetic structure of nemerteans of the "*Lineus ruber-viridis*" complex (Müller, 1774) from Arctic waters. *Polar Biol.* **2019**, *42*, 497–506. [CrossRef]
47. Hiebert, T.C.; Maslakova, S. Integrative taxonomy of the *Micrura alaskensis* Coe, 1901 species complex (Nemertea: Heteronemertea), with descriptions of a new genus *Maculaura* gen. nov. and four new species from the NE Pacific. *Zool. Sci.* **2015**, *32*, 615–637. [CrossRef] [PubMed]
48. Schwartz, M.L. Untying a Gordian knot of worms: Systematics and taxonomy of the Pilidiophora (phylum Nemertea). Ph.D. Thesis, George Washington University, Washington, DC, USA, 2009.
49. Chernyshev, A.V.; Polyakova, N.E.; Hiebert, T.C.; Maslakova, S.A. Evaluation of the taxonomic position of the genus *Carinina* (Nemertea: Palaeonemertea), with descriptions of two new species. *Invertebr. Syst.* **2021**, *35*, 245–260. [CrossRef]
50. Thollesson, M.; Norenburg, J.L. Ribbon worm relationships: A phylogeny of the phylum Nemertea. *Proc. R. Soc. Lond. B* **2003**, *270*, 407–414. [CrossRef] [PubMed]
51. Strand, M.; Sundberg, P. A DNA-based description of a new nemertean (phylum Nemertea) species. *Mar. Biol. Res.* **2011**, *7*, 63–70. [CrossRef]
52. Kajihara, H.; Abukawa, S.; Chernyshev, A.V. Exploring the basal topology of the heteronemertean tree of life: Establishment of a new family, along with turbotaxonomy of Valenciniidae (Nemertea: Pilidiophora: Heteronemertea). *Zool. J. Linn. Soc.* **2022**, *196*, 503–548. [CrossRef]
53. Hubrecht, A.A.W. Report on the Nemertea collected by H.M.S. Challenger during the years 1873-76. In *Report on the Scientific Results of the Voyage of H.M.S. Challenger during the Years 1873-76 under the Command of Captain George S. Nares and the Late Captain Frank Tourle Thomson, R.N, Zool*; Majesty: Edinburgh, UK, 1887; Volume 19, pp. 1–150.
54. Chernyshev, A.V. On the higher taxa of the Nemertea, with a review of the subclass Anopla. *Zool. Zh.* **1995**, *74*, 7–18.
55. Chernyshev, A.V. CLSM analysis of the phalloidin stained muscle system of the nemertean proboscis and rhynchocoel. *Zool. Sci.* **2015**, *32*, 547–560. [CrossRef]
56. Friedrich, H. Nemertini. In *The Zoology of Iceland*; Munksgaard: Copenhagen, Denmark, 1958; Volume 2, pp. 1–24.
57. Sundberg, P.; Kvist, S.; Strand, M. Evaluating the utility of single-locus DNA barcoding for the identification of ribbon worms (phylum Nemertea). *PLoS ONE* **2016**, *11*, e0155541. [CrossRef]
58. Berg, G. Studies on *Nipponnemertes* Friedrich (Nemertini, Hoplonemertini). II. Taxonomy of *Nipponnemertes pulcher* (Johnston) and some other species. *Zool. Scr.* **1985**, *14*, 239–246. [CrossRef]
59. Krämer, D.; Schmidt, C.; Podsiadlowski, L.; Beckers, P.; Horn, L.; von Döhren, J. Unravelling the *Lineus ruber/viridis* species complex (Nemertea, Heteronemertea). *Zool. Scr.* **2017**, *46*, 111–126. [CrossRef]

water MDPI

Article

Current Assessment of Water Quality and Biota Characteristics of the Pelagic Ecosystem of the Atlantic Sector of Antarctica: The Multidisciplinary Studies by the Institute of Biology of the Southern Seas

Natalia Mirzoeva *, Tatiana Polyakova, Ernest Samyshev, Tatiana Churilova, Vladimir Mukhanov, Alexandr Melnik, Vladislav Proskurnin, Evgeny Sakhon, Elena Skorokhod, Olga Chuzhikova-Proskurnina, Elena Chudinovskih, Natalia Minkina, Natalia Moiseeva, Victor Melnikov, Artem Paraskiv, Lidia Melnik and Tatiana Efimova

A. O. Kovalevsky Institute of Biology of the Southern Seas (IBSS), Russian Academy of Sciences, 299011 Sevastopol, Russia
* Correspondence: natmirz@mail.ru; Tel.: +7-(978)739-80-79

Abstract: Comprehensive studies of the ecosystem of the Atlantic sector of the Antarctic were carried out in the period between 2020 and 2022, during the 79th and 87th sea expeditions on the R/V "Akademik Mstislav Keldysh". Concentrations of soluble forms of 15 trace elements, except Mo, in surface waters of the Southern Ocean were matched the lower limit of their background concentrations in oceanic waters. A high spatial variability of chlorophyll *a*—an indicator of phytoplankton biomass, which is the main food object of the Antarctic krill—was revealed. In the Bransfield Strait, the abundance of bacterioplankton exhibited a weak relationship with water temperature and a pronounced dependence on chlorophyll fluorescence. It was determined, by using the bioluminescence method, that the largest concentrations of larvae and juveniles of krill were noted in the Bransfield Strait, on the shelf of the Antarctic Peninsula. Against the background of a decline in krill abundance in recent years, there has been an intensive development of *Salpa thompsoni*, the main food competitor of krill. New data on the distribution of mesoparasitic copepods on endemic fish of the Southern Ocean were also obtained, and their pathogenic effect on the body of fish hosts has been revealed.

Keywords: *Euphausia superba*; Salpa thompsoni; mesoparasitic copepods; bacterioplankton; bioluminescence; chlorophyll *a*; optically active components; trace elements; heavy metals; Atlantic sector of the Antarctic

Citation: Mirzoeva, N.; Polyakova, T.; Samyshev, E.; Churilova, T.; Mukhanov, V.; Melnik, A.; Proskurnin, V.; Sakhon, E.; Skorokhod, E.; Chuzhikova-Proskurnina, O.; et al. Current Assessment of Water Quality and Biota Characteristics of the Pelagic Ecosystem of the Atlantic Sector of Antarctica: The Multidisciplinary Studies by the Institute of Biology of the Southern Seas. *Water* **2022**, *14*, 4103. https://doi.org/10.3390/w14244103

Academic Editor: José Luis Sánchez-Lizaso

Received: 5 November 2022
Accepted: 12 December 2022
Published: 16 December 2022

1. Introduction

The Southern Ocean is one of the most productive and ecologically clean areas of the World Ocean. This area, with large-scale inventory of marine biological resources, is a fishing area (under international fishing agreements). Antarctic waters are differentiated by high biological productivity with relatively low biodiversity [1–3]. Whales, penguins, seals, seabirds, fish and numerous invertebrates coexist in the Southern Ocean ecosystem. These organisms survive predominantly by feeding on Antarctic krill (*Euphausia superba* Dana, 1852). Antarctic krill is an essential target of Antarctic fisheries, and the basis of the diet of its many consumers. Krill stocks in the Southern Ocean were, until recently, estimated at hundreds of millions of tons [2,4–8]. Now, assessment of the status of krill communities is one of the priority research areas in the Antarctic sector of the Atlantic Ocean.

According to international agreements, access to Antarctic resources is granted to those countries that carry out scientific research in this unique region of the planet [9,10]. The expediency of conducting research on bioresources in the Antarctic is due to the prospects for their development and the need to create a theoretical basis for fishing, primarily for Antarctic krill and several species of fish [2,3,11,12], as well as the impacts of climate change and anthropogenic action in the contemporary period [13–15]. At the same time, the methodological basis of biological research should be a multidisciplinary monitoring of the Antarctic ecosystem [16].

The 79th and 87th cruises of the research vessel (R/V) "Akademik Mstislav Keldysh" to the Atlantic sector of the Antarctic took place from 30 November 2019 to 8 May 2020, and from 7 December 2021 to 6 April 2022, respectively [17,18]. The organization and management of these expeditions was carried out by the P.P. Shirshov Institute of Oceanology of the Russian Academy of Sciences (RAS), whose scientific fleet includes the R/V "Akademik Mstislav Keldysh". These marine expeditions to study the natural complexes of the Antarctic waters were carried out within the framework of the international obligations of the Russian Federation as a party to the Antarctic Treaty [10], as well as the Convention on the Conservation of Antarctic Marine Living Resources [9]. During the cruise, some fundamental tasks, assigned to the scientists of Russia in a number of strategic documents, were addressed [19,20].

Research workers of six Departments of the A. O. Kovalevsky Institute of Biology of the Southern Seas (IBSS) RAS (Department of Radiation and Chemical Biology; Department of Marine Ecosystem Functioning; Department of Ecological Parasitology; Department of Plankton; Department of Biophysical Ecology; Scientific Research Center (SRC) "Geomatics") participated in study of the Atlantic sector of the Antarctic during the 79th and 87th cruises of the R/V Akademik Mstislav Keldysh. All specific scientific research activities were combined into multidisciplinary studies carried out at the level of the IBSS teams, and the research topics of all scientific organizations of Russia participating in these marine cruises were combined.

The aim of the scientific research of the IBSS team in 2020 and 2022 was to obtain new data for a current assessment of the ecological state of the Antarctic ecosystem in relation to the content of trace elements in sea water, including heavy metals. This was studied in order to determine the spatial, structural and functional characteristics of bacterio- and phytoplankton as the main food resources for Antarctic krill, as well as to understand the spatial and quantitative variability of zooplankton, primarily Antarctic krill. Additionally, the objective of parasitological research was to fill in significant gaps in the study of the parasite fauna of Antarctic animals.

To achieve the aim of this investigation, the following tasks were addressed: (1) determination and analysis of the concentrations of soluble forms of a number of trace elements (including heavy metals) in the surface waters of the Atlantic sector of the Antarctic in order to assess the current environmental situation in this area, as well as to identify possible sources of the input of the studied elements into the considered water area; (2) quantifying the total abundance of bacterioplankton and high nucleic acid (HNA) bacteria, and analyzing their relationship with chlorophyll fluorescence in the central part of the Bransfield Strait; (3) study of the variability of the spectral absorption of light by all optically active components and photosynthetic characteristics of phytoplankton; (4) observation and analysis of bioluminescence of Antarctic krill and salp (*Salpa thompsoni* Foxton, 1961) as important elements of the functioning of the pelagic community; (5) conducting studies of contemporary biological productivity, structure and spatial organization of Antarctic krill; (6) studying the pathogenic influence and distribution of mesoparasitic copepods in endemic bathypelagic fishes of the Southern Ocean.

Antarctic krill and Antarctic fish are valuable fishery resources for all mankind [9,10]. Therefore, this research has unconditional scientific, ecological and practical bioresource significance.

2. Materials and Methods

2.1. Trace Elements

The water samples for trace elements analysis were taken during the 87th cruise of the R/V "Akademik Mstislav Keldysh" ("AMK"), organized by the P.P. Shirshov Institute of

Oceanology of RAS. The Antarctic portion of the expedition was held in January–February 2022 (Figure 1).

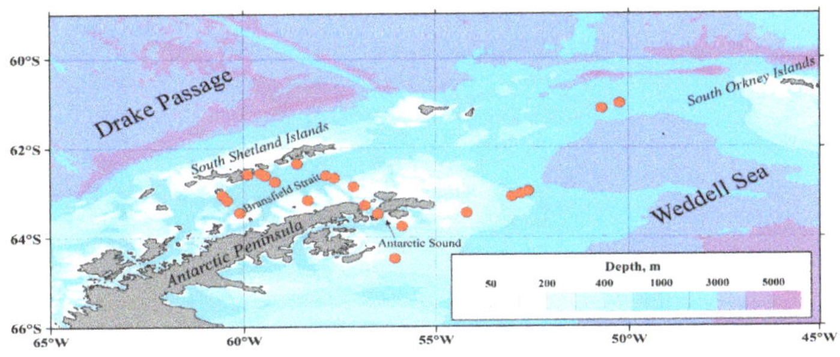

Figure 1. Location of water sampling sites for trace elements analysis in 87th cruise of R/V "Akademik Mstislav Keldysh".

The cruise undertook 3 transects across the Bransfield Strait in its eastern, central and western parts, a transect from Bransfield Strait through the Antarctic Sound to the NW part of the Weddell Sea (where the anomalous phytoplankton bloom observed at the time of sampling [21]), and a transect through the NW Weddell Sea in the NE direction. The results of this study were compared to those obtained in 2020 [22,23].

Surface water was sampled by an acid-cleaned plastic sampler directly from the shipboard on a rope. Seawater was vacuum-filtered through a 0.45 μm nitrate cellulose membrane immediately after sampling and acidified with high pure nitric acid to pH < 2. Acidified samples were stored in a refrigerator at 2–3 °C until they were treated in the ship's laboratory. The onboard sample treatment technique which was applied resulted in the extraction of selected trace elements with carbon tetrachloride (CTC or CCl_4) in the form of diethildithiocarbamate (DDC) complexes in 3 replicates for each sample, followed by re-extraction of elements by destruction of the complexes with concentrated nitric acid [23,24]. The set of measured elements was determined by their ability to forma complex with a DDC ion extractable with CTC, and included 15 elements: Be (beryllium), V (vanadium), Fe (iron), Co (cobalt), Ni (nickel), Cu (copper), Zn (zinc), As (arsenic), Se (selenium), Mo (molibdenium), Ag (silver), Cd (cadmium), Sb (antimony), Tl (thallium), and Pb (lead). Concentrated solutions of each sample were stored in a refrigerator during the cruise, and were then transported to the IBSS Center for Collective Use "Spectrometry and Chromatography", where concentrations of the elements were measured by the ICP-MS technique on a PlasmaQuant MS Elite (Analytik Jena AG) mass-spectrometer [25,26]. The spectrometer was calibrated using a standard solution: "Multi-element calibration standard IV-28, HNO_3/HF, 125 mL" (Inorganic Ventures). The mode of measurement with the mass-spectrometer included 7 replicates of 10 scans for each identified element from 10,000 to 100,000 μs, depending on its expected concentration. The calculation and registration of the measurement results were carried out in accordance with GOST R 56219-2014 and RD 52.10.243-92 [24,25]. The average relative determination error was not higher than ±10%.

2.2. *Bacterioplankton*

The data on bacterioplankton and chlorophyll *a* fluorescence were collected on January 21, 2020, over a southward transect across the central Bransfield Strait at 7 stations (st. 6587, 6590, 6591, 6592, 6593, 6594, and 6595) (Figure 2).

The length of the transect from Greenwich Island (South Shetland Islands) to the shelf of the Antarctic Peninsula was 93 km. Water samples were taken at various points, from 5–7 horizons from the surface to 190 m of depth, depending on the hydrological structure and fluorescence distribution. Bacterial abundances in the samples were estimated by flow cytometry. Counts were performed using a Beckman Coulter flow cytometer (Cytomics

FC 500, Beckman Coulter Inc., Brea, CA, USA) equipped with an air-cooled blue laser (15 mW, 488 nm) and a standard filter setup. Aliquots of 1mL water samples previously fixed with formaldehyde (2% final conc.) were stained with SYBR Green I (Molecular Probes Inc., Eugene, Oregon, USA), following the procedures described by [27]. SYBR-Green I fluorescence, at the green channel FL1 (525 nm), was considered proportional to intracellular nucleic acid content, and was interpreted as a measure of bacterial cell-specific metabolic activity [28]. Consequently, high nucleic acid (HNA) and low nucleic acid (LNA) bacteria were gated on the cytograms (Figure 3).

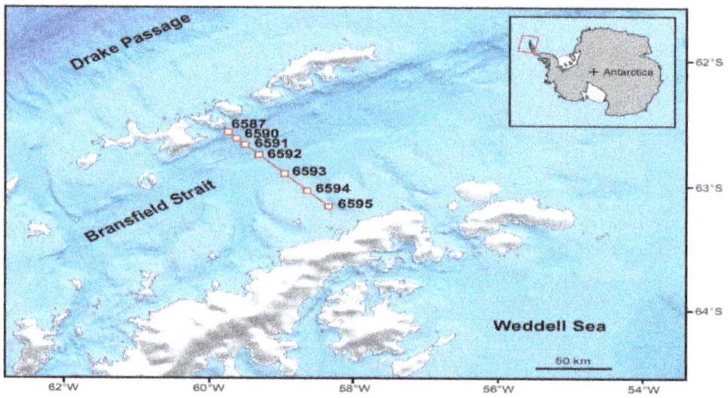

Figure 2. Locations of oceanographic stations in the Bransfield Strait (79th cruise of R/V "Akademik Mstislav Keldysh", January 2020) where bacterioplankton samples were collected at 5 to 7 depths in austral mid-summer.

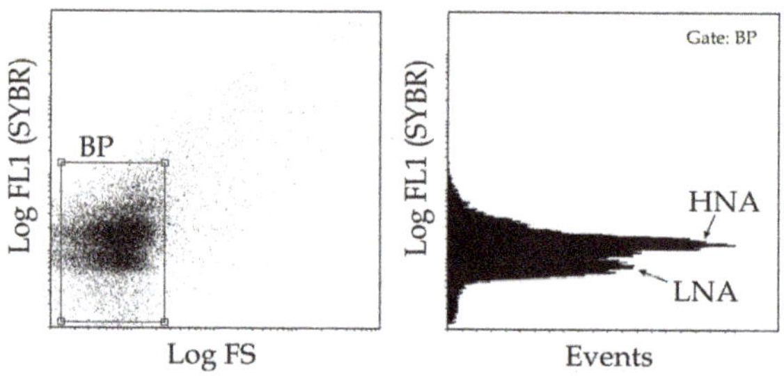

Figure 3. Gating total bacterioplankton (BP), HNA-, and LNA-bacteria in the space of direct light scattering (FS, cell size), and green fluorescence (FL1, SYBR).

Temperature, conductivity, and pressure data were collected using an SBE911 CTD attached to a Carrousel system with 24 5-L Niskin bottles for water sampling. Profile measurements of chlorophyll *a* fluorescence were conducted using a PUM-200 transparency meter equipped with a Minitracka-II fluorimeter (Chelsey Instruments, Molesey, UK).

2.3. Spectral Bio-Optical Properties and Productive Characteristics of Phytoplankton

Water samples were taken using 5-L bottles from different depths, which were chosen based on chlorophyll a fluorescence and temperature profiles. A map of the sampling scheme is shown in Figure 4.

Optical densities of the samples were measured with a dual-beam spectrophotometer, Lambda 35 (PerkinElmer), equipped with an integrating sphere. Chlorophyll *a* and phaeopigment concentrations were measured by the spectrophotometric method [29,30]. The light absorption of particles ($a_p(\lambda)$), phytoplankton ($a_{ph}(\lambda)$), non-algal particles ($a_{NAP}(\lambda)$), and colored dissolved organic matter ($a_{CDOM}(\lambda)$) were measured in accordance with NASA protocols [31]. For $a_{NAP}(\lambda)$ and $a_{ph}(\lambda)$ determination, pigments were bleached by the method used in [32], and β-correction was performed in accordance with [33]. The $a_{NAP}(\lambda)$

and $a_{CDOM}(\lambda)$ data were fitted by exponential function [34]. Slope coefficients (S_{NAP} and S_{CDOM}) were estimated for the wavelength domain of 400−700 nm and 350–500 nm, respectively. The maximum quantum yield of photosynthesis (ϕ_{max}) was calculated [35] based on the maximum quantum yield of PSII measured with PAM fluorimeter (Water PAM-II, Walz) [36,37]. The values of the quantum yield of photosynthesis (ϕ_z) at different depths were calculated based on ϕ_{max} and saturating light intensity (I_k) [38]. I_k was determined based on light curves of the relative electron transport rate, measured with a PAM fluorimeter.

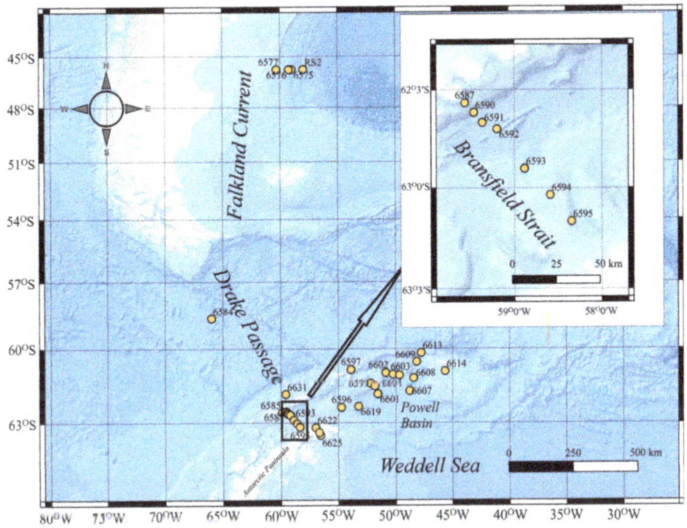

Figure 4. The map-scheme of sampling for scientific research of the spectral bio-optical properties and productive characteristics of phytoplankton.

2.4. Bioluminescence

The multiple sounding method was applied [39] using an autonomous hydrobiological system «Salpa» [40,41]. The «Salpa» was designed to study the intensity of the bioluminescence of organisms in the World Ocean's layer, at 0–250 m depths, using the method of multiple vertical sounding at a speed of up to 1.2 $m \cdot s^{-1}$. "Salpa" allows simultaneous measurements of bioluminescent potential, temperature, hydrostatic pressure, turbidity, electrical conductivity, and photosynthetically active radiation to be taken. Among the existing methods for measuring bioluminescence signals in pelagial environments (towing photometers, hanging them on a given horizon, installing them using special trusses on the bottom, etc.) in recent years, the method of sounding the water column has been recognized as the most promising and accurate. The essence of the applied method is that hydrobionts, which make the main contribution to the formation of the bioluminescent potential of the pelagic zone, generally flash with light when mechanically stimulated. The method of collecting and processing the data using the "Salpa-M" complex has previously been described in detail [40,41]. The bathyphotometer "Salpa" creates a standard level of mechanical excitation of bioluminescence, which makes it possible to achieve an accurate comparison of the measured values of the vertical structure of the bioluminescence field (BF) when performed in different regions and under various conditions. The data on bioluminescence were collected over a transect across the central Bransfield Strait at 9 stations (Figure 5).

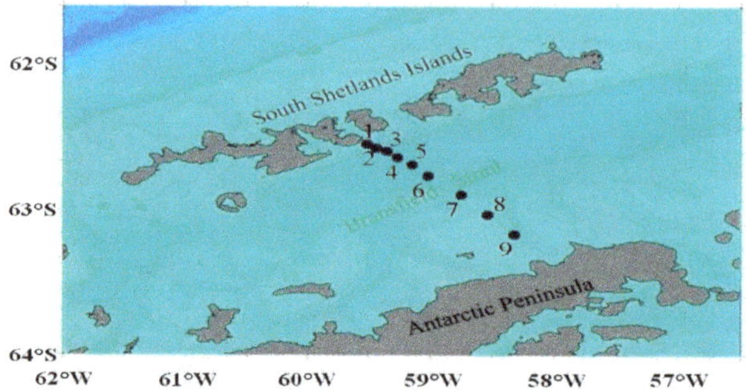

Figure 5. Location of bioluminescence stations in the Bransfield Strait.

2.5. Antarctic Plankton

The investigations on distribution krill and salps were studied during the 87th cruise of the R/V "Akademik Mstislav Keldysh" in January and February 2022 in the western zone of the Atlantic sector of Antarctica. A total of 31 tows were performed: 21 by the DSN plankton net and 10 by the Isaacs-Kidd trawl in the Samyshev–Aseev modification [42,43]. A sample of at least 300 individuals was selected for biological analysis. If there were fewer individuals in the catch, then it was analyzed completely. In cases where it was not possible to determine the total number of Antarctic krill and salps in the catch, the entire catch was weighed and, depending on the amount of krill and salps caught, a sample (from 100 mL to 1 L) was taken and extrapolated to the entire catch. A total of 3789 krill individuals were measured, and an analysis was performed for 2561 krill and 6421 salps.

2.6. Parasitological Studies of Antarctic Fish

Five specimens of the Antarctic deep-sea smelt *Bathylagus antarcticus* (Günther, 1878), infected with copepods, were caught in the Atlantic sector of Antarctica off the Powell basin, in the region of the northern submarine rises and on the shelf of the South Orkney Islands in February 2020. These were collected during the 79th cruise of the R/V "Akademik Mstislav Keldysh" Shirshov Institute of Oceanology of Russian Academy of Sciences (IO RAS). The fishes were caught with mid-water trawls at three stations: st. 6653 (62°26.9545′ S, 52°14.5092′ W), 2794–2882 m in depth; st. 6655 (62°28.7629′ S, 50°57.1162′ W), 730–838 m in depth; and st. 6690 (59°47.4053′ S, 50°25.5293′ W), approximately 1400 m in depth. Fish were immediately preserved in 70% ethanol. Copepods were collected from ethanol-fixed fish in the laboratory under a stereomicroscope with magnification 15–20. In the laboratory, copepods were carefully removed from the tissues of the fish, and then soaked in lactophenol for an entire day before dissection. Mesoparasitic copepods from all specimens were deposited in the Collection of Marine Parasites (CMP: 7 vouchers Nos.: 1370.Cr.40.v1; 1371.Cr.40.v2; 1372.Cr.40.v3; 1373.Cr.40.v4; 1374.Cr.40.v5; 1375.Cr.41.v1; 1376.Cr.41.v2) of the A.O. Kovalevsky Institute of Biology of the Southern Seas RAS (IBSS), Sevastopol [44].

3. Results

3.1. Trace Elements

Concentrations of all studied elements (Figures 6 and 7), except Mo, were generally low, matching the lower limit of their background concentrations in oceanic waters [16,23], or were even lower.

For some elements, the values obtained were below their detection limits: Be—<0.005, Co—<0.005, Ag—<0.01 and Tl—<0.001 $\mu g \cdot L^{-1}$. Concentrations of Se were extremely low, and slightly exceeded its detection limit (0.01 $\mu g \cdot L^{-1}$) only in a few samples from the southern and western parts of the Bransfield strait. Concentrations of V and Ni are not presented in figures, as they were distributed quite homogeneously within the studied region, and matched the background ranges for the elements reported for oceanic waters [16,23,45]. Ranges of the values measured in the actual study were 1.17 ÷ 1.79 $\mu g \cdot L^{-1}$ for V and 0.21 ÷ 0.39 for Ni.

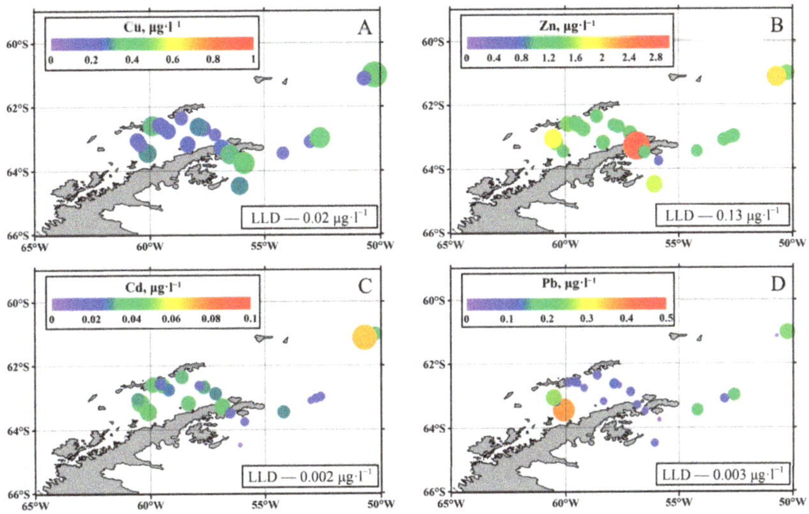

Figure 6. Concentrations of the main heavy metals ((**A**)—Cu; (**B**)—Zn; (**C**)—Cd; (**D**)—Pb) in surface waters of the studied area of the Antarctic region.

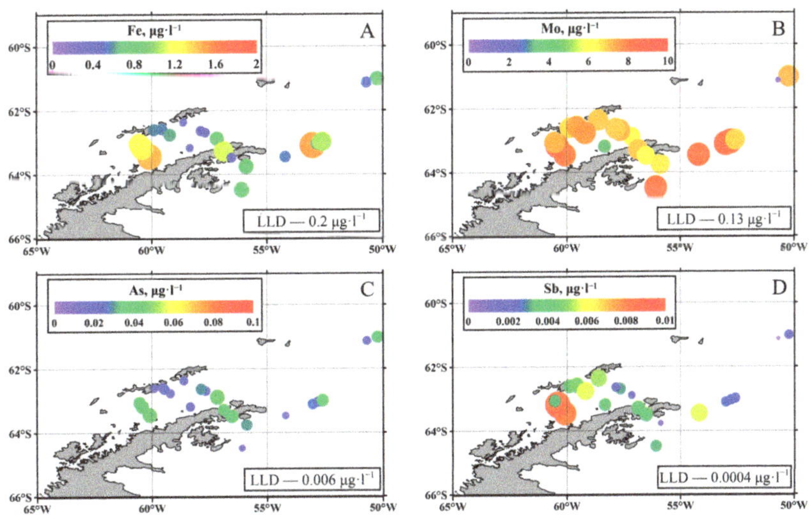

Figure 7. Concentrations of Fe (**A**), Mo (**B**), As (**C**), and Sb (**D**) in surface waters of the studied area of the Antarctic region.

3.2. Bacterioplankton–Phytoplankton Coupling and Bacterial Physiological Activity

In the Bransfield Strait, the bacterioplankton abundance varied from 0.14 to 1.03×10^6 cells mL^{-1} and averaged 0.36 ± 0.18 (SD) $\times 10^6$ cells mL^{-1}. The highest concentrations of bacteria were observed in the surface layer at st. 6593, and in the 20 m layer in the northern part of the strait (Figure 8).

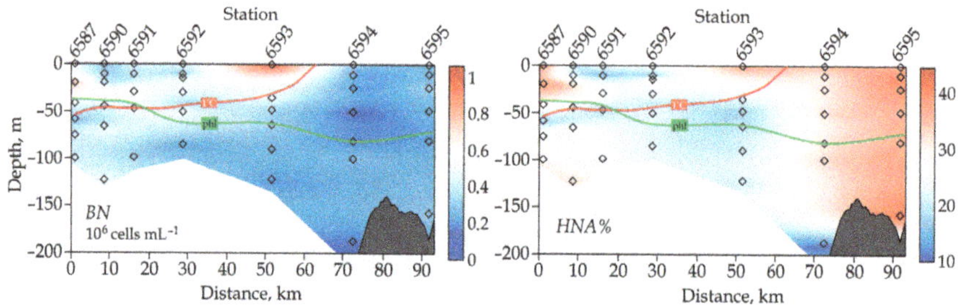

Figure 8. Vertical distribution of the total bacterioplankton (*BN*) and the fraction of HNA-bacteria (*HNA%)* along the transect in the Bransfield Strait during austral mid-summer, January 2020. Green and red lines represent the bottom of the photic layer (phl) and 1 °C isotherm (as a boundary between water masses), respectively.

The fraction of the HNA-bacteria was 28 ± 8% of the total abundance of bacterioplankton, and ranged from 10% to 43%. The highest HNA% values coincided with bacterioplankton peaks at st. 6587 and 6593, as well as in the southern part of the strait on the shelf of the Antarctic Peninsula (st. 6595), across the entire range of depth from the surface to the bottom (Figure 8).

The abundance of bacterioplankton exhibited a weak relationship with water temperature (R^2 = 0.29) and a stronger relationship (R^2 = 0.54) with chlorophyll fluorescence (Figure 9a,b). On the contrary, HNA% was not correlated with either temperature or chl*a* fluorescence (Figure 9c,d). Inclusion in the analysis of only physiologically active HNA-bacteria (Figure 9e,f) resulted in a significant weakening of the bacteria–phytoplankton relationship. On the contrary, (Figure 9g,h) LNA-bacteria showed a greater dependence on phytoplankton as the major source of organic matter.

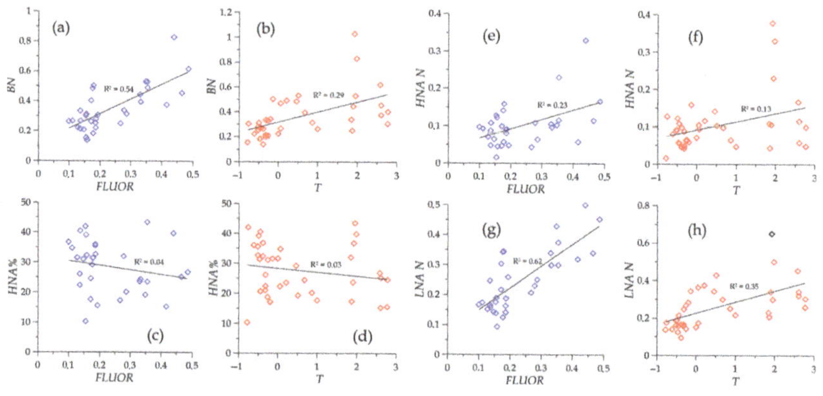

Figure 9. Relationships between bacterioplankton (*BN*, 10^6 cells mL^{-1}), HNA-bacteria (*HNA N*, 10^6 cells mL^{-1}), LNA-bacteria (*LNA N*, 10^6 cells mL^{-1}), the fraction of HNA-bacteria (*HNA*%), water temperature (*T*, °C), and chlorophyll *a* fluorescence (*FLUOR*, rel. u.).

3.3. Spectral Bio-Optical Properties and Phytoplankton Productive Characteristics

In the austral summer 2020 (from 11 January to 4 February) in the Atlantic sector of the Antarctic, a complex investigation of bio-optical parameters, including chlorophyll *a* and phaeopigments concentration; (TChl-*a*), $a_p(\lambda)$,$a_{NAP}(\lambda)$, $a_{ph}(\lambda)$,and $a_{CDOM}(\lambda)$; and phytoplankton photosynthetic characteristics (ϕ_{max} and I_k) (Figure 10) were carried out at 37 stations in the Drake Passage, Falkland Current, Bransfield Strait, and Powell Basin.

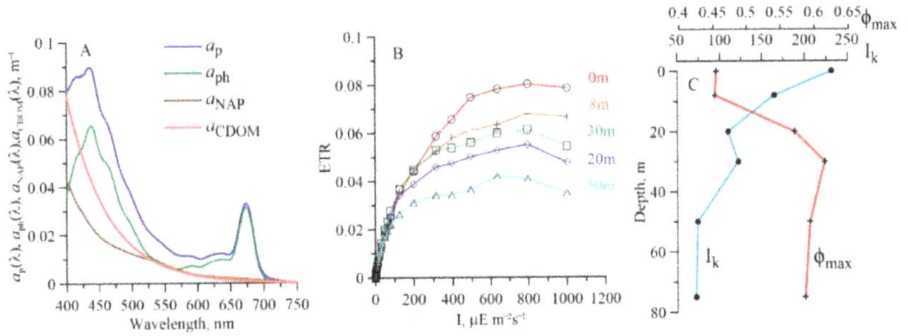

Figure 10. Examples of spectral light absorption coefficients of particles ($a_p(\lambda)$, m^{-1}), non-algal particles ($a_{NAP}(\lambda)$), phytoplankton $a_{ph}(\lambda)$, and colored dissolved organic matter($a_{CDOM}(\lambda)$) (**A**); Light curves of relative electron transport rate (ETR, rel. un.) (**B**); Vertical distribution of the saturating light intensity (I_k, $\mu E\ m^{-2}\ s^{-1}$, blue line) and maximum quantum yield of photosynthesis (ϕ_{max}, C quantum^{-1}, red line) (**C**).

Wide variability of the bio-optical properties (about two orders of magnitude) was observed (Figure 11). The surface TChl-*a* varied in a range from 0.20 mg m^{-3} (Drake Passage) to 4.4 mg m^{-3}(Powell Basin). The $a_{ph}(\lambda)$, $a_{NAP}(\lambda)$, and $a_{CDOM}(\lambda)$ at the wavelength of phytoplankton physiology (maximum of light absorbance at 438 nm) varied in the surface layer from 0.0049 to 0.29 m^{-1}, from 0.0046 to 0.034 m^{-1}, and from 0.0042 to 0.15 m^{-1}, correspondently (Figure 11).

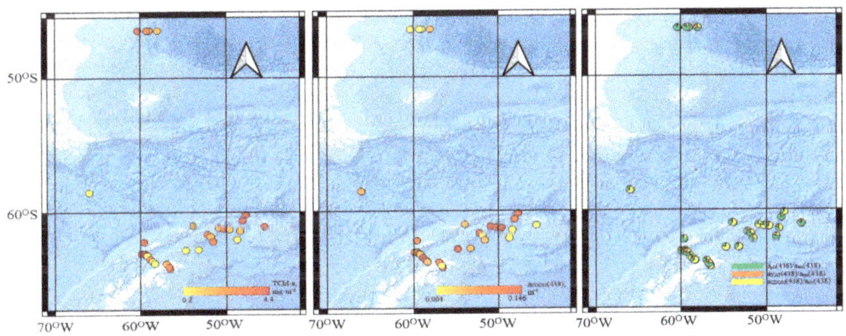

Figure 11. Map of the sum of chlorophyll *a* and phaeopigment concentration (TChl-*a*, left panel), light absorption coefficient of colored dissolved organic matter at 438 nm ($a_{CDOM}(438)$, middle panel), and contribution of the phytoplankton($a_{ph}(438)/a_{tot}(438)$), non-algal particles ($a_{NAP}(438)/a_{tot}(438)$), and CDOM ($a_{CDOM}(438)/a_{tot}(438)$) to total non-water absorption at 438 nm (left panel).

The $a_{ph}(\lambda)$ correlated with TChl-*a* ($a_{ph}(438) = 0.048 \times C_a^{1.2}$, $r^2 = 0.83$, $N = 126$; $a_{ph}(678) = 0.022 \times C_a^{1.2}$, $r^2 = 0.92$, $N = 126$). The power coefficients were >1. This reflects the increasing chlorophyll *a*-specific phytoplankton absorption coefficients ($a^*_{ph}(\lambda)$), with a rise of TChl-*a*. In high trophic waters (st. 6609), the $a_{ph}(\lambda)$, had a local maximum at 490 and 550 nm—the "signature" of phycobilin pigments.

As result of parameterization of non-algal particles (NAP) and colored dissolved organic matter (CDOM), absorption of S_{NAP} was equal to 0.011 ± 0.0017 nm^{-1}, and that of S_{CDOM} was 0.0151 ± 0.0016 nm^{-1}. The analysis showed that $a_{NAP}(\lambda)$ co-varied with both $a_{ph}(\lambda)$ and TChl-*a*, but $a_{CDOM}(\lambda)$ did not correlated with either $a_{ph}(\lambda)$ or (TChl-*a*). The contribution of phytoplankton and CDOM to the total non-water absorption at 438 nm varied significantly in the surface layer, from 7% to 88% and from 3% to 88%, respectively (Figure 11). The high phytoplankton and minimum CDOM contributions were observed in more trophic waters (TChl-*a* > 1 mg m^{-3}).

Analysis of the distribution of phytoplankton photosynthetic characteristics (ETR light curves, F_{max}, and I_k) revealed (Figure 8) that the I_k decreased from the surface to the bottom of the euphotic zone, and that F_{max} profiles showed a tendency to increase with depth.

3.4. Bioluminescence

Zooplankton sampling was carried out using a Bongo net [46], as well as via trawling using a Double squad net (DSN) and an Isaacs-Kidd trawl (IKMT). The Bongo net was a towed plankton net consisting of a frame with two metal rings, which had two filter cones with a mesh of 300 microns fixed onto them. The diameter of each frame ring was 60 cm [46]. DSN—a double square net with an inlet area of 1 m^2 and a 6 m long filter cone made of gas with a mesh size of 0.5 mm—was equipped with a water flow meter (Hydrobios, Germany) and a 24 kg wing-shaped depressor (Hydrobios, Germany). Oblique tows were carried out in layers, starting from 730 to 100 m, at vessel speeds from 2 to 3.1 knots. The towing depth was prompted by the pressure sensor readings of the Senti DT probe (Star Oddi, Iceland). An oblique tow was carried out in layers starting from 270 to 130 m, at a vessel speed of 1.7–2.3 knots. Fishing with the Bongo net showed the lowest level of effectiveness. The maximum number of specimens of *E. superba* and *S. thompsoni* were obtained using the IKMT trawl. The average size of *S. thompsoni* individuals was 30–50 mm. A large aggregation of *S. thompsoni* was registered while towing with the DSN, with an average individual size of 20–40 mm. The maximum number of salps in a catch did not exceed 400 individuals.

Aggregations of large mature krill were mainly present in oceanic areas distant from the coast, whereas larvae and juveniles were mainly concentrated in the shelf zone. The largest concentrations of larvae and juveniles of krill were noted in the Bransfield Strait on the shelf of the Antarctic Peninsula.

The maximum level of bioluminescence was registered at st.1, where it reached 272.3×10^{-12} W·cm^{-2}·L^{-1} at 166 m (Figure 12). The vertical structure of the bioluminescence at this station was single-maximum (Table 1), with an ocean surface temperature of 2.05 °C and salinity of 33.8 PSU.

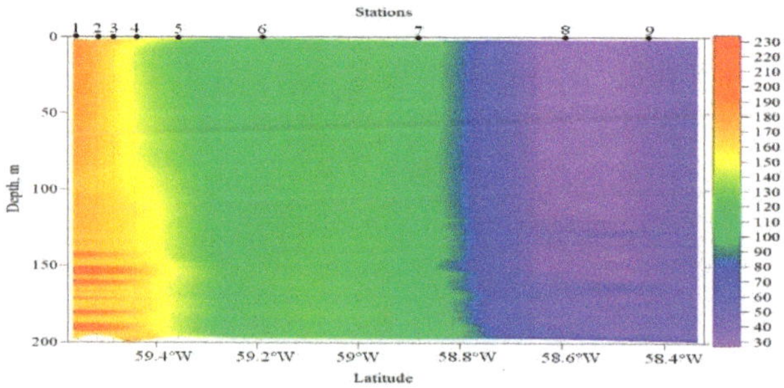

Figure 12. Vertical section of bioluminescence (10^{-12} W·cm^{-2}·L^{-1}) in the Bransfield Strait (Figure 5).

At stations 2, 3, 4, 5, 6, and 7, the average level of bioluminescence varied from 88×10^{-12} W·cm^{-2}·L^{-1} to 166×10^{-12} W·cm^{-2}·L^{-1}, the structure being evenly distributed across depth with slight peaks. At central stations of the section, a high-gradient zone was formed where the water temperature varied from 1.32 °C to 1.76 °C. The lowest level of the bioluminescent signal, with values of 18.69×10^{-12} W·cm^{-2}·L^{-1} and 17.42×10^{-12} W·cm^{-2}·L^{-1} throughout the study area, were recorded at stations 8 and 9, respectively. At the same time, the coldest ocean surface temperature layer, −0.006 °C, was also recorded at station 9 (Table 1).

In the area of the archipelago of the South Shetland Islands, where the flow of water from the southern periphery of the Antarctic Circumpolar Current enters, the accumulations of salps and a low number of krill of older age groups were recorded. The distribution of *E. superba* and *S. thompsoni* coincided with the structure of the BF in the studied region (Figure 12).

Table 1. Mean of bioluminescence, temperature, and salinity at several stations in the region.

Station Coordinates	Station Number	Max. Biolum.	Max. Level (m) of Biolum.	Biolum. Means	Temp, °C	Salinity, PSU
62°33.48′ S 059°33.85′ W	1	272.3	166	149.57	2.05	33.8
62°34.96′ S 059°31.92′ W	2	159.4	176	88.4	1.9	33.8
62°35.93′ S 059°29.42′ W	3	169.87	179	165.84	1.76	33.85
62°36.89′ S 059°27.30′ W	4	190.8	198	166.3	1.88	33.9
62°40.01′ S 059°22.01′ W	5	115.67	113	100.72	1.32	34.11
62°44.86′ S 059°11.89′ W	6	110	179	107.81	1.38	34.1
62°54.04′ S 058°53.23′ W	7	109	111	101.59	0.68	34.18
63°02.14′ S 058°35.66′ W	8	81.4	128	18.69	0.07	34.19
63°10.01′ S 058°20.05′ W	9	55.93	152	17.42	0.07	34.19

3.5. Antarctic Krill

Our research showed that krill abundance ranged from 0 to 537 ind·1000 m^{-3}, and biomass ranged from 0 to 331.8 g·1000 m^{-3}. The maximum catch by the Isaacs-Kidd trawl was 33 kg. Significant numbers of large, immature krill were found southeast of James Ross Island, where krill numbers averaged 357 ind·1000 m^{-3}. The third-largest catch was in the Bransfield Strait, with a krill count of 321 ind·1000 m^{-3} (Figure 13).

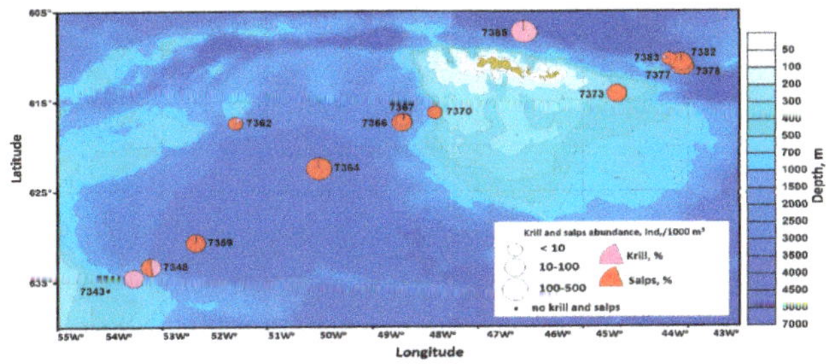

Figure 13. Spatial distribution of the number of accumulations (ind·1000 m^{-3}) of Antarctic krill and salps (%) in the Powell Basin and the area of the South Orkney Islands, according to results from the 87th cruise of R/V "Akademik Mstislav Keldysh" in 2022.

The bulk of the catch consisted of juvenile fish.

3.6. Parasitological Research

Five specimens of sphyriid and two specimens of pennilid copepods were identified on five exemplars of *Bathylagus antarcticus*. Only one species of copepod parasitized on infected fish. Five individuals of mature egg-bearing copepod females were found on three specimens of *B. antarcticus*, which morphologically corresponded to the *Paeonocanthus antarcticensis* (Hewitt, 1965) (Figure 14A–D). In the cephalothorax and part of the neck, copepods of this species penetrated deep into the muscles, eventually reaching the abdominal cavity and penetrating into the livers of the fish.

In the two other specimens of *B. antarcticus*, two mature, egg-bearing female copepods were found, which, according to their morphometric character, correspond to the diagnosis of the genus *Sarcotretes* Jungersen (Figure 14E–G). Among the four valid species of copepods of this genus, the species *S. scopeli* Jungersen is the closest to the studied individuals. It should be noted that the studied female copepods, *Sarcotretes* sp. from *B. antarcticus*, were gravid, with smaller body sizes than the *S. scopeli* copepods parasitizing *Bathylagus* sp. from the Southern Ocean (13.4 mm vs. 21.15 mm sensu [47]), as well as than myctophid fishes caught in southwestern Greenland, the Scotian Shelf, and off New Jersey (13.4 mm vs. 16–25 mm sensu [48]).

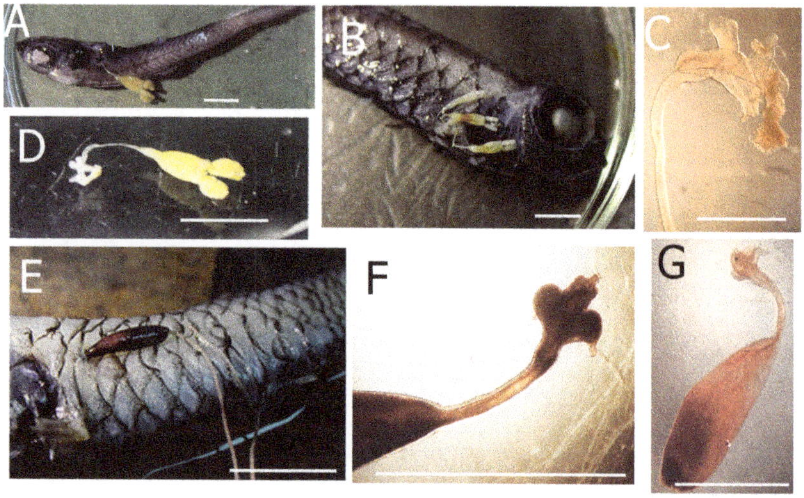

Figure 14. *Paeonocanthus antarcticensis* (Hewitt) (**A–D**) female and *Sarcotretes* sp. (**E–G**) female on *Bathylagus antarcticus* Günther. (**A**,**B**)—Attached copepods on fish body, cephalothorax, and neck, embedded in the body muscles; (**C**)—cephalothorax copepods after being soaked in lactophenol; (**D**)—live copepod with egg sac, after extraction from the fish; (**E**)—attached copepod with implanted cephalothorax in fish heart; (**F**)—live copepod cephalothorax after extraction from the fish; (**G**)—whole copepod after being soaked in lactophenol. Scalebars: (**A**,**B**,**D**,**E**)—1 cm, (**F**, **G**)—0.5 cm, (**C**)—0.2 cm.

Observed differences are likely related to the host variability of these copepods; however, further studies of a larger number of *Sarcotretes* specimens from *Bathylagus* in Antarctica are needed to confirm this statement. The cephalothorax of discovered copepods of this genus penetrated through all the organs and muscles of the fish, and directly into the cavity of the heart.

4. Discussion

4.1. Trace Elements

The measured concentrations of Zn, Cd, Pb (Figure 6), Fe, As, and Sb (Figure 7) were generally higher in samples from the western section in the Bransfield Strait, with a slight decrease in their content eastward through the Strait. This trend is highly likely to reflect the input of the enhanced content of these elements from the west of the Strait with their further dilution with Weddell Sea waters, which generally have lower concentrations of the studied metals and metalloids [22,23]. This concept is in good agreement with studies of water mass transport through the Strait [18,21,49]. The most evident trends were obtained for Fe, Pb, and Sb. Further analyses of elements' distribution within the entire studied region indicates that internal sources of Fe (Figure 7A) and Zn (Figure 6B) were likely to be present in the Weddell Sea, as enhanced values were observed for both elements in the north west part of the Sea. Perhaps this is due to the influx of elements into the sea environment during the melting of polar glaciers [16,50]. The input of considerable quantities of Fe in suspended form, sourced from the shore lands of islands and the Antarctic continent, caused leaching further toward the solution, and then assimilation with phytoplankton species was observed for the South Scotia Ridge region (NW Weddell Sea) [51]. The almost homogeneous distributions of Mo (Figure 7B) and V in the studied area were quite typical for oceanic ecosystems [52,53], since these elements reveal conservative behavior in oxygenated marine waters [45]. The average values of Mo (6.76 $\mu g \cdot L^{-1}$) and V (1.52 $\mu g\ L^{-1}$) were in agreement with those reported for sea waters of the World Ocean [45,52,53]. The lack of Mo in two samples (Figure 7B) may be explained by local hypoxia due to the commencing storms, causing the mixing of highly oxygenated surface waters with less oxygenated underlying water layers, since Mo has been shown to be sensitive to decreases in oxygen concentration [53]. The further analysis of the particulate material sampled at the same stations of the cruise may clarify this point.

A comparison of the results obtained in 2020 [22] and 2022 on the concentration of trace elements in the surface waters of the Bransfield Strait showed an increase in the concentrations of Mo and V in 2022 by 2.4 and 2.3 times, respectively. Additionally, in 2022, an excess of the average concentrations of Mo and V in the sea water of the Bransfield Strait by 1.6–1.7 times, respectively [54], was noted.

4.2. Bacterioplankton–Phytoplankton Coupling and Bacterial Physiological Activity

We have revealed that the highest bacterioplankton abundances can be found in the northern part of the strait, where the water masses dominate with Bellingshausen Sea influence, water temperatures of >1 °C, and salinities of <34.1 psu [55,56]. The low boundary of the water mass is well-marked by the 1 °C isotherm (Figure 9). High Chl-*a* concentrations, found at the same stations, agreed well with the earlier long-term observations: well-stratified waters off the South Shetland Islands are associated with a shallower photic layer and higher Chl-*a* concentrations at or near the surface [57]. Thus, the Bransfield Strait bacterioplankton demonstrated good coupling with phytoplankton, the major supplier of organic carbon in the Southern Ocean, under the conditions of negligible input of all ochthonous organic matter to the coastal waters [58]. According to our results, the fraction of HNA bacteria was relatively small compared with other regions of the World Ocean. This agreed well with other data [59,60] and supported the hypothesis about the important role of other mechanisms (such as bacterivory and phage infection) in controlling Antarctic bacterioplankton [61–63]. Our attempt to reveal a stronger relationship between HNA bacteria and phytoplankton was unsuccessful, indicating that in the Antarctic waters, bacterial abundance may be closely controlled by loss processes rather than resource supply [64].

4.3. Spectral Bio-Optical Properties and Phytoplankton Productive Characteristics

The investigated water region had high spatial heterogeneity in the distribution of the bio-optical parameters caused by dynamics and hydrological structure of the waters [65,66]. The water dynamics resulted in variability of the main environmental factors—light, temperature, and nutrients supply—controlling the abilities of phytoplankton and its photosynthetic characteristics [67,68]. The revealed links between $a_{ph}(\lambda)$ and TChl-*a* differed from their relationship in the Antarctic waters [69] and World Ocean [70] by the higher power coefficients (>1). This reflected the increasing levels of $a^*_{ph}(\lambda)$, with a rise in TChl-*a* caused by a dominance of small-celled species of phytoplankton in warmer (surface temperature >3 °C) and trophic waters (TChl-*a*> 1 mg m^{-3}). The "signature" of phycobilin pigments on the $a_{ph}(\lambda)$ justified an increasing of cryptophyte abundance in phytoplankton and a shift in dominating taxon to Cryptophyte [71]. The collected bio-optical data was the basis for a validation of the standard satellite products, for parameterization of absorption by all optically active components, and for refining the developed three bands algorithm [72] for Antarctic waters. The measured bio-optical properties and quantum efficiency of photosynthesis will be used for the calculation of primary production (PP) by a full spectral approach, using spectral downwelling irradiance ($E_d(\lambda)$), $a^*_{ph}(\lambda)$, and ϕ_{max}, I_k) [73]. The spectral PP takes into account the effect of global warming on the species and size structure of phytoplankton, with its state caused by adaptation to ambient light and nutrient availability (via $a^*_{ph}(\lambda)$, and ϕ_{max}, I_k) [36,70,74], as well as the effect of ice melting on the variability of the NAP and CDOM content (via $E_d(\lambda)$) [75].

4.4. Study of Bioluminescence

Thanks to the use of hydrobiological complexes and the method of multiple soundings, there was an opportunity to register the fields and the structure of accumulations of Antarctic krill, salps, and other luminous organisms in the 0–200 m layer in Antarctic waters. It was stated that one of the main features of the vertical structure of Antarctic bioluminescence fields (BFs) is their stratification, determined both by the parameters of the pelagic community and by the characteristics of the water masses. The depth of the layer or layers of maximum luminescence intensity (MLIL) and their number are also important characteristics of the BF. During the study period, the MLILs recorded at almost all stations were below the isobath of 100 m. Intense outbreaks of bioluminescence in the area of the South Shetland Islands archipelago (with a low abundance of krill) can be associated with an increase in the abundance of salps, which are capable of generating outbreaks of such potential. A rapid increase in the number of salps was likely due to their rapid asexual reproduction in ice-free spring waters [2]. The increase in the intensity of the bioluminescence field in the studied waters was likely due to an improvement in the supply of nutrients to the waters and a significant increase in the content of digestible organic matter. Moving to the south, the intensity of bioluminescence was noticeably decreasing, which was likely due to the horizontal and vertical structure of the waters. A significant decrease in temperature and an increase in salinity have a substantial impact on the distribution of luminous organisms [76]. Low temperatures seem to have a negative effect on the development of salps; therefore, they are not found in the southern regions. The largest concentration of larvae and juveniles of krill, which, unlike accumulations of salps, cannot produce a powerful bioluminescent signal, was registered on the shelf of the Antarctic Peninsula [76].

4.5. Antarctic Krill and Salpa Thompsoni

Against the background of a decline in Antarctic krill abundance in recent years, there has been an intensive development of *Salpa thompsoni*, the main food competitor of Antarctic krill. Catastrophic outbreaks of this species due to climate change, observed since 1975 and up to the present time, raise reasonable concerns about possible adverse changes in the pelagic ecosystem of the Southern Ocean [77–83]. These jelly planktonic organisms change the structure of trophic chains, which affects the ecology of many animals, including fish, birds, and mammals. According to our experimental determinations of energetic metabolism [84], the dietary requirements of *S. thompsoni* are up to two orders of magnitude (depending on the size of tunicates) greater than those of Antarctic krill [85]. The mechanism of salps penetration to high latitudes is based on the meridional gyres exchange through the Antarctic Circumpolar Current front [86]. Under the conditions of large-scale atmospheric processes, this mechanism is universal for all sectors of Antarctica [2]. It appears to be key to the awareness of the expected serious changes in the Antarctic ecosystem under the influence of global climate change, displacement of polar fronts, and main continental margin currents, and it is most pronounced in the Atlantic sector [87]. During the period of the IBSS surveys in the Atlantic sector, the R/V "Ernst Krenkel" (1998) and Horizon (2002) recorded potentially commercial-scale accumulations of krill, with stocks estimated at millions of tons [85]. In the time of the 87th expedition of RV "Akademik Mstislav Keldysh" in 2022, it was discovered that the krill fields had been replaced by accumulations of salps. Their numbers ranged from 0 to 202 ind·1000 m^{-3} and from 0 to 72 g·1000 m^{-3}, respectively. It is thought that the possible catastrophic consequences for the Antarctic ecosystem will involve three main factors [2]—mechanical removal from the base of krill habitat; suppression of its population development due to inhibition of reproduction in the crustacean, whose spawn develops on the polluted bottom of shallow waters; and through food competition with salps.

4.6. Parasitological Research

Parasitic copepods are common fish parasites, and can have pathogenic effects on the host, sometimes causing their death. This can result in economic problems in the fishing industry [88–91]. Currently, there is limited information about ecto- and mesoparasitic crustaceans from fish of the Southern Ocean [47,92,93]. According to [93,94], only nine species of copepods, from seven genera and four families, have been identified from fish from this region between 1966 to 2015. In the Southern Ocean, only two species of sphyriid and pennilid copepods have previously been found in the bathypelagic fish *Bathylagus* Günther. In *B. antarcticus,* in the Pacific Sector of Antarctica, D'Urville Sea (65° S, 139°59.6′ E), the copepod *Paeonocanthus antarcticensis* from the family Sphyriidae Wilson was observed [92]. In the Antarctic, *Sarcotretes* copepods were mainly recorded in the gills of mesopelagic fishes (Myctophidae) [93], and the only record of *S. scopeli* was recorded in a bathypelagic fish, *Bathylagus* sp., in the Indian Ocean Sector of Antarctica, Davis Sea, and Lützow-Holm Bay [47,48]. Before our studies, there had been no repeated findings of copepods of the *Paeonocanthus* and *Sarcotretes* genera in bathypelagic fishes in Antarctic waters. The discovery of copepods of these two genera in endemic bathypelagic fish of the Southern Ocean in a new area, the Atlantic sector of Antarctica off the Powell basin, significantly expands our know ledge about their range and the species composition of their fish hosts. According to [90], representatives of the genus *Sarcotretes* can penetrate into the liver when penetrating the cephalothorax and part of the neck into the body cavity of the fish, which leads to a decrease in their volume in comparison with uninfected fish. In addition, V.N. Kazachenko [90] noted the penetration of cephalothorax copepods of this genus into the abdominal cavity near the intestines of fish, which led to a decrease in the intestinal lumen and, as a result, a disruption in the functioning of this organ. During our study, we found a more pathogenic influence of copepods of the genus *Sarcotretes,* namely the penetration of the cephalothorax and part of the neck by *Sarcotretes* sp., into the heart cavity of *Bathylagus antarcticus,* leading to the parasite feeding on the blood of the fish. We also noted a deep penetration of the cephalothorax of another copepod species, *Paeonocanthus antarcticensis,* into the liver of *B. antarcticus*. In this casem the muscles at the site of penetration were very loose, and the liver was slightly deformed. The formation of tumors at the sites of penetration of copepods of both species was not found in *Bathylagus*.

5. Conclusions

In general, the study of trace element distribution in surface waters of the Atlantic Sector of Antarctica in 2022 ensured that the studied water area was the reference ecological background region in terms of content of trace elements, including heavy metals. However, comparison of the results obtained in 2020 [22] and 2022 on the concentration of trace elements in the surface water of the Bransfield Strait showed an increase in the concentrations of Mo and V in 2022 by 2.4 and 2.3 times, respectively. Such an increase did not result in excess of any maximum permissible concentrations, but led to the average concentrations of Mo and V in the sea water exceeding the recommended "target values" (considered absolutely safe for a marine environment) of these elements by 1.6–1.7 times [49]. This testifies to the need to continue monitoring chemoecological studies in this region of the Southern Ocean, in order to identify the sources of trace elements entering the studied area.

In Antarctica, the primary production of phytoplankton determines not only the matter and energy flow up to the higher trophic levels, including Antarctic krill, but also the dependence of heterotrophic bacteria on algae-produced labile dissolved organic matter. In 2020 the highest abundance of bacterioplankton (up to 1.03 × 106 cells mL^{-1}) was observed in the northern part of the Bransfield Strait in the warmer and less salty water mass under Bellingshausen Sea influence, just where the phytoplankton bloom was registered. A lower fraction of physiologically active HNA bacteria (28 ± 8% on average) and their weaker coupling with phytoplankton indicated that the Bransfield Strait bacterioplankton seemed to be closely controlled by predatory and viral mortality rather than organic carbon supply by phytoplankton. The variability in the bio-optical properties of the Southern Ocean was assessed in the austral summer of 2020. The high spatial heterogeneity in the distribution of light absorption by all optically active components was also revealed. The collected bio-optical data present the scientific basis for the development of regional bio-optical algorithms for operative assessment of the current state of the Southern Ocean's pelagic ecosystem, as well as forecasting changes due to global warming. Satellite data provide primary production estimates with high spatial and temporal resolution. The combination of in situ data and data calculated using satellite algorithms on the abundance of phytoplankton (the concentration of chlorophyll *a* is used as a biomass marker) and its production activity will make it possible to further study the patterns of formation of primary production and food supply of krill in contemporary conditions, as well as the basis of established relationships, to predict possible changes due to global warming.

The findings of mesoparasitic copepods of two genera (*Paeonocanthus* and *Sarcotretes*) in bathypelagic fish endemic to the Southern Ocean in a new area, the Atlantic sector of Antarctica off the Powell basin, significantly expands our knowledge about their geographical distribution and the species composition of their definitional fish hosts—*Bathylagus spp.* The conducted research and the obtained results are aimed at supporting management decisions regarding the exploitation and conservation of the unique ecological and biological resources of Antarctica. This research is also the basis for further research by the IBSS in the Atlantic sector of the Antarctic.

Author Contributions: Conceptualization, N.M. (Natalia Mirzoeva), E.S. (Ernest Samyshev), T.C., V.M. (Vladimir Mukhanov), T.P. and A.M.; methodology and validation, N.M. (Natalia Minkina), V.P., E.S. (Evgeny Sakhon), E.S. (Elena Skorokhod), N.M. (Natalia Moiseeva), E.C., O.C.-P., A.P. and T.E.; investigation, T.P., V.M. (Vladimir Mukhanov), N.M. (Natalia Moiseeva), V.P., O.C.-P., E.S. (Elena Skorokhod), A.P., E.C., E.S. (Evgeny Sakhon) and A.M.; formal analysis and writing—original draft preparation, N.M. (Natalia Mirzoeva), E.S. (Ernest Samyshev), T.C., V.M. (Vladimir Mukhanov), T.P., A.M., V.M. (Victor Melnikov), V.P., N.M. (Natalia Minkina), N.M. (Natalia Moiseeva), A.P. and L.M.; writing—review and editing, N.M. (Natalia Mirzoeva), T.P., E.S. (Ernest Samyshev), T.C., V.M. (Vladimir Mukhanov), V.P. and A.M. All authors have read and agreed to the published version of the manuscript.

Funding: This study was conducted within the framework of the Russian state task No. 121090800137-6 "Comprehensive studies of the current state of the ecosystem of the Atlantic sector of Antarctica".

Informed Consent Statement: Not applicable.

Data Availability Statement: All data used in this study are available upon request from the corresponding author.

Acknowledgments: The authors are sincerely grateful to the crew of the R/V "Academician Mstislav Keldysh" for their help with the fieldwork. Special thanks to the head of the expedition, Morozov E. G., and the deputy head of the expedition, Molodtsova T. N. (IO RAS) for supporting and organizing trawling operations. The authors thank Bitiutskii D. G., Leading Researcher, and Usachev S. I., Chief specialist, (AZNIIRH) for their assistance in sampling processing for all participants of the sea expedition. The authors are sincerely grateful to the director of the FRC IBSS, Gorbunov R. V. for the organization of these scientific researches.

Conflicts of Interest: The authors declare no conflict of interest.

References

1. Laws, R.M. Ecology of the Southern Ocean. *Am. J. Sci.* **1985**, *73*, 26–40.
2. Samyshev, E.Z. *Antarctic Krill and the Structure of Planktonic Community in Its Distribution Area*, 2nd ed.; Nauka: Moscow, Russia, 1991; pp. 1–168. (In Russian)
3. Samyshev, E.Z. The structure and functioning of Antarctic marine coastal ecosystems in the conditions of natural and anthropogenic contaminations. *Mar. Ekol. J.* **2011**, *10*, 3–25.
4. Everson, I. Role of Krill in Marine Food Webs, the Southern Ocean. In *Krill: Biology, Ecology and Fisheries*; Everson, I., Ed.; Fisheries and Aquatic Resources, Series 6; Blackwell Science: Oxford, UK; pp. 194–201.
5. Bykov, V.P.; Bykova, V.M.; Krivosheina, L.L.; Golovkova, G.N.; Shust, K.V.; Shevtsov, V.V.; Kartintsev, A.V.; Ezhova, E.A. *Antarctic Krill: Manual*; Bykova, V.M., Ed.; Russian Federal Research Institute for Fisheries and Oceanography (VNIRO): Moscow, Russia, 2001; pp. 1–207. (In Russian)
6. Atkinson, A.; Siegel, V.; Pakhomov, E.A.; Jessopp, M.J.; Loeb, V. A reappraisal of the total biomass and annual production of Antarctic krill. *Deep. Sea Res. Part I Oceanogr. Res. Pap.* **2009**, *56*, 727–740. [CrossRef]
7. Siege, V. *Biology and Ecology of Antarctic Krill*; Advances in Polar Ecology Book Series; Springer International Publishing: Cham, Switzerland, 2016; Volume 1, pp. 1–441. [CrossRef]
8. Milhalev, Y. *Whales of the Southern Ocean. Biology, Whaling and Perspectives of Population Recovery*; Advances in Polar Ecology Book Series; Book 5; Springer: Cham, Switzerland, 2019; pp. 1–408.

9. Commission for the Conservation of Antarctic Marine Living Resources (CCAMLR). Available online: https://www.ccamlr.org/ (accessed on 11 August 2021).
10. Antarctic Treaty. International Treaty of 1 December 1959 (with an Additional Protocol of 4 October 1991). Available online: http://docs.cntd.ru/document/1901494 (accessed on 11 August 2021). (In Russian)
11. Samyshev, E.Z.; Bibik, V.A.; Savich, M.S.; Grishin, A.N.; Okanev, O.A.; Alekseenko, V.R. On the issue of the state of the Antarctic krill population and the pelagic ecosystem in the area of the Scotia Sea. *Bull. Ukr. Antarct. Cent.* **1997**, *1*, 132–136.
12. Duhamel, G.; Hulley, P.-A.; Causse, R.; Koubbi, P.; Vacchi, M.; Pruvost, P.; Vigetta, S.; Irisson, J.-O.; Mormede, S.; Belchier, M.; et al. Chapter 7: Biogeographic patterns of fish. In *Biogeographic Atlas of the Southern Ocean*; de Broyer, C., Koubbi, P., Raymond, B., d'Udekemd'Acoz, C., van de Putte, A., Griffiths, H.I., David, B., Grant, S., Gutt, J., Held, C., et al., Eds.; Scientific Committee on Antarctic Research: Cambridge, UK, 2014; pp. 328–362.
13. Kennicutt, M.C.; Chown, S.L.; Cassano, J.J.; Liggett, D.; Peck, L.S.; Massom, R.; Rintoul, S.R.; Storey, J.; Vaughan, D.G.; Wilson, T.J.; et al. A roadmap for Antarctic and Southern Ocean science for the next two decades and beyond. *Antarct. Sci.* **2015**, *27*, 3–18. [CrossRef]
14. Parker, E.; Jones, C.D.; Arana, P.M.; Alegría, N.A.; Sarralde, R.; Gallardo, F.; Phillips, A.J.; Williams, B.W.; Dornburg, A. Infestation dynamics between parasitic Antarctic fish leeches (Piscicolidae) and their crocodile icefish hosts (Channichthyidae). *Polar Biol.* **2020**, *43*, 665–677. [CrossRef]
15. Florindo, F.; Siegert, M.; De Santis, L.; Naish, T.R. Chapter 1: Antarctic climate evolution. In *Antarctic Climate Evolution*, 2nd ed.; Florindo, F., Siegert, M., De Santis, L., Naish, T.R., Eds.; Elsevier: Amsterdam, The Netherlands, 2022; pp. 1–7. [CrossRef]
16. Samyshev, E.Z.; Minkina, N.I. Coastal Ecosystem Contamination by Heavy Metals as an Indicator of Climate Change in Antarctica. *J. Comput. Theor. Nanosci.* **2019**, *16*, 228–236. [CrossRef]
17. Morozov, E.G.; Spiridonov, V.A.; Molodtsova, T.N.; Frey, D.I.; Demidova, T.A.; Flint, M.V. Research of the ecosystem of the Atlantic sector of the Antarctic (79th cruise of the research vessel "Akademik Mstislav Keldysh"). *Oceanology* **2020**, *60*, 823–825. [CrossRef]
18. Morozov, E.G. Brief results of the Antarctic expedition 2021–2022—87th cruise of RV "Akademik Mstislav Keldysh". *Oceanol. Res.* **2022**, *50*, 126–128. [CrossRef]
19. Strategy of Scientific and Technological Development of the Russian Federation. Decree of the President of the Russian Federation No. 642 of 1 December 2016, as Amended and Supplemented on 15 March 2021. Available online: https://www.sudact.ru/law/ukaz-prezidenta-rf-ot-1122016-n-642/ (accessed on 4 November 2022).
20. Strategy for the Development of Maritime Activities of the Russian Federation until 2030. Decree of the Government of the Russian Federation of 30 August 2019, No. 1930-r. Available online: https://www.zakonrf.info/rasporiazhenie-pravitelstvo-rf-1930-r-30082019/ (accessed on 4 November 2022).
21. Morozov, E.G.; Frey, D.I.; Krechik, V.A.; Polukhin, A.A.; Sapozhnikov, P.V. Water Masses, Currents, and Phytoplankton in the Bransfield Strait in January 2020. In *Antarctic Peninsula Region of the Southern Ocean*; Morozov, E.G., Flint, M.V., Spiridonov, V.A., Eds.; Advances in Polar Ecology Book Series; Springer: Berlin/Heidelberg, Germany, 2021; Volume 6, pp. 55–64. [CrossRef]
22. Mirzoeva, N.Y.; Tereshchenko, N.N.; Paraskiv, A.A.; Proskurnin, V.Y.; Morozov, E.G. Heavy metals in surface water of the Atlantic sector of the Antarctic during the 79th cruise of the research vessel "AkademikMstislavKeldysh". *Mar. Biol. J.* **2020**, *5*, 56–68. [CrossRef]
23. Paraskiv, A.A.; Mirzoeva, N.Y.; Tereshchenko, N.N.; Proskurnin, V.Y.; Sidorov, I.G.; Arkhipova, S.I.; Morozov, E.G. Heavy metals and anthropogenic radionuclides in the region of the Antarctic Peninsula. In *Antarctic Peninsula Region of the Southern Ocean*; Morozov, E.G., Flint, M.V., Spiridonov, V.A., Eds.; Advances in Polar Ecology Book Series; Springer: Berlin/Heidelberg, Germany, 2021; Volume 6, pp. 433–455. [CrossRef]
24. *Guide to the Chemical Analysis of Sea Waters: Guide Document RD 52.10.243-92*; Gidrometeoizdat: Saint Petersburg, Russia, 1993; 264p. Available online: https://meganorm.ru/Data2/1/4293815/4293815261.pdf (accessed on 14 September 2020). (In Russian)
25. *GOST R 56219-2014*; Water. Determination of 62 Elements by Inductively Coupled Plasma Mass Spectrometry (ICP-MS) Method: Data Vvedeniya01.01.2016. Standartinform, Moscow, Russia, 2015. 36p. Available online: https://docs.cntd.ru/document/1200 11543 (accessed on 5 November 2022). (In Russian)
26. Analytik Jena AG. *Inductively Coupled Plasma Mass Spectrometers (IPS-MS): PlasmaQuant MS and PlasmaQuant MS Elite. User Manual*; Edition 12/2014; Analytik Jena AG: Überlingen, Germany, 2014; 143p. (In Russian)
27. Gasol, J.M.; Del Giorgio, P.A. Using flow cytometry for counting natural planktonic bacteria and understanding the structure of planktonic bacterial communities. *Sci. Mar.* **2000**, *64*, 197–224. [CrossRef]
28. Servais, P.; Casamayor, E.O.; Courties, C.; Catala, P.; Parthuisot, N.; Lebaron, P. Activity and diversity of bacterial cells with high and low nucleic acid content. *Aquat. Microb. Ecol.* **2003**, *33*, 41–51. [CrossRef]
29. Jeffrey, S.W.; Humphrey, G.F. New spectrophotometric equations for determining chlorophylls a,b,c1 and c2 in higher plants, algae and natural phytoplankton. *Biochem. Physiol. Pflanz.* **1975**, *167*, 191–194. [CrossRef]
30. Lorenzen, C.J. Determination of chlorophyll and pheopigments: Spectrophotometric equations. *Limnol. Oceanogr.* **1967**, *12*, 343–346. [CrossRef]
31. Neeley, A.R.; Mannino, A. *Inherent Optical Property Measurements and Protocols: Absorption Coefficient*; Dartmouth, N.S., Ed.; International Ocean Colour Coordinating Group (IOCCG): Dartmouth, NS, Canada, 2018. [CrossRef]

32. Kishino, M.; Takahashi, N.; Okami, N.; Ichimura, S. Estimation of the spectral absorption coefficients of phytoplankton in the sea. *Bull. Mar. Sci.* **1985**, *37*, 634–642.
33. Mitchell, B.G. Algorithms for determining the absorption coefficient for aquatic particulates using the quantitative filter technique. In Proceedings of the 1990 Technical Symposium on Optics, Electro-Optics, and Sensors—Ocean Optics X, Orlando, FL, USA, 16–20 April 1990; Volume 1302, pp. 137–148. [CrossRef]
34. Babin, M.; Stramski, D.; Ferrari, G.M.; Claustre, H.; Bricaud, A.; Obolensky, G.; Hoepffner, N. Variations in the light absorption coefficients of phytoplankton, nonalgal particles, and dissolved organic matter in coastal waters around Europe. *J. Geophys. Res.* **2003**, *108*, 3211. [CrossRef]
35. Bjorkman, O.; Demmig, B. Photon yield of O_2 evolution and chlorophyll fluorescence characteristics at 77K among vascular plants of diverse origins. *Planta* **1987**, *170*, 489–504. [CrossRef]
36. Schreiber, U.; Bilger, W.; Neubauer, C. Chlorophyll fluorescence as a nonintrusive indicator for rapid assessment of in vivo photosynthesis. In *Ecophysiology of Photosynthesis*; Springer: Berlin/Heidelberg, Germany, 1995; pp. 49–70. [CrossRef]
37. Ke, B. *Photosynthesis: Photobiochemistry and Photobiophysics*; Springer Science and Business Media: Berlin, Germany, 2001; pp. 1–792.
38. Marra, J.; Trees, C.C.; Bidigare, R.R.; Barber, R.T. Pigment absorption and quantum yields in the Arabian Sea. *Deep. Sea Res. Part II Top. Stud. Oceanogr.* **2000**, *47*, 1279–1299. [CrossRef]
39. Gitelson, I.I.; Levin, L.A.; Utyushev, K.N.; Cherepanov, O.A.; Chugunov, Y.N. *Ocean Bioluniinescence*; Gidrometeoizdat: Saint-Petersburg, Russia, 1992; pp. 1–283. (In Russian)
40. Melnik, A.; Melnik, L.; Mashukova, O.; Melnikov, V. Field studies of bioluminescence in the Antarctic sector of the Atlantic Ocean in 2002 and 2020. *Luminescence* **2021**, *36*, 1910–1921. [CrossRef]
41. Hydrobiological System "Salpa". Available online: http://ecodevice.com.ru/ecodevice-catalogue/complex-salpa-gbf (accessed on 29 November 2022).
42. Bouchard, C.; Mollard, S.; Suzuki, K.; Robert, D.; Fortier, L. Contrasting the Early Life Histories of Sympatric Arctic Gadids *Boreogadus Saida* and *Arctogadus Glacialis* in the Canadian Beaufort Sea. *Polar Biol.* **2016**, *39*, 1005–1022. [CrossRef]
43. Kobyliansky, S.G.; Orlov, A.M.; Gordeeva, N.V. Composition of Deepsea Pelagic Ichthyocenes of the Southern Atlantic, from Waters of the Range of the Mid-Atlantic and Walvis Ridges. *J. Ichthyol.* **2010**, *50*, 932–50949. [CrossRef]
44. Dmitrieva, E.V.; Lyakh, A.M.; Kornyychuk, Y.M.; Polyakova, T.A.; Popyuk, M.P. IMBR Collection of Marine Parasites: The Collection of Marine Parasites Maintained by the Institute of Marine Biological Research. 2015. Available online: http://www.marineparasites.org (accessed on 11 March 2022).
45. Morris, A.W. Dissolved molybdenum and vanadium in the northeast Atlantic Ocean. *Deep. Sea Res. Oceanogr. Abstr.* **1975**, *22*, 49–54. [CrossRef]
46. Bongo Plankton Net. Available online: https://www.nhbs.com/bongo-plankton-net (accessed on 20 August 2022).
47. Kabata, Z.; Gusev, A.V. Parasitic Copepoda of fishes from the collection of the Zoological Institute in Leningrad. *J. Linn. Soc. (Zool.)* **1966**, *46*, 155–207. [CrossRef]
48. Hogans, W.E. Review of Sarcotretes Jungersen, 1911 (Copepoda: Pennellidae) from midwater and demersal fishes in the north Atlantic Ocean. *Can. J. Zool.* **1988**, *66*, 1371–1375. [CrossRef]
49. Morozov, E.G. Currents in Bransfeld Strait. *Dokl. Earth Sci.* **2007**, *415*, 984–986. [CrossRef]
50. Gulin, S.B.; Stokozov, N.A. ^{137}Cs concentrations in Atlantic and Western Antarctic surface waters: Results of the 7th Ukrainian Antarctic Expedition, 2002. *J. Environ. Radioact.* **2005**, *83*, 1–7. [CrossRef]
51. Sanchez, N.; Reiss, C.S.; Holm-Hansen, O.; Hewes, C.D.; Bizsel, K.C.; Ardelan, M.V. Weddell-Scotia confluence effect on the iron distribution in waters surrounding the South Shetland (Antarctic Peninsula) and South Orkney (Scotia Sea) Islands during the Austral summer in 2007 and 2008. *Front. Mar. Sci.* **2019**, *6*, 771. [CrossRef]
52. Collier, R.W. Molybdenum in the Northeast Pacific Ocean. *Limnol. Oceanogr.* **1985**, *30*, 1351–1354. [CrossRef]
53. Emerson, S.R.; Huested, S.S. Ocean anoxia and the concentrations of molybdenum and vanadium in seawater. *Mar. Chem.* **1991**, *34*, 177–196. [CrossRef]
54. Warmer, H.; van Dokkum, R. *Water Pollution Control in the Netherlands Policy and Practice2001*; RIZA Report 2002.009; Rijksinstituut voor Integraal Zoetwater Beheer en Afvalwater Behandeling: Dordrecht, The Netherlands, 2002; 76 .
55. Tokarczyk, R. Classification of water masses in the Bransfield Strait and southern part of the Drake Passage using a method of statistical multidimensional analysis. *Polar Res.* **1987**, *8*, 333–366.
56. Polukhin, A.A.; Morozov, E.G.; Tishchenko, P.P.; Frey, D.I.; Artemiev, V.A.; Borisenko, G.V.; Vidnichuk, A.V.; Marina, E.N.; Medvedev, E.V.; Popov, O.S.; et al. Water structure in the Bransfield Strait (Antarctica) in January 2020: Hydrophysical, optical and hydrochemical features. *Oceanology* **2021**, *61*, 724–736. [CrossRef]
57. Holm-Hansen, O.; Mitchell, B.G.; Hewes, C.D.; Karl, D.M. Phytoplankton blooms in the vicinity of Palmer Station, Antarctica. *Polar Biol.* **1989**, *10*, 49–57. [CrossRef]
58. Morán, X.A.G.; Estrada, M.; Gasol, J.M.; Pedrós-Alió, C. Dissolved primary production and the strength of phytoplankton–bacterioplankton coupling in contrasting marine regions. *Microb. Ecol.* **2002**, *44*, 217–223. [CrossRef]
59. Davidson, A.T.; Thomson, P.G.; Westwood, K.; van den Enden, R. Estimation of bacterioplankton activity in Tasmanian coastal waters and between Tasmania and Antarctica using stains. *Aquat. Microb. Ecol.* **2004**, *37*, 33–45. [CrossRef]
60. Pearce, I.; Davidson, A.T.; Bell, E.M.; Wright, S. Seasonal changes in the concentration and metabolic activity of bacteria and viruses at an Antarctic coastal site. *Aquat. Microb. Ecol.* **2007**, *47*, 11–23. [CrossRef]

61. Guixa-Boixereu, N.; Vaqué, D.; Gasol, J.M.; Sanchez-Cámara, J.; Pedrós-Alió, C. Viral distribution and activity in Antarctic waters. *Deep. Sea Res. Part II Top. Stud. Oceanogr.* **2002**, *49*, 827–845. [CrossRef]
62. Vaqué, D.; Guixa-Boixereu, M.; Gasol, J.M.; Pedrós-Alió, C. Distribution of microbial biomass and importance of protists in regulating prokaryotic assemblages in three areas close to the Antarctic Peninsula in spring and summer 1995/96. *Deep. Sea Res. Part II Top. Stud. Oceanogr.* **2002**, *49*, 847–867. [CrossRef]
63. Duarte, C.M.; Agustí, S.; Vaqué, D.; Agawin, N.S.R.; Felipe, J.; Casamayor, E.O.; Gasol, J.M. Experimental test of bacteria–phytoplankton coupling in the Southern Ocean. *Limnol. Oceanogr.* **2005**, *50*, 1844–1854. [CrossRef]
64. Ortega-Retuerta, E.; Reche, I.; Pulido-Villena, E.; Agustí, S.; Duarte, C.M. Exploring the relationship between active bacterioplankton and phytoplankton in the Southern Ocean. *Aquat. Microb. Ecol.* **2008**, *52*, 99–106. [CrossRef]
65. Azaneu, M.; Heywood, K.J.; Queste, B.Y.; Thompson, A.F. Variability of the Antarctic slope current system in the northwestern Weddell Sea. *J. Phys. Oceanogr.* **2017**, *47*, 2977–2997. [CrossRef]
66. Collares, L.; Mata, M.M.; Kerr, R.; Arigony-Neto, J.; Barbat, M.M. Iceberg drift and ocean circulation in the Northwestern Weddell Sea, Antarctica. *Deep. Sea Res. Part II Top. Stud. Oceanogr.* **2018**, *149*, 10–24. [CrossRef]
67. Kerr, R.; Mata, M.M.; Mendes, C.R.B.; Secchi, E.R. Northern Antarctic Peninsula: A marine climate hotspot of rapid changes on ecosystems and ocean dynamic. *Deep. Sea Res. Part II Top. Stud. Oceanogr.* **2018**, *149*, 4–9. [CrossRef]
68. Ducklow, H.W.; Stukel, M.R.; Eveleth, R.; Doney, S.C.; Jickells, T.; Schofield, O.; Baker, A.R.; Brindle, J.; Chance, R.; Cassar, N. Spring–summer net community production, new production, particle export and related water column biogeochemical processes in the marginal sea ice zone of the Western Antarctic Peninsula 2012–2014. *Philos. Trans. R. Soc. A* **2018**, *376*, 20170177. [CrossRef]
69. Ferreira, A.; Ciotti, Á.M.; Mendes, C.R.B.; Uitz, J.; Bricaud, A. Phytoplankton light absorption and the package effect in relation to photosynthetic and photoprotective pigments in the northern tip of Antarctic Peninsula. *J. Geophys. Res. Ocean.* **2017**, *122*, 7344–7363. [CrossRef]
70. Bricaud, A.; Babin, M.; Morel, A.; Claustre, H. Variability in the chlorophyll-specific absorption coefficients of natural phytoplankton: Analysis and parameterization. *J. Geophys. Res.* **1995**, *100*, 13321–13332. [CrossRef]
71. Mendes, C.R.B.; Tavano, V.M.; Leal, M.C.; de Souza, M.S.; Brotas, V.; Garcia, C.A.E. Shifts in the dominance between diatoms and cryptophytes during three late summers in the Bransfield Strait (Antarctic Peninsula). *Polar Biol.* **2013**, *36*, 537–547. [CrossRef]
72. Suslin, V.; Churilova, T.A. Regional algorithm for separating light absorption by chlorophyll-a and coloured detrital matter in the Black Sea, using 480–560 nm bands from ocean colour scanners. *Int. J. Remote Sens.* **2016**, *37*, 4380–4400. [CrossRef]
73. Churilova, T.; Suslin, V.; Krivenko, O.; Efimova, T.; Moiseeva, N. Spectral Approach to Assessment of Phytoplankton Photosynthesis Rate in the Black Sea Based on Satellite Information: Methodological Aspects of the Regional Model Development. *J. Sib. Fed. Univ. Biol.* **2016**, *9*, 367–384. (In Russian) [CrossRef]
74. Morel, A.; Bricaud, A. Theoretical results concerning light absorption in a discrete medium and application to specific absorption of phytoplankton. *Deep-Sea Res.* **1981**, *28*, 1375–1393. [CrossRef]
75. Churilova, T.; Suslin, V.; Efimova, T.; Moiseeva, N.; Skorokhod, E. Effect of dissolved and suspended matter on light in the sea and phytoplankton light absorbance capacity in coastal waters of the Black Sea. *Fundam. Prikl. Gidrofiz.* **2020**, *13*, 42–53. [CrossRef]
76. Bitiutskii, D.G.; Samyshev, E.Z.; Minkina, N.I.; Melnikov, V.V.; Chudinovskih, E.S.; Usachev, S.I.; Salyuk, P.A.; Serebrennikov, A.N.; Zuev, O.A.; Orlov, A.M. Distribution and Demography of Antarctic Krill and Salps in the Atlantic Sector of the Southern Ocean during Austral Summer 2021–2022. *Water* **2022**, *14*, 3812. [CrossRef]
77. Pakhomov, E.A.; Froneman, P.W.; Perissinoto, R. Salp/krill interactions in the Southern Ocean: Spatial segregation and implications for the carbon flux. *Deep Sea Res.* **2002**, *2*, 1881–1907. [CrossRef]
78. Atkinson, A.; Siegel, V.; Pakhomov, E.A.; Rothery, P. Long-term decline in krill stock and increase in salps within the Southern Ocean. *Nature* **2004**, *432*, 100–103. [CrossRef] [PubMed]
79. Atkinson, A.; Hill, S.L.; Pakhomov, E.; Siegel, V.; Anadon, R.; Chiba, S.; Daly, K.L.; Downie, R.; Fielding, S.; Fretwell, P.; et al. KRILLBASE: A circumpolar database of Antarctic krill and salp numerical densities, 1926–2016. *Earth Sci. Data Syst.* **2017**, *9*, 193–2107. [CrossRef]
80. Bombosch, A. Euphausia Superba or Salpa Thompsoni—Who Is Going to Win? 2008. Available online: https://www.ir.canterbury.ac.nz/bitstream/handle/10092/13921/Bombosch.pdf?sequence=1/ (accessed on 14 December 2022).
81. Groeneveld, Y.; Berger, U.; Henschke, N.; Pakhomov, E.; Reiss, C.; Meyer, B. Blooms of a key grazer in the Southern Ocean—An individual-based model of *Salpa thompsoni*. *Prog. Oceanogr.* **2020**, *185*, 102339. [CrossRef]
82. Luo, J.Y.; Stock, C.A.; Henschke, N.; Dunne, J.P.; O'Brien, T.D. Global ecological and biogeochemical impacts of pelagic tunicates. *Prog. Oceanogr.* **2022**, *205*, 102822. [CrossRef]
83. Johnston, N.M.; Murphy, E.J.; Atkinson, A.; Andrew, J.; Constable, A.J.; Cotté, C.; Cox, M.; Daly, K.L.; Driscoll, R.; Flores, H.; et al. Status, Change and Futures of Zooplankton in the Southern Ocean. *Front. Ecol. Evol.* **2022**, *9*, 624692. [CrossRef]
84. Minkina, N.I.; Samyshev, E.Z.; Pakhomov, E.A.; Melnikov, V.V. Temporal and satial variability of energy exchange in Antarctic salps. *Res. Sq.* **2022**, submitted. [CrossRef]
85. Samyshev, E.Z. Conclusion on the state of krill population and pelagic ecosystem in the western region of the Atlantic part of Antarctica in the pre-winter period of 1998. *Bull. Ukr. Antarct. Cent.* **2000**, *3*, 231–236.
86. Lomakin, P.D.; Samyshev, E.Z. Oceanographic conditions in the area of the South Shetland Islands in March–April 1997, 1998 and their influence on the distribution of krill and salp. *Oceanology* **2004**, *44*, 882–891.

87. Artamonov, Y.V. Seasonal Variability of Large-Scale Water Circulation and Thermohaline Fronts of the South Atlantic. Ph.D. Thesis, Marine Hydrophysical Institute of the National Academy of Sciences of Ukraine, Sevastopol, Ukraine, 2005; 363p.
88. Kabata, Z. *Csustacea as Enemies of Fishes: Diseaes of Fishes*; Snieszko, S.F., Axelrod, H.R., Eds.; TFH PublicationsInc.: Jersey City, NJ, USA, 1970; pp. 1–171.
89. Kabata, Z. Diseases caused by metazoans: Crustaceans. In *Diseases of Marine Animals*; Kinne, O., Ed.; Biologische Anstalt Helgoland Jahresbericht: Hamburg, Germany, 1984; Volume 4, Part 1; pp. 321–399.
90. Kazachenko, V.N. Influence of parasitic copepods on fish. *Izv. TINRO* **2008**, *154*, 204–213.
91. Kazachenko, V.N.; Kovaleva, N.N.; Nguyen, V.T.; Ngo, H.D. Three new species and one new genus of parasitic copepods (Crustacea: Copepoda) from fishes of the South China Sea. *Russ. J. Mar. Biol.* **2017**, *43*, 264–269. [CrossRef]
92. Ho, J.S.; Kim, I.H.; Nagasawa, K.; Saruwatari, T. *Paeonocanthusantarcticensis* (Hewitt, 1965): A rare copepod parasite of bathypelagic fish, *Bathylagusantarcticus* Günther, from the Antarctic Ocean. *Zool. Sci.* **2003**, *20*, 1315–1321. [CrossRef] [PubMed]
93. Cherel, Y.; Boxshall, G.A. *Sarcotretes* (Copepoda: Pennellidae) parasitizing myctophid fishes in the Southern Ocean: New information from seabird diet. *J. Parasitol.* **2004**, *90*, 1288–1292. [CrossRef] [PubMed]
94. Polyakova, T.A.; Gordeev, I.I. Parasites as an inseparable part of Antarctic and Subantarctic marine biodiversity. In *Antarctic Peninsula Region of the Southern Ocean*; Morozov, E.G., Flint, M.V., Spiridonov, V.A., Eds.; Advances in Polar EcologyBook Series; Springer: Berlin/Heidelberg, Germany, 2021; Volume 6, pp. 321–354. [CrossRef]

water

MDPI

Article

Age Structure and Spatial Distribution of *Euphausia superba* Larvae off the Antarctic Peninsula, Southern Ocean

Valentina V. Kasyan

Laboratory of Systematics and Morphology, A.V. Zhirmunsky National Scientific Center of Marine Biology, Far Eastern Branch, Russian Academy of Sciences (NSCMB FEB RAS), Vladivostok 690041, Russia; valentina-k@yandex.ru

Abstract: The Antarctic krill, *Euphausia superba* Dana, 1850, is a species forming high biomass and, therefore, playing a major role in the Antarctic marine food web. The age structure and patterns of spatial distribution of *E. superba* larvae in the waters of the Bransfield Strait (Antarctic Sound, Powell Basin), and off the South Orkney Islands, were studied based on data collected through a research survey in January and February 2022. Eggs and larvae (naupliar, calyptopis, and furcilia stages) of *E. superba* were found in these regions. Eggs and nauplii were concentrated in the southern, deep-sea part of the Antarctic Sound and over the northeastern and southwestern slopes of the Powell Basin, while calyptopis and furcilia larvae were concentrated north of the South Orkney Islands. The larvae abundance increased in an easterly direction. Four groups of communities comprising krill larvae at different development stages were identified. These groups were located in two subregions with the border between them running off the South Orkney Islands. The distribution and abundance of *E. superba* larvae showed a clear relationship with environmental conditions, in particular with a combination of such factors as sea surface temperature and chlorophyll *a* concentration.

Keywords: Antarctic krill *Euphausia superba*; eggs; nauplii; calyptopis; furcilia; abundance; distribution; Atlantic sector; Southern Ocean

Citation: Kasyan, V.V. Age Structure and Spatial Distribution of *Euphausia superba* Larvae off the Antarctic Peninsula, Southern Ocean. *Water* **2022**, *14*, 3196. https://doi.org/10.3390/w14203196

Academic Editor: Morozov Eugene

Received: 2 September 2022
Accepted: 8 October 2022
Published: 11 October 2022

1. Introduction

The Antarctic krill, *Euphausia superba* Dana, 1850, (hereafter referred to as krill) is a major link between primary producers and vertebrates (seals, whales, and seabirds) in the food web of the Southern Ocean [1]. It is also an important target species of commercial fisheries [2,3]. Over the past 25 years, the population density of krill in the Atlantic sector has markedly decreased [4–6], and therefore, recovery processes in this species attract increasing research attention. Although pelagic communities in the Atlantic sector of the Southern Ocean have been quite thoroughly studied to date [7–10], there is still little information on the distribution of early development stages of *E. superba* in this region. Previous studies described the relationship between spatial variations in larval aggregations and concentration of adult krill [11,12], the distribution of life-history stages and overlaps of their ranges, and also their relationships with environmental dynamics [13–15]. Krill development passes through a series of successive stages. Fertilized eggs descend from the surface layer to a depth of 1000 m, where nauplii hatch and begin rising back to the surface. During the ascent, the development stages of krill larvae sequentially change from nauplius to metanauplius, calyptopis, and furcilia [11,14].

Significant warming has been observed in the Southern Ocean in recent decades: since 1955, the temperature of the upper ocean layers west of the Antarctic Peninsula has increased by more than 1 °C [16]. The reduction in the sea ice extent and proportion of multi-year sea ice is still ongoing [17,18]. The continuing climate change in the Southern Ocean [19,20] makes pelagic communities adapt to new conditions [21,22]. Since krill larvae are extremely sensitive to environmental factors [23,24], study of their responses to variations in environmental factors is of particular relevance to address the conservation of the species and provide sustainability of krill fishery in the future. In the present study, we assessed the current spatial distribution of krill larvae in the waters off the Antarctic Peninsula (Atlantic sector of the Southern Ocean) and identified the main environmental factors responsible for the distribution and density of various krill larvae stages. The results of the study are expected to be of high practical value for monitoring the current status of the changing Antarctic marine ecosystem.

2. Materials and Methods

The material for this study was collected in January and February 2022 at 23 integrated stations. The study area included the Bransfield Strait, the Antarctic Sound, the Powell Basin of the Weddell Sea, and the waters off the South Orkney Islands. Sampling was carried out using the opening/closing Multinet system [25] (0.25 m^2 aperture) equipped with five 150-µm mesh nets, a WP-2 net [26] with a mesh size of 150 µm, and a Bongo net (505-µm mesh and 0.6 m mouth diameter) [27] (Figure 1). The tows from a 200-m depth to the surface were performed obliquely. The vertical, stratified sampling tows from the bottom (maximum 1000 m) to depths of 500, 500–200, 200–100, 100–50, and 50–0 m were made at a velocity of 0.5–1 m/s. The Bongo net was equipped with a water flow counter (Hydrobios, Germany). A total of 120 plankton samples were collected. Fresh samples, immediately after being hauled aboard and brought to the laboratory, were viewed in a Bogorov counting chamber under a SZX7 binocular microscope (Olympus, Japan) and then fixed with a 4% formalin solution. The development stages of *E. superba* were identified using special identification keys [11,28]. Hydrological measurements of water temperature (T, °C) and salinity (S, psu) were carried out using an OCEAN SEVEN 320 Plus CTD multiparameter probe (Idronaut, Italy), lowered from the surface to the bottom. Chlorophyll *a* (Chl *a*) profiles were measured with a calibrated fluorometer (Seapoint Sensors Inc., Exeter, NH, USA) mounted on the CTD probe. Temperature and salinity within the upper 50 m horizon were referred to as "surface". Mean values of Chl *a* concentration were calculated for the upper 100 m of the water column, which was the estimated average depth of the euphotic zone.

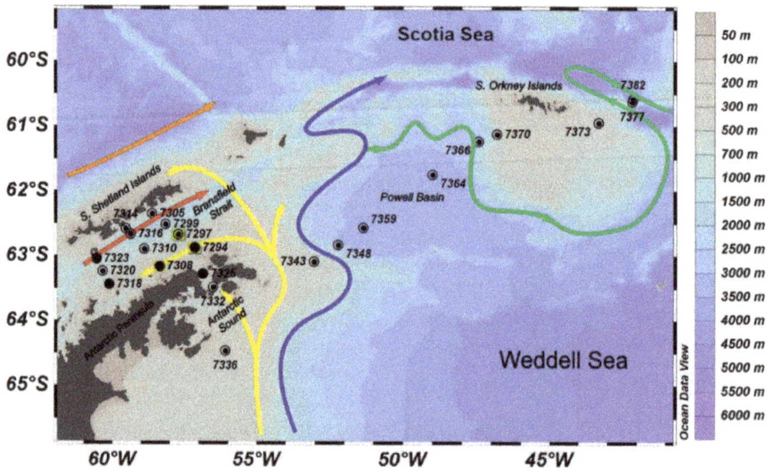

Figure 1. The sampling stations and the major currents in the study area. Orange line is the Antarctic Circumpolar Current (ACC); red line, the Bransfield Current (BC); yellow line, the Antarctic Coastal Current (ACoC); blue line, the Antarctic Slope Front (ASF); green line, the Weddell Front (WF). Plankton nets used at the stations are designated as follows: ◉—Multinet, ●—Bongo, and •—WP-2.

Statistical analysis was carried out using the PRIMER v. 6 software package [29]. For data processing, a matrix of larvae similarity of samples was calculated on the basis of the Bray–Curtis dissimilarity index. Number of larvae at different development stages per 1 m^3 of water was used as initial data. To assess the reliability of clustering results, a SIMPROF permutation test was performed (with a number of repetitions of 999, $p = 0.05$). In addition, effects of environmental parameters (water temperature, salinity, and Chl *a*) on the age structure of krill larvae were analyzed by BEST (BIOENV analysis) relating the larvae abundance matrix to the Euclidean distance matrix of environmental parameters. The BEST procedure selects the subset of abiotic variables that maximizes the rank correlation (ρ) between the biotic and abiotic (dis)similarity matrices [30]. The Spearman's rank correlation analysis was performed to show correlations between abundance and hydrological variables in order to identify the variables most responsible for their distribution between the stations (run in PAST ver. 4.05). The maps of the sampling area and spatial distribution of krill larvae abundance were composed using the Ocean Data View v. 5 software.

3. Results

The highest sea surface temperature (SST) values (with a maximum of 1.8 °C) were recorded from the waters northeast of the South Orkney Islands and in the Bransfield Strait off the South Shetland Islands; medium values were in the middle of the Bransfield Strait and in the Powell Basin; the lowest values, in the Antarctic Sound and in the waters off the Antarctic Peninsula, where one extreme temperature minimum (stn. 7336) was recorded (Figure 2A). The highest sea surface salinity (SSS) values were recorded from the areas in the middle of the Bransfield Strait and in the Antarctic Sound; medium values were in the Powell Basin and northeast of the South Orkney Islands; the lowest values were in the Bransfield Strait off the South Shetland Islands and off the Antarctic Peninsula (Figure 2B). The distribution of mean Chl *a* concentrations is shown in Figure 2C.

The age structure and abundance of krill larvae differed between the waters off the Antarctic Peninsula and off the South Orkney Islands coast. The maxima of abundance of krill eggs and larvae (2300 ind./m^3) were recorded from three areas: the Antarctic Sound, the Powell Basin, and the waters northeast of the South Orkney Islands. The minimum abundance values of krill eggs and larvae (1.3 ind./m^3) were in the Bransfield Strait and off the Antarctic Peninsula, with their concentration increasing 3–5-fold from the coast to the middle of the Bransfield Strait (Figure 3A). Krill eggs and nauplii were concentrated in three areas: the southern Antarctic Sound and over the southwestern and northeastern slopes of the Powell Basin (Figure 3A). Krill larvae in calyptopis and furcilia stages were concentrated at a maximum abundance northeast of the South Orkney Islands. In general, the krill larvae abundance increased in an easterly direction. In the waters off the Antarctic Peninsula and in the Powell Basin, the major proportion (%) of krill larvae was formed by eggs and nauplii, while east of this area, in the waters off the South Orkney Islands, calyptopis and furcilia larvae accounted for a maximum proportion (Figure 3B).

The age structure and vertical distribution pattern of *E. superba* larvae are shown in Figure 4. The maximum values of average abundance of krill eggs and nauplii were observed in the relatively warm surface (0–50 m) layer in the warm Antarctic Surface Water (AASW) (Figure 4F). Relatively high concentrations of krill eggs and nauplii were recorded from the deep-sea (500–800 m) layer in the cool and saline Transitional Zonal Water with Weddell Sea influence (TWW) and Shelf Water (SW) (Figure 4D,E). The maximum concentrations of calyptopis and furcilia larvae were recorded from the warmer 0–100 m layer in the AASW (Figure 4G). Highly abundant krill eggs and larvae were observed in the surface water, where they co-occurred with blooming phytoplankton (visually in samples).

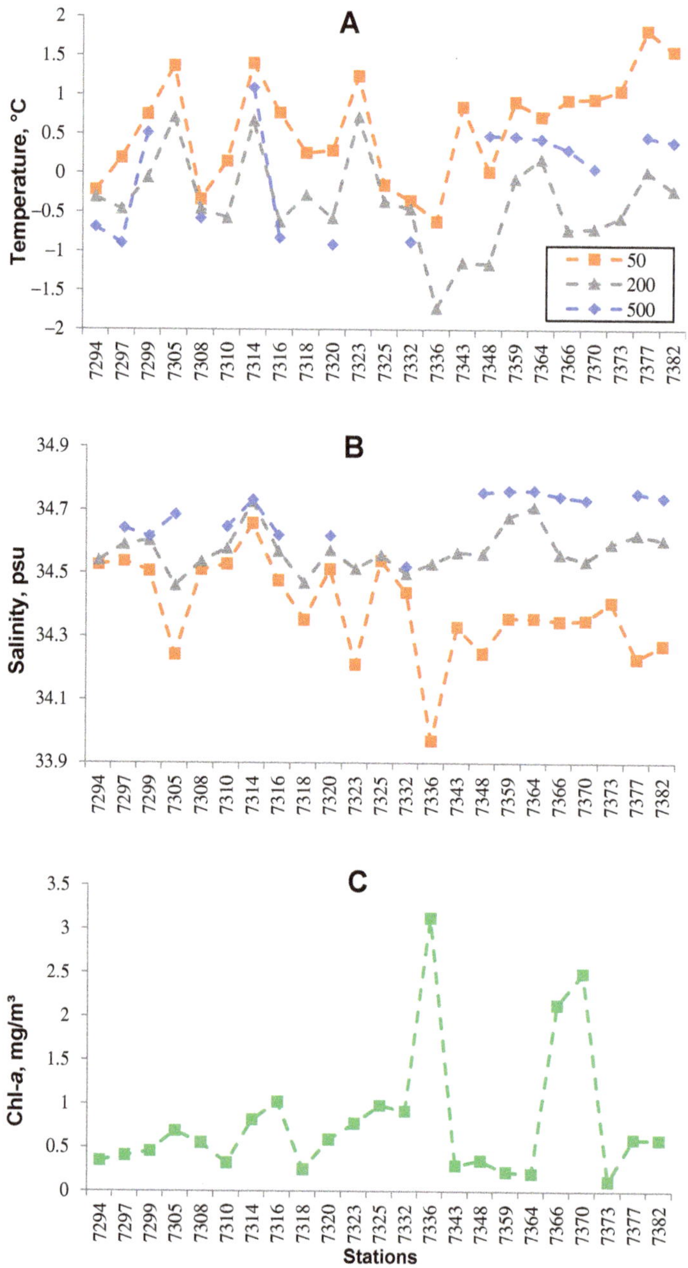

Figure 2. Mean values of water temperature (**A**), salinity (**B**) at the surface (to 50 m), in the 200-m layer, and in the 500-m layers, and also mean values of chlorophyll *a* (**C**) concentrations (mg/m^3) at stations sampled during January and February 2022.

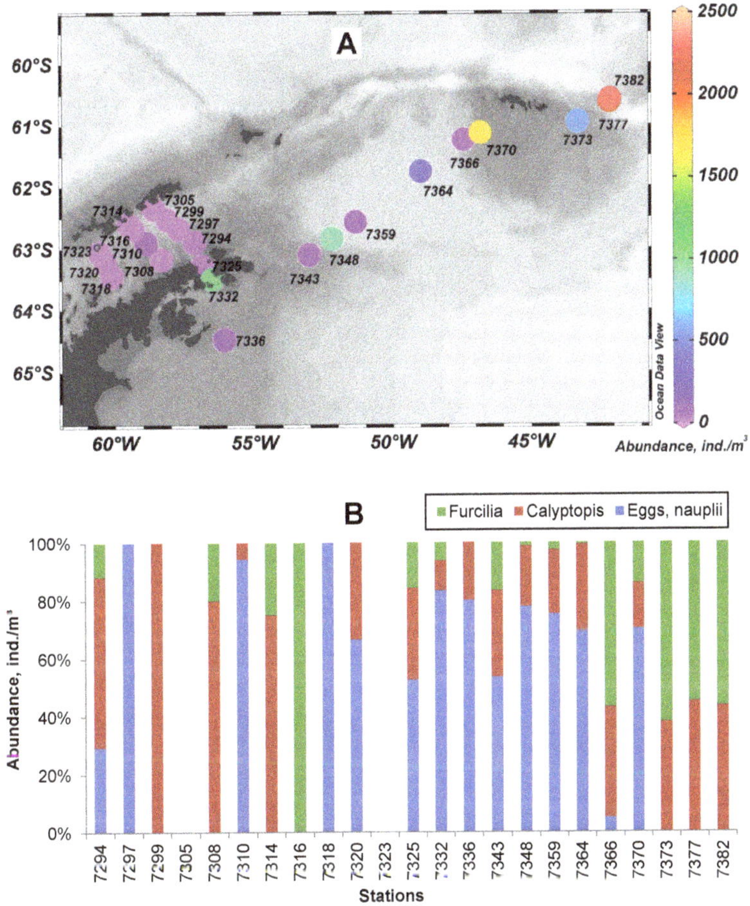

Figure 3. Spatial distribution of total abundance ((**A**) ind./m^3) and contribution ((**B**) %) of *Euphausia superba* eggs and larvae at different development stages in January and February 2022.

Four significant groups of stations were identified (SIMPROF R = 0.73, p = 0.05%) (Figure 5A). Group A was mainly confined to the waters of the Bransfield Strait and Powell Basin (Figure 5B), characterized by high values of SSS and Chl *a* concentration. Group B tended to the waters off the South Orkney Islands that had high values of SST, water temperature at depths of 200 and 500 m, and salinity at 200 and 500 m. Group C was mainly located in the waters of the Antarctic Sound and Powell Basin with low temperatures at 200 m, a low SSS, and a high Chl *a* concentration; Group D was confined to the Bransfield Strait (Figure 5B) with low values of SST, temperature at 200 and 500 m, a low SSS, salinity at 200 and 500 m, and a low mean Chl *a* concentration in 0–100 m.

Differences in the spatial dominance of eggs and larvae at different development stages were observed between the identified groups of stations. The groups A, C, and D were dominated by krill eggs and nauplii (first subregion); the group B, by larvae of calyptopis and furcilia stages (second subregion). The border between the subregions ran near the South Orkney coast.

The Spearman's rank correlation analysis indicated that SST was the most important factor determining the distribution of the different development stages of krill larvae ($\rho = 0.307$, $p < 0.05$). A higher correlation coefficient was obtained with a combination of SST and Chl *a* concentration ($\rho = 0.369$, $p < 0.05$) (Table 1). The abundance of krill eggs and nauplii positively correlated with Chl *a* concentration and salinity at 500 m and negatively correlated with the other factors. The abundance of both calyptopis and furcilia larvae significantly positively correlated with SST and temperature at 500 m, as well as with salinity at 200 and 500 m, and negatively correlated with SSS, temperature at 200 m, and Chl *a* concentration. The abundance of furcilia larvae proved to be closely related with that of calyptopis larvae (Table 2).

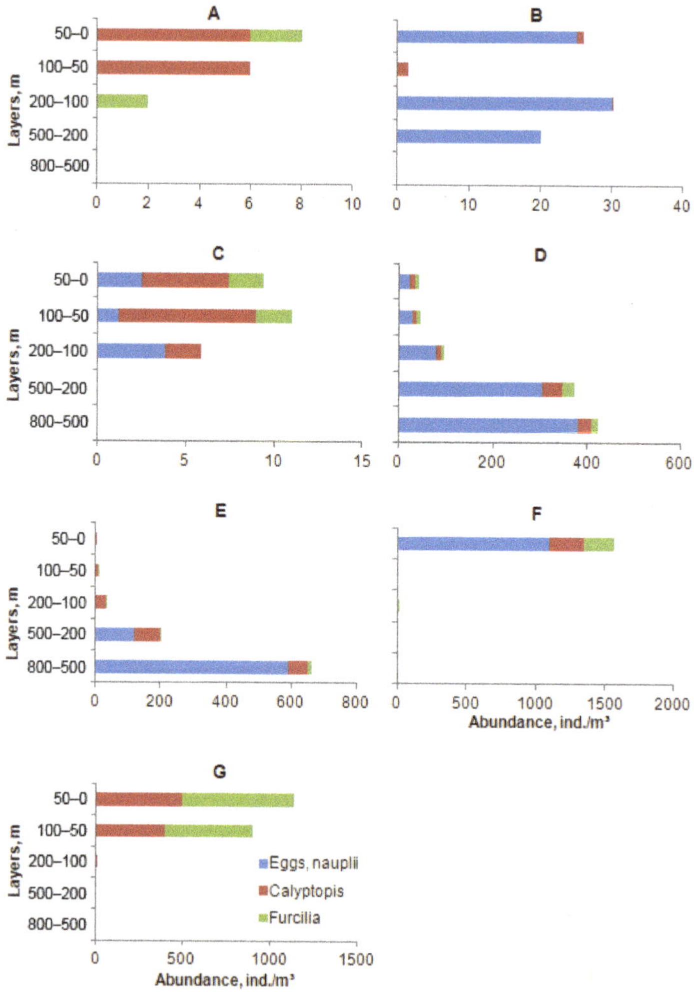

Figure 4. Vertical distribution of average abundances of *Euphausia superba* eggs and larvae (ind./m^3) in the study area: (**A**) in the Bransfield Strait off the South Shetland Islands; (**B**) in the middle of the Bransfield Strait; (**C**) in the Bransfield Strait off the Antarctic Peninsula; (**D**) in the deep-sea waters of the Antarctic Sound; (**E**) over the southwestern slope of the Powell Basin; (**F**) over the northeastern slope of the Powell Basin; (**G**) northeast of the South Orkney Islands.

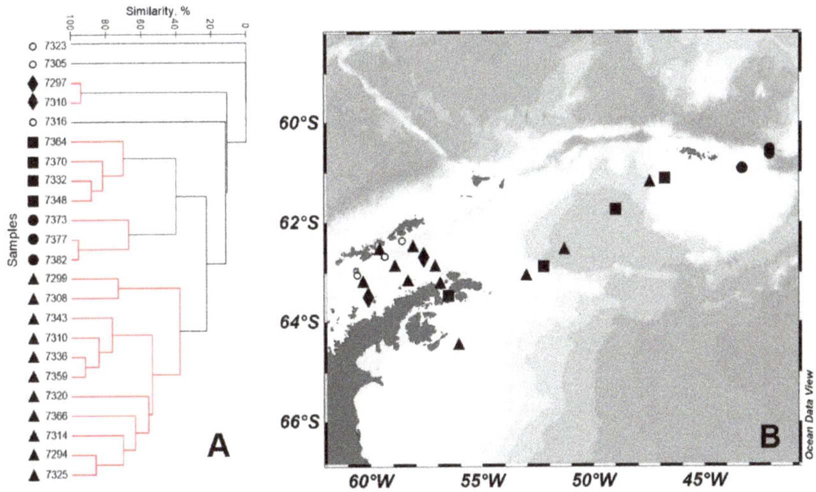

Figure 5. Dendrogram of stations resulting from a cluster analysis based on the abundances of *Euphausia superba* eggs and larvae at different development stages (**A**); geographical distributions of the identified groups of stations (**B**). Groups are as follows: ▲—A, •—B, ■—C, ♦—D, ○—no groups.

Table 1. Correlation coefficients inferred from the BEST analysis of relationships between the distribution and age structure of *Euphausia superba* and the environmental parameters such as water temperature (T, °C), salinity (S, psu), and chlorophyll *a* concentration (Chl *a*, mg/m^3).

Environmental Factors	ρ	Mean ± SD			
		Group A	Group B	Group C	Group D
SST, °C	0.307	0.36 ± 0.65	1.48 ± 0.39	0.34 ± 0.61	0.22 ± 0.04
200 m T, °C	0.096	−0.48 ± 0.61	−0.25 ± 0.30	−0.54 ± 0.56	−0.37 ± 0.13
500 m T, °C	0.089	0.03 ± 0.75	0.43 ± 0.05	0.02 ± 0.63	−0.90 ± 0.07
SSS, psu	0.101	34.43 ± 0.18	34.30 ± 0.09	34.35 ± 0.08	34.44 ± 0.13
200 m S, psu	0.093	34.59 ± 0.06	34.61 ± 0.02	34.58 ± 0.09	34.53 ± 0.09
500 m S, psu	0.081	34.69 ± 0.07	34.75 ± 0.01	34.69 ± 0.12	34.64 ± 0.03
Chl a, mg/m^3	0.163	0.89 ± 0.91	0.43 ± 0.27	0.99 ± 1.05	0.33 ± 0.11
BEST combination of factors (SST and Chl *a*)	0.369				

Table 2. Spearman's rank correlations between the abundances of different development stages of *Euphausia superba* and the environmental factors ($p < 0.05$).

	SST	SSS	200 m T	200 m S	500 m T	500 m S	Chl *a*	E and N	C	F
E and N	−0.38	−0.16	−0.44	−0.02	−0.08	0.29	0.36	1		
C	0.22	−0.32	−0.20	0.37	0.41	0.57	−0.09	0.58	1	
F	0.33	−0.18	−0.16	0.26	0.23	0.44	−0.06	0.32	0.86	1

Notes: T is water temperature (°C); S, water salinity (psu); Chl *a*, chlorophyll *a* concentration (mg/m^3); E and N, eggs and nauplii; C, calyptopis; F, furcilia.

4. Discussion and Conclusions

The pattern of the major currents in the study area was described earlier in the literature [31–33]. In the Bransfield Strait, a multidirectional (two jets) system of currents is observed: the continuation of the cold Antarctic Coastal Current (ACoC) flowing southwest and carrying the relatively cool and saline Transitional Zonal Water with Weddell Sea influence (TWW), and the warm Bransfield Current (BC) directed southeast along the South Shetland Islands and carrying the worm Transitional Zonal Water with Bellingshausen

Sea influence (TBW) [34–36]. The Weddell Surface Water spreading zone is located in the Antarctic Sound [37], where the northward TWW is flowing in the southern, deep-sea part of the strait. A current of the very cold (close to freezing) temperatures (–1.9 °C), Shelf Water (SW) is running south of the Antarctic Sound [38]. There are three northward currents in the Powell Basin: the ACoC, the Antarctic Slope Front (ASF), and the Weddell Front (WF), where the surface layer is occupied by the warm Antarctic Surface Water (AASW), with the Warm Deep Water (WDW) located below [39]. In the region off the South Orkney Islands, the Antarctic Circumpolar Current (ACC) and the Weddell Sea water interact [40].

The results we obtained characterize the known patterns of spatial distribution of krill eggs and larvae, which have not changed in general. However, the average value of krill larvae abundance in our study was 3–5-fold lower than those reported by other researchers [13,14,41,42]. It is probable that the low abundance of krill eggs and larvae that we observed in 2022 coincided in time with a period between high krill abundance waves. Effects of various currents determine the pattern of distribution of krill eggs and larvae that are transported by the longitudinal component of the currents from deep horizons of deep-sea waters to the divergence zones, where local density maxima of krill juveniles are formed. The krill distribution is associated with the action of vortex currents near the surface in the mixing zone of the Weddell Sea and ACC waters and in the gyres and shadow zones off the South Orkney Islands [11,43]. In our study, the concentration of krill eggs and larvae increased 5-fold in the direction from the coast towards the middle of the Bransfield Strait, in the zone where the warm BC water is mixed with the cold and saline ACoC water.

We observed a close relationship of distribution and abundance of krill larvae at different development stages with a certain combination of abiotic environmental factors, in particular, SST and Chl *a* concentration, and also found that larvae tended to the areas characterized by such a combination. The circulation of water masses within the region is also an important factor affecting the dynamics of this relationship. The study area was influenced by the warm and saline water of the ACC southern jet in the west [44] and the colder and freshened water of the Weddell Sea and the associated fronts in the east [45]. In the Antarctic Sound and over the Powell Basin slope, influenced by the cold (-0.5 to -1.5 °C), saline (34.5 psu), and chlorophyll *a*-rich (up to 2.5 mg/m^3) waters carried by the ACoC, we recorded the maximum concentrations of krill eggs and nauplii having the Weddell Sea origin. This distribution can be explained by the transport of eggs and nauplii from the shelf to the slope in a northeasterly direction through the Weddell Gyre [46]. As in our studies, Gao et al. [13] reported an abundance of krill larvae at early development stages in areas with high Chl *a* concentration. The timing of spawning and conditions for subsequent larval development (temperature and food type) have important ecological implications for the success of recruitment in the krill populations [47]. The presence of krill eggs in the study area indicated a late spawning season in 2022, which probably occurred in the Weddell Sea in early February. Thus, the surface layer of the warm and freshened AASW off the South Orkney Islands was dominated by calyptopis and furcilia larvae that are indicators of spring/early summer krill spawning [48,49]. There is a great deal of uncertainty as to whether krill larvae survive the upcoming winter due to late spawning, which may eventually become a cause of low krill recruitment in the following year, since the growth and survival rates of krill larvae decrease at sub-zero water temperatures [23]. As the distribution of maximum concentrations of *E. superba* larvae in the study area shows, the age of larvae increases in the west to east direction. An assumption can be made that the early spawning in krill occurred east of the Antarctic Peninsula, in the waters of the South Orkney Islands.

On the basis of the distribution pattern of krill larvae, the study area can be divided into two subregions. In the waters between the Antarctic Peninsula and southwest of the South Orkney Islands, the age structure of krill larvae was mainly composed of eggs and nauplii, while northeast of the South Orkney Islands, it was dominated by calyptopis and furcilia. Aggregations of krill larvae found in the waters of the South Orkney Islands were more mature and more abundant than those of larvae in the shelf waters off the

Antarctic Peninsula. As in our study, Makarov and Menshenina [50], Gao et al. [13], and Spiridonov et al. [42] found that krill eggs and early larvae are present in the waters of the South Orkney Islands, and in summer these can be transported by the western branch of the Weddell Gyre [46] and be mixed with the group of late larvae transported by the water of the southern ACC jet [51]. In previous studies, eggs and larvae of Antarctic krill were found in the northwestern Weddell Sea [15,43,46], over the shelf slope of the Antarctic Peninsula [14,41,42,45,49], in the waters of the South Orkney Islands, and in the Scotia Sea [11,13,50] during the southern summer until late autumn and the early southern winter (May and June) [48,52]; differences in the development stages of krill larvae were often recorded from these regions.

Krill eggs and larvae are known to inhabit different depths and drift in various directions and, therefore, their aggregations are differentiated in space [43]. According to our observations, krill eggs and nauplii tended to deep-sea layers (up to 800 m), while calyptopis and furcilia larvae tended to the surface layers. The distribution and abundance of krill eggs and nauplii positively correlated with high Chl *a* concentrations, and with salinity at 500 m, whereas the abundance of krill calyptopis and furcilia larvae positively correlated with above-zero SST. In addition to above-zero water temperatures, a high phytoplankton concentration can also contribute to the development of krill larvae [23,47,53,54]. We observed krill eggs and nauplii in the surface (0–50 m) layer, above the layer of high phytoplankton concentration (visually in samples) 50–100 m, whereas the distribution of calyptopis and furcilia larvae found in the 0–50 and 50–100 m layers overlapped with the maximum microalgae concentration (visually in samples).

The climate change that has been observed in the Southern Ocean in recent decades may be both part of a natural systemic process [7,55,56] and a negative trend for the Southern Ocean ecosystem [22,57,58]. The Antarctic krill, *E. superba*, needs further study at all life-history stages for monitoring the actual status of the changing Southern Ocean ecosystem and its biological resources.

Funding: This research was supported by the Russian State Assignment no. 122072000067-9.

Data Availability Statement: Not applicable.

Acknowledgments: We are grateful to the administration of the Shirshov Institute of Oceanology, Russian Academy of Sciences for the organization of the expedition and field operations, to the plankton and benthos research teams for collecting and providing plankton samples, to the hydrology research team for providing data on water temperature, salinity, and chlorophyll *a* concentration, and also to the captain and the crew of the R/V *Akademik Mstislav Keldysh* for their comprehensive assistance. Special thanks are due to E. Shvetsov for his help in improving the English. The authors thank two anonymous reviewers for their valuable comments and suggestions which allowed for considerable improvement of the manuscript.

Conflicts of Interest: The author declares no conflict of interest.

References

1. Laws, R. Seals and whales of the Southern Ocean. *Philos. Trans. R. Soc. Lond. B Biol. Sci.* **1977**, *279*, 81–96. [CrossRef]
2. Kawaguchi, S.; Nicol, S. Learning about Antarctic krill from the fishery. *Antar. Sci.* **2007**, *19*, 219–230. [CrossRef]
3. Santa Cruz, F.; Ernst, B.; Arata, J.A.; Parada, C. Spatial and temporal dynamics of the Antarctic krill fishery in fishing hotspots in the Bransfield Strait and South Shetland Islands. *Fish. Res.* **2018**, *208*, 157–166. [CrossRef]
4. Atkinson, A.; Siegel, V.; Pakhomov, E.; Rothery, P. Long-term decline in krill stock and increase in salps within the Southern Ocean. *Nature* **2004**, *432*, 100–103. [CrossRef] [PubMed]
5. Atkinson, A.; Hill, S.L.; Pakhomov, E.A.; Siegel, V.; Reiss, C.S.; Loeb, V.J.; Steinberg, D.K.; Schmidt, K.; Tarling, G.A.; Gerrish, L.; et al. Krill (*Euphausia superba*) distribution contracts southward during rapid regional warming. *Nat. Clim. Change* **2019**, *9*, 142–147. [CrossRef]
6. Yang, G.; Atkinson, A.; Hill, S.L.; Guglielmo, L.; Granata, A.; Li, C. Changing circumpolar distributions and isoscapes of Antarctic krill: Indo-Pacific habitat refuges counter long-term degradation of the Atlantic sector. *Limnol. Oceanogr.* **2020**, *66*, 272–287. [CrossRef]
7. Voronina, N.M. Comparative abundance and distribution of major filter-feeders in the Antarctic pelagic zone. *J. Mar. Sys.* **1998**, *17*, 375–390. [CrossRef]

8. Hewes, C.D.; Reiss, C.S.; Holm-Hansen, O. A quantitative analysis of sources for summertime phytoplankton variability over 18 years in the South Shetland Islands (Antarctica) region. *Deep-Sea Res. II* **2009**, *56*, 1230–1241. [CrossRef]
9. Garcia, M.D.; Dutto, M.S.; Chazarreta, C.J.; Berasategui, A.A.; Schloss, I.R.; Hoffmeyer, M.S. Micro- and mesozooplankton successions in an Antarctic coastal environment during a warm year. *PLoS ONE* **2020**, *15*, e0232614. [CrossRef] [PubMed]
10. Pakhomov, E.; Pshenichnov, L.; Krot, A.; Paramonov, V.; Slypko, I.; Zabroda, P. Zooplankton distribution and community structure in the Pacific and Atlantic Sectors of the Southern Ocean during austral summer 2017–2018: A Pilot Study Conducted from Ukrainian Long-Liners. *J. Mar. Sci. Eng.* **2020**, *8*, 488. [CrossRef]
11. Makarov, R.R. Larvae of *Euphausia superba* Dana in plankton from the Sea of Scotia. *Tr. Vses. Nauchno-Issled. Ryb. Okeanogr.* **1974**, *99*, 84–103.
12. Shulenberger, E.; Wormuth, J.H.; Loeb, V.J. A large swarm of *Euphausia superba*: Overview of path structure and composition. *J. Crustacean Biol.* **1984**, *4*, 75–95. [CrossRef]
13. Gao, Q.; Xu, Z.; Huang, H.; Chen, X.; Feng, C.; Li, L. Geographical distribution and age composition of *Euphausia superba* larvae (Crustacea: Euphausiacea) in the South Shetland Islands region and southern Scotia Sea in relation to environmental conditions. *Acta Oceanol. Sin.* **2013**, *32*, 59–67. [CrossRef]
14. Sologub, D.O. Distribution and age structure of euphausiids larvae at the Western Antarctic Peninsula (Subarea 48.1). *Trudy VNIRO* **2015**, *154*, 3–15.
15. Yakovenko, V.A.; Spiridonov, V.A.; Gorbatenko, K.M.; Shadrin, N.V.; Samyshev, E.Z.; Minkina, N.I. Macro- and mesozooplankton in the Powell Basin (Antarctica): Species composition and distribution of abundance and biomass in February 2020. In *Antarctic Peninsula Region of the Southern Ocean*; Morozov, E.G., Flint, M.V., Spiridonov, V.A., Eds.; Springer: Cham, Switzerland, 2021; Volume 6, pp. 131–141. [CrossRef]
16. Meredith, M.P.; King, J.C. Rapid climate change in the ocean west of the Antarctic Peninsula during the second half of the 20th century. *Geophys. Res. Let.* **2005**, *32*, L19604. [CrossRef]
17. Stammerjohn, S.E.; Martinson, D.G.; Smith, R.C.; Yuan, X.; Rind, D. Trends in Antarctic annual sea ice retreat and advance and their relation to El Nino-Southern Oscillation and Southern Annular Mode variability. *J. Geophys. Res Ocean.* **2008**, *113*, C03S90. [CrossRef]
18. Meredith, M.; Sommerkorn, S.; Cassotta, C.; Derksen, A.; Ekaykin, A.; Hollowed, G.; Kofinas, A.; Mackintosh, J.; Melbourne-Thomas, M.M.C.; Muelbert, G.; et al. Polar Regions. In *IPCC Special Report on the Ocean and Cryosphere in a Changing Climate*; Pörtner, H.-O., Roberts, D.C., Masson-Delmotte, V., Zhai, P., Tignor, M., Poloczanska, E., Mintenbeck, K., Alegría, A., Nicolai, M., Okem, A., et al., Eds.; Cambridge University Press: Cambridge, UK; New York, NY, USA, 2019; pp. 203–320. [CrossRef]
19. Anisimov, O.A.; Vaughan, D.G.; Callaghan, T.V.; Furgal, C.; Marchant, H.; Prowse, T.D.; Vilhjalmsson, H.; Walsh, J.E. Polar regions (Arctic and Antarctic). In *Climate Change 2007: Impacts, Adaptation and Vulnerability*; Contribution of Working Group II to the Fourth Assessment Report of the Intergovernmental Panel on Climate Change; Parry, M.L., Canziani, O.F., Palutikof, P.J., van der Linden, P., Hanson, C.E., Eds.; Cambridge University Press: Cambridge, UK, 2007; pp. 653–685.
20. Xiao, C. Changes in antarctic climate system: Past, present and future. *Adv. Clim. Change Res.* **2008**, *4*, 1–7.
21. Behrenfeld, M.J. Climate-mediated dance of the plankton. *Nat. Clim. Change* **2014**, *4*, 880–887. [CrossRef]
22. Johnston, N.M.; Murphy, E.J.; Atkinson, A.; Constable, A.J.; Cotte, C.; Cox, M.; Daly, K.L.; Driscoll, R.; Flores, H.; Halfter, S.; et al. Status, change, and futures of zooplankton in the Southern Ocean. *Front. Ecol. Evol.* **2022**, *9*, 624692. [CrossRef]
23. Ross, R.M.; Quetin, L.B.; Kirsch, E. Effect of temperature on developmental times and survival of early larval stages of *Euphausia superba* Dana. *J. Exp. Mar. Bio. Ecol.* **1988**, *121*, 55–71. [CrossRef]
24. Perry, F.A.; Kawaguchi, S.; Atkinson, A.; Sailley, S.F.; Tarling, G.A.; Mayor, D.J.; Lucas, C.H.; King, R.; Cooper, A. Temperature–induced hatch failure and nauplii malformation in Antarctic krill. *Front. Mar. Sci.* **2020**, *7*, 501. [CrossRef]
25. Weikert, H.; John, H.-C. Experiences with a modified be multiple opening–closing plankton net. *J. Plankton Res.* **1981**, *3*, 167–176. [CrossRef]
26. Tranter, D.J. Reviews on zooplankton sampling methods. In *Zooplankton Sampling*; Part I; Tranter, D.J., Ed.; UNESCO: Paris, France, 1968; pp. 11–144.
27. Smith, P.E.; Richardson, S. Standard techniques for pelagic fish egg and larva surveys. *FAO Fish. Tech. Pap.* **1977**, *175*, 27–73.
28. Pertsova, K.N. Larvae of euphausiids of the Antarctic. In *Ecology and biogeography of plankton: Proceedings of the Institute of Oceanology RAS*; Nauka: Moscow, Russia, 1976; Volume 105, pp. 147–170.
29. Clarke, K.R.; Gorley, R.N. *PRIMER v6: User Manual/Tutorial*; PRIMER–E Ltd.: Plymouth, UK, 2006; p. 192.
30. Clarke, K.R.; Ainsworth, M. A method of linking multivariate community structure to environmental variables. *Mar. Ecol. Prog. Ser.* **1993**, *92*, 205–219. [CrossRef]
31. Heywood, K.J.; Garabato, A.C.N.; Stevens, D.P.; Muench, R.D. On the fate of the Antarctic Slope Front and the origin of the Weddell Front. *J. Geophys. Res.* **2004**, *109*, C06021. [CrossRef]
32. Thompson, A.F.; Heywood, K.J.; Thorpe, S.E.; Renner, A.H.H.; Trasvica, A. Surface circulation at the tip of the Antarctic Peninsula from drifters. *J. Phys. Oceanog.* **2009**, *39*, 3–26. [CrossRef]
33. Morozov, E.G.; Flint, M.V.; Spiridonov, V.A. *Antarctic Peninsula region of the Southern Ocean*; Advances in Polar Ecology; Springer: Cham, Switzerland, 2021; Volume 6, p. 433. [CrossRef]

34. Zhou, X.Q.; Zhu, G.P.; Hu, S. Influence of tides on mass transport in the Bransfield Strait and the adjacent areas. *Antarctic Polar Sci.* **2020**, *23*, 100506. [CrossRef]
35. Fedotova, A.A.; Kashin, S.V. Interannual variations of water mass properties in the central basin of the Bransfield Strait. In *Antarctic Peninsula Region of the Southern Ocean*; Morozov, E.G., Flint, M.V., Spiridonov, V.A., Eds.; Springer: Cham, Switzerland, 2021; Volume 6, pp. 131–141. [CrossRef]
36. Krechik, V.A.; Frey, D.I.; Morozov, E.G. Peculiarities of water circulation in the central part of the Bransfield Strait in January 2020. *Dokl. Earth Sci.* **2021**, *496*, 92–95. [CrossRef]
37. Tokarczyk, R. Classification of water masses in the Bransfield Strait and southern part of the Drake Passage using a method of statistical multidimensional analysis. *Polish Pol. Res.* **1987**, *8*, 333–366.
38. Krek, A.V.; Krek, E.V.; Krechik, V.A. The Circulation and Mixing Zone in the Antarctic Sound in February 2020. In *Antarctic Peninsula Region of the Southern Ocean*; Morozov, E.G., Flint, M.V., Spiridonov, V.A., Eds.; Springer: Cham, Switzerland, 2021; Volume 6, pp. 83–100. [CrossRef]
39. Fedotova, A.A.; Stepanova, S.V. Water mass transformation in the Powell Basin. In *Antarctic Peninsula Region of the Southern Ocean*; Morozov, E.G., Flint, M.V., Spiridonov, V.A., Eds.; Springer: Cham, Switzerland, 2021; Volume 6, pp. 115–129. [CrossRef]
40. Deacon, G.E.R. A general account of the hydrology of the South Atlantic Ocean. *Discov. Rep.* **1933**, *7*, 171–238.
41. Kittel, W.; Jazdzewski, K. Studies on the larval stages of *Euphausia superba* Dana (Crustacea, Euphausiacea) in the southern Drake Passage and in the Bransfield Strait in February and March 1981 during the BIOMASS-FIBEX expedition. *Pol. Polar Res.* **1982**, *3*, 273–280.
42. Spiridonov, V.A.; Zalota, A.K.; Yakovenko, V.A.; Gorbatenko, K.M. Composition of population and transport of juveniles of Antarctic krill in Powell Basin region (northwestern Weddell Sea) in January 2020. *Trudy VNIRO* **2020**, *181*, 33–51. [CrossRef]
43. Voronina, N.M. *Pelagic Ecosystems of the Southern Ocean*; Nauka: Moscow, Russia, 1984; p. 206.
44. Orsi, A.H.; Whitworth, T.; Nowlin, W.D., Jr. On the meridional extent and fronts of the Antarctic Circumpolar Current. *Deep Sea Res. I* **1995**, *42*, 641–673. [CrossRef]
45. Morozov, E.G.; Krechik, V.A.; Frey, D.I.; Polukhin, A.A.; Artemiev, V.A.; Kasyan, V.V.; Sapozhnikov, P.V.; Mukhametianov, R.Z. Frontal Zone Between Relatively Warm and Cold Waters in the Northern Weddell Sea. In *Antarctic Peninsula Region of the Southern Ocean*; Morozov, E.G., Flint, M.V., Spiridonov, V.A., Eds.; Springer: Cham, Switzerland, 2021; Volume 6, pp. 31–51. [CrossRef]
46. Melnikov, I.; Spiridonov, V. Antarctic krill under perennial sea ice in the western Weddell Sea. *Antarct. Sci.* **1996**, *8*, 323–429. [CrossRef]
47. Ross, R.M.; Quetin, L.B.; Baker, K.S.; Vernet, M.; Smith, R.C. Growth limitation in young *Euphausia superba* under field conditions. *Limnol. Oceanogr.* **2000**, *45*, 31–43. [CrossRef]
48. Hempel, I.; Hempel, G. Field observations on the developmental ascent of larval *Euphausia superba* (Crustacea). *Polar Biol.* **1986**, *6*, 123–126. [CrossRef]
49. Spiridonov, V.A. Spatial and temporal variability in reproductive timing of Antarctic krill (*Euphausia superba*). *Polar Biol.* **1995**, *15*, 161–174. [CrossRef]
50. Makarov, R.R.; Menshenina, L.L. On the study of distribution and phenology of euphausiid larvae in the Scotia Sea. *Antarctica* **1988**, *28*, 166–174.
51. Hofmann, E.E.; Klinck, J.M.; Locarnini, R.A.; Fach, B. Krill transport in the Scotia Sea and environs. *Antarct. Sci.* **1998**, *10*, 406–415. [CrossRef]
52. Siegel, V.; Watkins, J.L. Distribution, Biomass and Demography of Antarctic Krill, *Euphausia superba*. In *Biology and Ecology of Antarctic Krill*; Advances in Polar Ecology; Siegel, V., Ed.; Springer: Cham, Switzerland, 2016; pp. 21–100. [CrossRef]
53. Meyer, B.; Atkinson, A.; Blume, B.; Bathmann, U.V. Feeding and energy budgets of larval Antarctic krill *Euphausia superba* in summer. *Mar. Ecol. Prog. Ser.* **2003**, *57*, 167–177. [CrossRef]
54. Schmidt, K.; Atkinson, A.; Venables, H.J.; Pond, D.W. Early spawning of Antarctic krill in the Scotia Sea is fuelled by "superfluous" feeding on non–ice associated phytoplankton blooms. *Deep. Sea Res. Part II Top. Stud. Oceanogr.* **2012**, *59–60*, 159–172. [CrossRef]
55. Atkinson, A.; Ward, P.; Hunt, B.P.V.; Pakhomov, E.A.; Hosie, G.W. An overview of Southern Ocean zooplankton data: Abundance, biomass, feeding and functional relationships. *CCAMLR Sci.* **2012**, *19*, 171–218.
56. McBride, M.M.; Dalpadado, P.; Drinkwater, K.F.; Godo, O.R.; Hobday, A.J.; Hollowed, A.B.; Kristiansen, T.; Murphy, E.J.; Ressler, P.H.; Subbey, S.; et al. Krill, climate, and contrasting future scenarios for Arctic and Antarctic fisheries. *ICES J. Mar. Sci.* **2014**, *71*, 1934–1955. [CrossRef]
57. Hays, G.C.; Richardson, A.J.; Robinson, C. Climate change and marine plankton. *Trends Ecol. Evol.* **2005**, *20*, 337–344. [CrossRef]
58. Constable, A.J.; Melbourne-Thomas, J.; Corney, S.P.; Arrigo, K.R.; Barbraud, C.; Barnes, D.K.A.; Bindoff, N.L.; Boyd, P.W.; Brandt, A.; Costa, D.P.; et al. Climate change and Southern Ocean ecosystems I: How changes in physical habitats directly affect marine biota. *Glob. Change Biol.* **2014**, *20*, 3004–3025. [CrossRef]

water

MDPI

Article

Distribution and Demography of Antarctic Krill and Salps in the Atlantic Sector of the Southern Ocean during Austral Summer 2021–2022

Dmitrii G. Bitiutskii [1,2], Ernest Z. Samyshev [3], Natalia I. Minkina [3], Victor V. Melnikov [3,*], Elena S. Chudinovskih [3], Sergei I. Usachev [1], Pavel A. Salyuk [4], Alexander N. Serebrennikov [5], Oleg A. Zuev [6] and Alexei M. Orlov [6,7]

1 Sector of the World Ocean, Azov-Black Sea Branch of the Russian Federal Research Institute of Fisheries and Oceanography ("AzNIIRKH"), 344002 Rostov-on-Don, Russia
2 Environmental Biochemistry Laboratory, Institute of Biology of the Karelian Research Centre of the Russian Academy of Sciences (IB KarRC RAS), 185910 Petrozavodsk, Russia
3 A.O. Kovalevsky Institute of Biology of the Southern Seas of RAS, 299002 Sevastopol, Russia
4 V.I. Il'ichev Pacific Oceanological Institute Far Eastern Branch Russian Academy of Sciences, 690041 Vladivostok, Russia
5 Institute of Natural and Technical Systems, Russian Academy of Sciences, 299011 Sevastopol, Russia
6 Shirshov Institute of Oceanology, Russian Academy of Sciences, 117997 Moscow, Russia
7 Department of Ichthyology and Hydrobiology, Tomsk State University, 634050 Tomsk, Russia
* Correspondence: sevlin@rambler.ru; Tel.: +7-978-8292940

Citation: Bitiutskii, D.G.; Samyshev, E.Z.; Minkina, N.I.; Melnikov, V.V.; Chudinovskih, E.S.; Usachev, S.I.; Salyuk, P.A.; Serebrennikov, A.N.; Zuev, O.A.; Orlov, A.M. Distribution and Demography of Antarctic Krill and Salps in the Atlantic Sector of the Southern Ocean during Austral Summer 2021–2022. *Water* 2022, *14*, 3812. https://doi.org/10.3390/w14233812

Academic Editor: Marina Marcella Manca

Received: 12 October 2022
Accepted: 17 November 2022
Published: 23 November 2022

Abstract: The study aimed to investigate krill (*Euphausia superba*) and salp (*Salpa thompsoni*) populations in the Atlantic sector of the Southern Ocean in January and February 2022. Samples were obtained to measure the abundance, biomass and distribution patterns of krill and salp. Sex differences and feeding habits of the Antarctic krill were determined. The dependence of the physiological state of the studied aquatic organisms on changes in environmental parameters was analyzed. Current data on the association of the dynamics of hydrometeorological parameters and processes with the distribution of chlorophyll a, krill, and salp were obtained. It was established that, at numerous stations, the biomass of salps prevailed over krill. The result indicates the replacement of the Antarctic krill populations by gelatinous zooplankton. The obtained results allow assessment of the biological resource potential in the studied region based on the analysis of the samples collected.

Keywords: chlorophyll *a*; Euphausiacea; *Euphausia superba*; *Salpa thompsoni*; planktonic tunicate; climate change; feeding competition

1. Introduction

The Antarctic krill *Euphausia superba* Dana, 1850 is one of the most abundant marine species of the Euphausiacea. It constitutes staple food for a wide range of animals and plays an important role in the functioning of Antarctic ecosystems and the Antarctic food web [1,2]. It has circumpolar distribution stretching from the coastal zone of the continental shelf to the northern boundary of the Antarctic Convergence. Due to its huge biomass and high biological value, it has become one of the important resources of the world's fisheries [3,4].

The maximum abundance of krill is observed in the Atlantic sector of the Southern Ocean (ASO), where its dense aggregations are formed in mesoscale gyres near seamounts and islands [5–7]. Over the past 80 years, krill abundance in the ASO has almost halved [8]. The extermination of whales in the 20th century did not lead to the expected increase in krill abundance: large baleen whales were replaced by penguins, fish, cephalopods, crabeater seals and small whales (Minke). The number of krill consumed by them exceeds the total diet of all baleen whales in the past by more than twice [9]. Current assessments of the distribution and abundance of Antarctic krill in the ASO show that the prevailing environmental factors are primary production, population of predators (including the restoration of the population of whales and seals), surface water temperature, ice conditions, sharp climatic fluctuations, etc. At the same time, the impact of Antarctic krill fishing is considered to be minimal [10].

During summer, the adult part of the krill population is located in the surface water layer, usually above the seasonal pycnocline, at an average depth of about 50 m. They have an irregular spatial distribution, which can be explained by their active behavior. Krill often forms surface patches and can search for more favorable environmental conditions. Water dynamics can also have both a direct impact (cumulative effect of gyres) and an indirect one (for example, through changing of the phytoplankton abundance). The confluence of positive factors contributing to the formation of extensive and stable krill aggregations is most likely to happen in coastal/near-ice and near-island areas, where the topography slows down the system of gyres. In the oceanic zone, such aggregations are rare and exist for a short time until they disintegrate due to the frequent changes in hydrophysical conditions. Small aggregations of animals, represented by individual flocks, predominate here. The krill areas in the continental seas and adjacent waters are largely formed by the Coastal Antarctic Current and the circulation system by orographic and hydrological factors [11,12].

Despite the high fecundity of krill, its reproduction success depends on the possibilities of breeding in shallow waters, at the bottom of which krill eggs develop [12]. Several authors, using the 41-year-old (1976–2016) KRILLBASE-abundance [13] and KRILLBASE-length-frequency [14] databases, showed that the distributions of eggs, krill nauplii, and metanauplii had maximum intensity and success of spawning on the shelf and above the shelf slope. Association of the krill breeding zone with shallow waters clearly corresponds to the idea that the core of the population area is confined to the circulation of the Coastal Antarctic Current and the Weddell Sea.

Therefore, interannual fluctuations in the abundance of krill are mainly explained by its reproduction success (when spawning in shallow waters) and the intensity of the mechanical transfer of larvae and mature organisms away from spawning grounds by the currents during the intensification of meridional processes The latter also determines the abundance of other plankton (including phyto- and mesozooplankton) [12].

Krill abundance has been declining since the 1970s due to the penetration of salps into the high latitudes of the Southern Hemisphere [15–22]. Invasion of gelatinous planktonic animals, *Salpa thompsoni* Foxton, 1966, into the krill's core areas to the south of 60° S is followed by a dramatic increase in its abundance. The latter causes legitimate concern since the feeding rate of tunicates is high, and the diet composition of these species, i.e., salps, makes them feeding competitors of both krill and mesozooplankton [10,23]. Compared to other tunicates (ascidians, pyrosomes, doliolids), which draw food-containing water into the body using cilia, the salps have the functions of locomotion and food consumption combined. Salps swim and feed uninterruptedly pushing water through the esophagus with contractions of the muscles of the tunic and not experiencing saturation, which makes them a "biological pump". They have a "mucus" net that captures even tiny food particles which form food boluses with the help of slime [24]. This way of feeding makes the salp indiscriminate filter feeders. Salps are also well known for their fast growth rate among metazoans [25–27]. This way of feeding and a complex reproductive cycle also help salps increase their abundance [28]. Changes in the biomass and distribution of *S. thompsoni* are now considered a potential determinant of the future structure and functioning of the Antarctic and Southern Ocean ecosystems [29].

The purpose of our research was to obtain current data on the impact of changes in wind patterns, hydrological structure and water dynamics on the distribution of chlorophyll *a* fields and the abundance and biomass of krill and salps; we also aimed to assess the resource potential of the region based on biological data, including size, sex composition and examination of the physiological state of the animals.

2. Materials and Methods

Oceanographic studies were carried out in cruise 87 of the R/V Akademik Mstislav Keldysh in January–February 2022 [30] in the areas of the Bransfield Strait, Antarctic Sound, and the Powell Basin of the Weddell Sea, near the South Orkney, James Ross and Shishkov islands (latitude 65°–60° S, longitude 62°–41° W).

2.1. Remote Sensing Methods

Surface wind (SW) was studied using CCMP OCW vector data available with a 6-h temporal resolution and a spatial resolution of 0.25° × 0.25° [31]. Based on these data, averaged maps of the absolute dynamic topography (ADT) and absolute dynamic velocity (ADV) of the geostrophic current for the period from 15 January 2022 to 15 February 2022 were constructed. In the course of the work, daily maps of ocean surface temperature (OST) taken from the CMEMS with a spatial resolution of 0.05 × 0.05° [32] were also analyzed.

To analyze the spatial distribution of chlorophyll a, a map of satellite estimates of chlorophyll a concentration was drawn based on sea colour measurements from 15 January to 15 February 2022. To exclude omissions, use was made of the combined data of level 3 from satellite radiometers MODIS-Aqua, MODIS-Terra, and VIIRS-SNPP [33] with chlorophyll a concentrations from the chlor_a product, which uses the colour index algorithm (CI) [34] for low chlor_a values (<0.15 mg/m^3) and the OCx algorithm [35] for more productive waters. To build a map in each pixel of a regular 4-km grid, the median values of the chlorophyll a concentration between 15 January and 15 February 2022, were calculated.

2.2. CTD Measurements and Sampling

Hydrology. Our CTD data were obtained using an Idronaut OCEAN SEVEN 320Plus manufactured by Idronaut, Italy. The sounding complex was equipped with: a high-precision pressure sensor (PA-10X) with a measurement range of 0–100 Mpa, an accuracy of 0.01% and a resolution of 0.002%; two redundant temperature sensors with a measurement range from −5 °C to 45 °C, an accuracy of 0.001 °C, and a resolution of 0.0001 °C; two redundant conductivity sensors with a measurement range of 0–7 siemens/m, an accuracy of 0.0001 siemens/m, a resolution of 0.00001 siemens/m. The currents were measured using a TRDI Workhouse Monitor (Lowered Acoustic Doppler Current Profiler, LADCP) submersible acoustic Doppler profiler with a frequency of 300 kHz paired with a Shipborne Acoustic Doppler Current Profiler (LADCP) TRDI Ocean Surveyor-75 with a frequency of 75 kHz. Both profilers are manufactured by Teledyne Technologies Inc., 1049 Camino Dos Rios Thousand Oaks, CA 91360, USA. The obtained data were processed using the LDEO Software version IX.10 [36]. Additionally, tidal forces were taken into account and calculated using the software described in [37].

Chlorophyll a. Concentrations of chlorophyll *a* were measured on board by a standard spectrophotometric method based on the analysis of the absorption spectra of chlorophyll *a* extract in acetone [38]. Sea water was sampled at the stations by Niskin bathometers mounted on a rosette. Instrumental measurements of maximum chlorophyll *a* concentrations of about 6 mg/m^3 were performed on 28–29 January at 63.5–64.5° S, 55–56.5° W. Detailed information on the spatial distribution of the chlorophyll *a* concentrations was collected throughout the voyage by a continuous-flow laser fluorometer [39]; the intensity of chlorophyll *a* fluorescence was measured in motion with an interval of 60 s, which corresponded to approximately 300 m at a vessel speed of 10 knots. The seawater sampling depth was 5 m; the values of chlorophyll *a* fluorescence intensity were converted into chlorophyll *a* concentrations according to standard definitions data (38 points, R^2 = 0.9).

Macrozooplankton. Sampling was carried out in the late January and the first half of February 2022 in the western region of the ASO (Bransfield Strait (BS) and Antarctic Sound (AS), near the South Orkney Islands (SOI), the Powell Basin of the Weddell Sea (PB), near the islands of James Ross (JR) and Shishkov (SH) (Figure 1). The material was obtained from 34 hauls with the following fishing gear: double square net (DSN)—21 hauls, Isaak-Kidd mid-water trawl modified by Samyshev-Aseev (IKMT-SA)—10 hauls and 3 catches with a standard Bongo net.

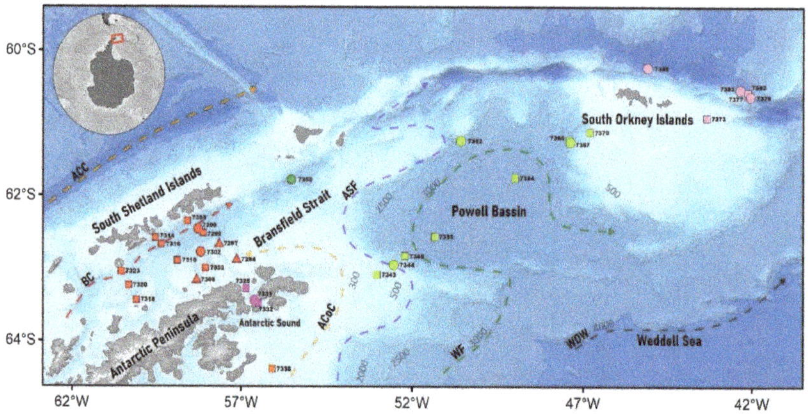

Figure 1. Map of the Antarctic krill and salp sample collection area with indication of currents (by: Heywood et al., 2004; Thompson et al., 2018). All stations were numerically marked. ACC—Antarctic Circumpolar Current, BC—Bransfield Current, ACoC—Antarctic Coastal Current, ASF—Antarctic Shelf Front, WF—Weddell Front, WDW—Weddell Deep Water. Fishing gear: •—IKMT-SA, ■—DSN, ▲—Bongo net. Area: •—BS, •—AS, •—PB, •—SOI, •—JR, •—SH (designations according to Table 1).

Table 1. Number of krill and salp specimens studied by areas: Bransfield Strait (BS), Antarctic Sound (AS), South Orkney Islands Area (SOI), Powell Basin of the Weddell Sea (PB), James Ross Island area (JR), Shishkov Island area (SH).

Parameter	**Number of individuals**						
	BS	**AS**	**SOI**	**PB**	**JR**	**SH**	**Total**
			Antarctic Krill				
Mass measurements	1140	548	733	743	500	125	3789
Biological analysis	734	330	529	543	300	125	2561
			Salps				
Mass measurements	1549	-	2433	2136	-	303	6421

DSN—a double square net with an inlet area of 1 m^2 and a 6 m long filter cone made of gas with a mesh size of 0.5 mm [40], net was equipped with a water flow meter (Hydrobios, Germany) and a 24 kg wing-shaped depressor (Hydrobios, Germany). Oblique tows were carried out in layers starting from 730 to 100 m at vessel speeds from 2 to 3.1 knots. The towing depth was prompted by the pressure sensor readings of the Senti DT probe (Star Oddi, Iceland).

IKMT-SA—Isaacs-Kidd trawl modified by Samyshev-Aseev—is a non-closing trawl with a mouth area of 6 m^2, the net part is 25 m long and made out of a knotless net with a mesh of 6 mm and an insert at the trawl end of a nylon sieve № 15 (0.67 mm) [41]. Oblique tows were carried out in layers starting from 1900 to 440 m, at a vessel speed of 2.6–3.3 knots.

The Bongo net was a towed plankton net consisting of a frame with two metal rings, which had two filter cones with a mesh of 300 microns fixed on them. The diameter of each frame ring was 60 cm [42]. Oblique tow was carried out in layers starting from 270 to 130 m, at a vessel speed of 1.7–2.3 knots.

It should be noted that krill was obtained from catches using various types of gear (Bongo net, DSN, IKMT-SA) with fishing depths ranging from 100 to 1900 m. As Antarctic krill can live at depths down to 800 m and the main concentrations of krill are in the depth range of 0–200 m, some measurements of the abundance and biomass of krill obtained from fishing with IKMT-SA at greater depths may be underestimates.

After trawling, each sample was weighed to an accuracy of 1 g and analyzed for composition. When the samples were large, the entire catch was weighed and a subsample (100–300 g) was taken out of it; the results of quantifying and weighing of such a subsample were extrapolated to the entire catch.

For biological analysis, a subsample of at least 300 specimens was taken. If there were fewer specimens in the catch, then it was analyzed in its entirety. For the calculation of biomass and abundance of Antarctic krill and salp aggregations, the data on each haul, tow depth and volume of filtered water per unit volume (1000 m^3) were taken into consideration [43]. Krill samples were processed following the generally accepted procedures and CCAMLR recommendations [44,45].

Biological analysis of freshly caught krill included measurements of the standard length from the outer edge of the eye to the end of the telson with an accuracy of 1 mm using a laminated paper scale [46]. Sex and maturity stage were determined under a UlabWF20X binocular (TM ULAB, China) with ×30 magnification or using a digital USB microscope ADSM 301 (Shenzhen Andonstar Technology Co., Ltd., Shenzhen, Guangdong, China) with digital zoom up to x4 according to the method of Makarov and Denys [47].

Stomach fullness (from 0 to 4 units, where 0 is empty, 4 is full), intestinal fullness (from 0 to 4 units, where 0 is empty, 4 is full), and liver colour were studied [48]. Each type of measurement was performed by the same person to minimize variability of results [49]. Krill specimens were weighed individually on a 20-g equal-arm mechanical scale VSM-20 (JSC Nizhny Tagil Medico-Instrumental Plant, Russia) with an accuracy of 10 mg (Table 1).

All sampled salps were fixed in 4–6% formalin for subsequent laboratory testing. Then a preliminary analysis of the selected samples was carried out by measuring the length of the salps with laminated graph paper by size groups with a step of 5 mm. Salps were weighed individually and in groups on a 20-g equal-arm mechanical scale VSM-20 (JSC Nizhny Tagil Medico-Instrumental Plant, Russia) with an accuracy of 10 mg (Table. 1).

3. Results

3.1. Hydrology

For the entire study area (latitude 66°–60° S, longitude 55°–40° W), we drew maps of ocean surface temperature (OST) with the surface wind (Figure 2) and absolute dynamic topography with geostrophic current velocities (Figure 3). In the tested area the heavy southeast wind (8 m/s) forked into the east and southwest winds (shown by arrows). The latter swirled in the cyclonic northwest direction (Figure 2). The swirling of the geostrophic current (shown by arrows) in the cyclonic direction was also observed (Figure 3). The wind and geostrophic vorticity led to a water rise in the cyclone centre with a simultaneous sea level lowering (the blue area in Figure 3). The lowering of the sea level occurred due to a higher density of deep waters compared to surface waters. The data in Figures 2 and 3 are averaged over the period from 15 January 2022 to 15 February 2022. Thus, the cyclonic vorticity, which led to the rise of deep waters during this period, was one of the reasons for the increased content of chlorophyll *a* (approximately at latitude 64° S and longitude 48° W).

CTD data for the upper 200 m layer show a spatial temperature distribution typical for the warm season (Figure 4). Positive temperature anomalies were recorded in the upper 50 m layer in most of the BS area and the coastal zones, while elsewhere a latitudinal dependence with a decrease in temperature values to the south was observed. At depths of 100–200 m, temperatures were below 0 °C, except for a narrow coastal strip along the South Shetland Islands associated with a current in the BS. Local temperature minimums were registered at the southernmost stations and on the southwestern slope of the Powell Basin.

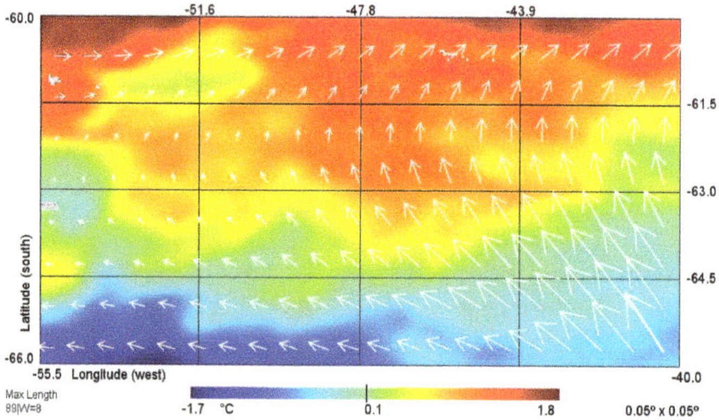

Figure 2. Spatial distribution of OST and SW (arrows) according to satellite measurement data (explanations in the text).

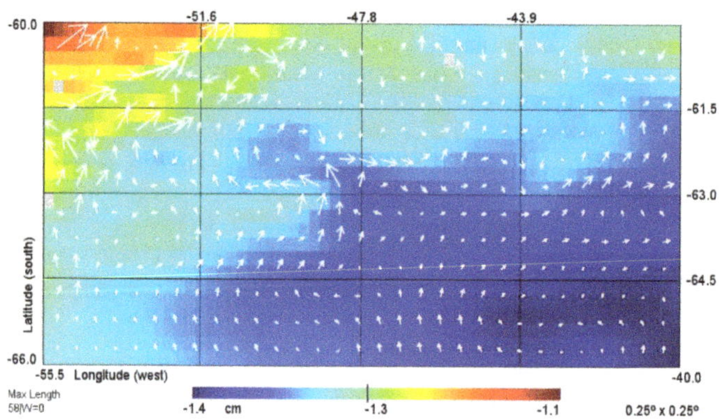

Figure 3. Spatial distribution of ADT and ADV (arrows) according to satellite measurement data (explanations in the text).

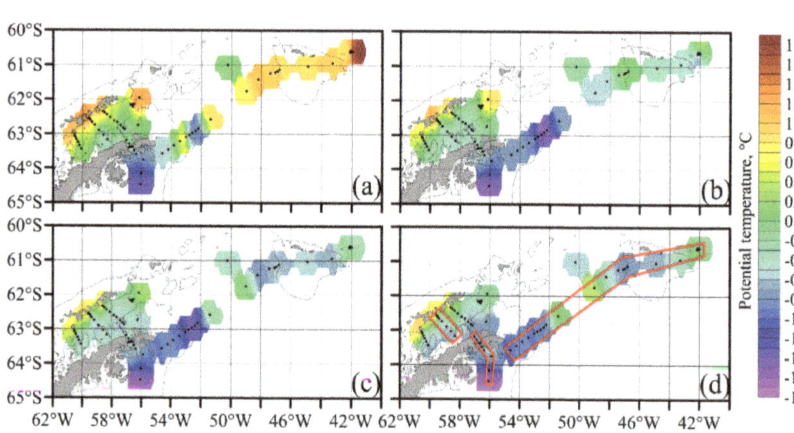

Figure 4. Spatial distribution of potential temperature on the horizons 50 (**a**), 100 (**b**), 150 (**c**) and 200 (**d**) m. The 500 m isobath is marked in grey. Stations are marked with black dots, sections are marked with red.

Most of the Bransfield Strait (Figure 5a,b) was occupied by the colder and saltier Transitional Zonal Water with Weddell Sea influence (TWW), which spread from north to south. Transitional Zonal Water with Bellingshausen Sea influence (TBW) and with the warm Bransfield current spread as a narrow jet close to the South Shetland Islands, reaching the middle of the strait only in the upper 50 m. The tentative boundary between these water masses runs along the 0 °C isotherms.

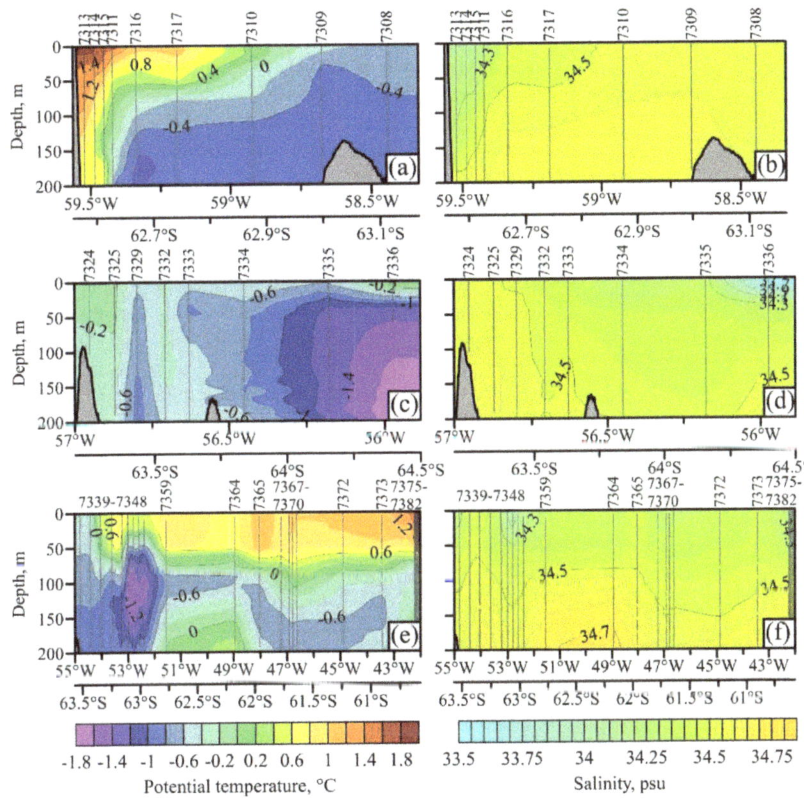

Figure 5. Potential temperature (**a**,**c**,**e**) and salinity (**b**,**d**,**f**) for three sections: the Bransfield Strait (**a**,**b**), the Antarctica Strait (**c**,**d**) and the Powell Basin (**e**,**f**). Station numbers are at the top, station positions are marked with grey lines. Seabed topography is shown on GEBCO2021 database. (https://www.gebco.net, accessed on 22 November 2022).

In the Antarctic Sound (Figure 5c,d), a decrease in temperature and salinity from north to south was observed. The northern shallow part of the strait was filled with the waters of the Bransfield Strait characterized by higher temperatures with a maximum of −0.10 °C in the upper 60 m layer and higher water salinity with a maximum of 34.55 parts per 1000 in the bottom layer. The deeper part of the strait was influenced by the waters from the Weddell Sea and by some freshening due to active ice melting.

The freshening was most pronounced in the upper 50 m layer (up to 34.29). Notably, st. 7336 to the south of the Antarctic Sound had outstanding minimum salinity values (33.55) and positive temperature values (0.10 °C) in the near-surface layer as well as greater gradients of thermohaline characteristics (at a depth of 200 m the values were −1.82 °C and 34.58).

The upper 200 m layer in the Powell Basin (Figure 5e,f) contained two water masses: the surface layer from 20 to 100 m in thickness was occupied by the Antarctic Surface

Water (AASW) formed during summer heating, while the underlying Cold Intermediate Water (CIL) between 50 and 150 m in thickness was the result of winter convection. In the AASW, the temperature increased from south to north from 0.50 °C to 2 °C; no latitudinal dependence was noted in the salinity distribution. In the CIL, the minimum temperatures were observed at the slope stations and reached −1.69 °C; the salinity increased smoothly with depth.

Strong near-surface currents observed in some subareas had a significant impact on the hydrological structure and water dynamics. The two-jet system of currents in the Bransfield Strait corresponded to the thermohaline structure: a powerful narrow jet to the northeast is the warm Bransfield Current (BC), while a weak wide jet to the southwest is a continuation of the Antarctic Coastal Current (ACoC). The highest ACoC velocities of 0.4 m/s were recorded on the shelf to the northeast of Joinville Island. A constant water flow from the Weddell Sea to the Bransfield Strait through the Antarctic Sound was not recorded; a cyclonic gyre was present in the southern part of the Strait.

3.2. *Chlorophyll a*

The highest chlorophyll *a* concentrations (more than 3 mg/m^3) were observed in the northern central and in the northwestern parts of the Weddell Sea (Figure 6). In the northern central part, increased values were recorded along the ice edge on 15–20 January; the chlorophyll *a* concentrations visible from the satellite gradually decreased to the values of about 1 mg/m^3 two weeks later. In the northwestern part of the Weddell Sea, on the shelf, high chlorophyll *a* concentrations were observed throughout the entire research period, both in the presence of an ice edge until 25 January, and later, until 15 February, when it was replaced by moving broken ice and icebergs.

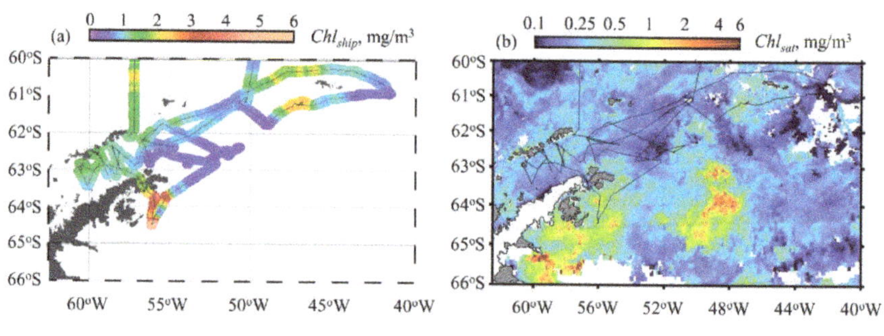

Figure 6. Spatial distribution of chlorophyll-a concentration (mg/m^3). (**a**) Flow-through fluorimetric measurements from 5 m depth while the vessel is under way; (**b**) median values of the combined satellite estimates from 15 January to 15 February 2022.

Average chlorophyll *a* concentrations (2–3 mg/m^3 according to ship data and about 1 mg/m^3 according to satellite data) were registered on the shelf of the South Orkney Islands in the direction of the southwest (7 February 2022) and northwest (12 February 2022). Relatively high values of chlorophyll *a* concentration (1–2 mg/m^3 according to ship measurements and 0.5–1 mg/m^3 according to satellite data) were observed in the Bransfield Strait closer to the South Shetland Islands.

The central part of the Powell Basin and the area to the east of the South Orkney Islands had very low chlorophyll *a* concentrations of less than 0.5 mg/m^3 and can, therefore, be classified as oligotrophic waters.

3.3. *Macrozooplankton*

Krill. The relative abundance of Antarctic krill varied from 0 to 537 individuals/1000 m^3. Its maximum values were recorded on the shelf to the north of the South Orkney Islands (Figure 1), where the IKMT-SA catch was ~33 kg. At this station, the trawling depth was maximal −1900 m—while the acoustic equipment of the vessel registered krill accumulations at a depth of about 50–100 m.

To the southeast of James Ross Island a large spot of chlorophyll *a* was discovered, which was investigated with a DSN at a separate station. At a depth of 210 m, the relative abundance of krill represented by large immature individuals was 357 ind./1000 m^3. The third largest catch was in the Bransfield Strait with a relative krill abundance of 321 ind./1000 m^3. The main part of this catch consisted of juveniles. The relative biomass of Antarctic krill ranged from 0 to 331.8 g/1000 m^3. The two catches with the highest biomass were off the north of the South Orkney Islands and to the southeast of James Ross Island (331.8 and 207.5 g/1000 m^3, respectively), similarly to the catches with the highest abundance. The third largest biomass value was recorded in the Bransfield Strait (Figure 1)—73.4 g/1000 m^3.

Size and weight composition of krill catches Large-sized krill with a length of more than 45 mm prevailed near Shishkov Island, in the Antarctic Sound and in the area of the South Orkney Islands (50.7, 51.6, 80.8%, respectively). Medium-sized krill (35–45 mm) dominated the Powell Basin and off James Ross Island (51.3, 74.2%, respectively), and were also present in significant numbers in the South Orkney Islands area (48.6%). The minimum proportion of small-sized krill (<35 mm) was observed in the catches off the South Orkney Islands—0.7%, off the Shishkov Island—0.8%, in the Antarctic Sound—7.9%, in the Powell Basin—14.4%, off the James Ross Island—18.8%. In the Bransfield Strait, the proportion of small-sized krill was the highest and amounted to 80.7%. In the Bransfield Strait, 27 mm long krill prevailed among juveniles; adult lengths were 43 mm among females and 42 mm among males. In the Antarctic Sound, the individuals were 27, 46 and 50 mm long, respectively. In the area of James Ross Island, juveniles were dominated by 37 mm long individuals; such large juveniles were also found in the Powell Basin, in the Brasfield Strait, and near the South Orkney Islands. Both females and males in the area of James Ross Island were dominated by 40 mm long individuals. In the Powell Basin, 33 mm long krill typified the major part of the juveniles, adult females were mostly 45 mm long, and males were 40 mm long. In the waters near Shishkov Island, juveniles were represented by single specimens, whereas 48 mm long adult females and 47 mm long males predominated (Figure 7). The length-weight ratio of krill individuals was similar in all studied areas, all significant standard deviations being within the margin of error of individual measurements (Figure 8).

The results of cluster analysis allowed us to identify four krill groups according to their size composition (Figure 9). A violin-shaped graph was used to visualize the distribution of krill size data in various clusters. The main part of small-sized krill in cluster S was represented by juveniles 21–30 mm long; cluster M1 included medium-sized krill with body length of 35–41 mm, while medium-sized krill also prevailed in cluster M2, large individuals, 39–53 mm long, were also present in significant numbers; cluster L mainly included large-sized, 47–54 mm long, krill.

Small-sized krill (cluster S) were concentrated mainly in the Bransfield Strait near the shelf of the Antarctic Peninsula. Large krill (cluster L) were recorded in the deep part of the Bransfield Strait, in the area of Shishkov Island, in the northern part of the Powell Basin and the northeast of the South Shetland Islands. Large and medium-sized krill (cluster M2) were concentrated mainly in the Antarctic Sound, in the southwest of the Powell Basin and north of the South Orkney Islands. Medium-sized krill (cluster M1) were registered in the Antarctic Sound, in the southwest of the Powell Basin, south of James Ross Island, and in the Bransfield Strait.

Krill sex composition. The predominance of juveniles was noted only in the Bransfield Strait, whereas juveniles were also present in significant numbers in the catches off James Ross Island and in the Powell Basin (34.3%, 25.41%, respectively). Females largely predominated in the Powell Basin, where the female to male ratio was 2.6:1.0. Females also prevailed in the Antarctic Sound with a gender ratio of 1.3:1.0. The ratio of males and females near the South Orkney Islands was equal. Males prevailed slightly in the waters near James Ross Island at 1.3:1.0 and to a large extent near Shishkov Island, at 4.1:1.0 (Figures 10 and 11).

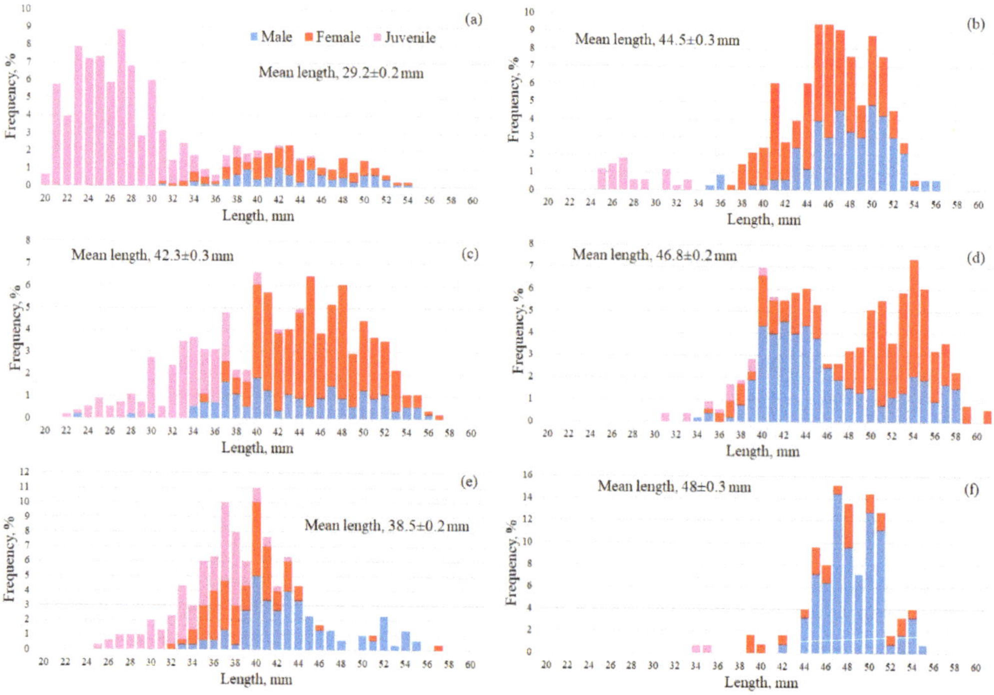

Figure 7. Size composition of Antarctic krill males, females, and juveniles in different areas: **(a)**—Bransfield Strait, **(b)**—Antarctic Sound, **(c)**—Powell Basin of the Weddell Sea, **(d)**—South Orkney Islands Area, **(e)**—James Ross Island area, **(f)**—Shishkov Island area.

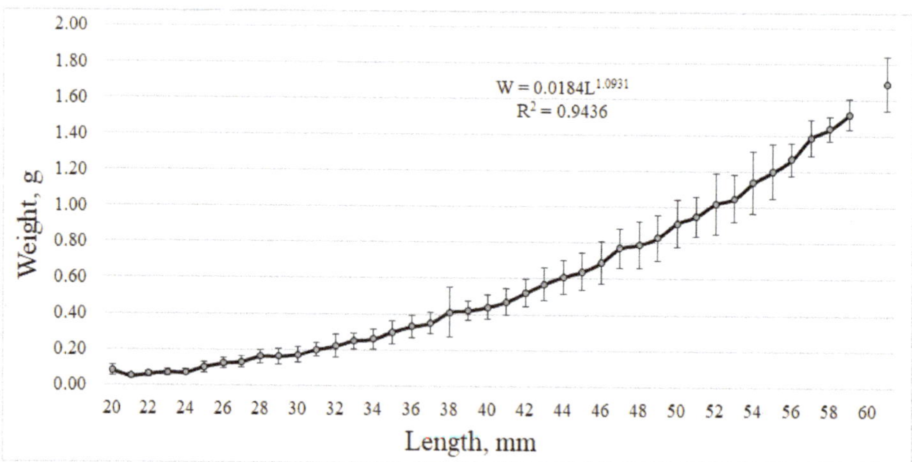

Figure 8. Relationship between the length and body weight of Antarctic krill in catches.

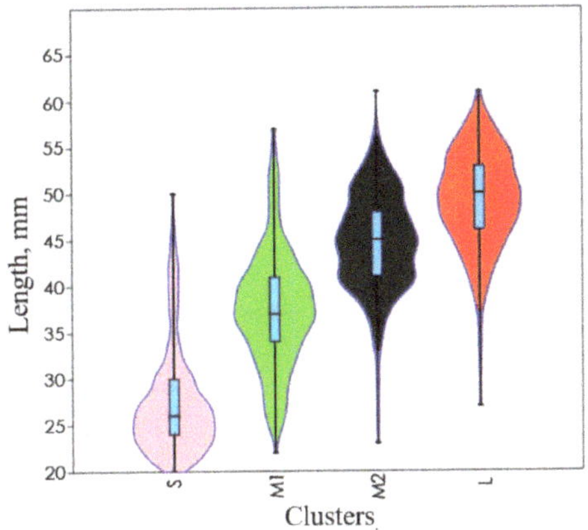

Figure 9. Krill distribution by size clusters (—interquartile range, —95% confidence intervals, |- median).

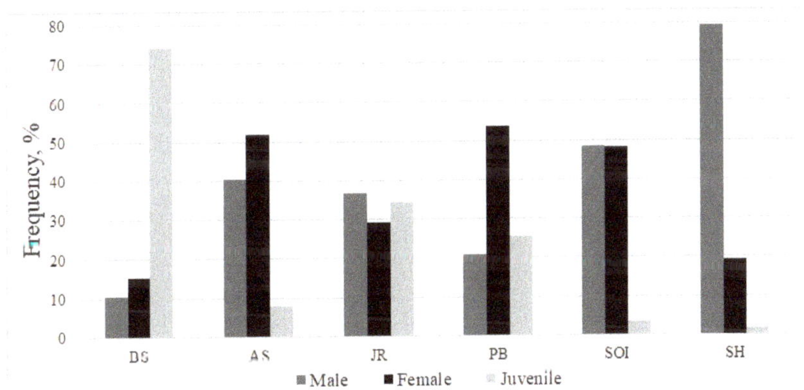

Figure 10. Ratio of Antarctic krill females, males, and juveniles in different areas (designations according to Table 1).

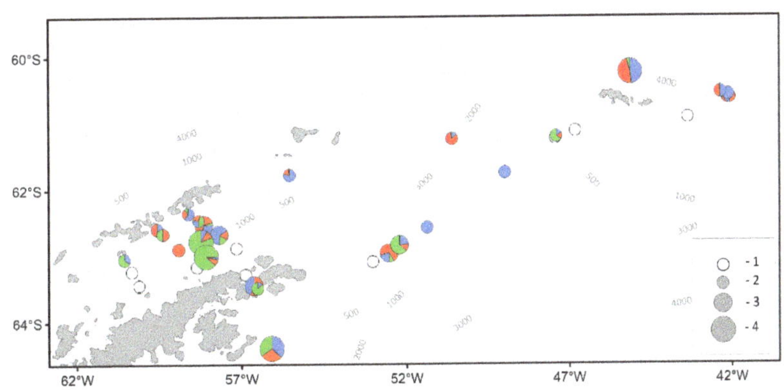

Figure 11. Spatial distribution of Antarctic krill relative abundance (ind./1000 m^3) and sex ratio (%) in ASO: •—males, •—females, •—juveniles. Relative abundance (ind./1000 m^3): 1–0; 2–< 10; 3–11-100; 4–101-400.

Females which had completed spawning at the stage of maturity IIIE were absent in almost all catches; a small number of them (1.2%) were found near the South Orkney Islands, however. In this area, mostly females at stages IIIB and IIIC were registered—31.0% and 34.5%, respectively. In the Bransfield Strait, both mature pre-spawning females at stage IIIB (35.5%) and immature females with a fully formed thelyum at stage IIB (34.6%) prevailed.

In the Antarctic Sound, females at stages IIIB and IIIC predominated, at 39.8% and 41.0% of the total, respectively. In the Powell Basin and near Shishkov Island, females at stage IIIB also largely dominated, as 54.8 and 58.3%, respectively. In the James Ross Island area, 74.7% of females were at stage IIB (Figure 12).

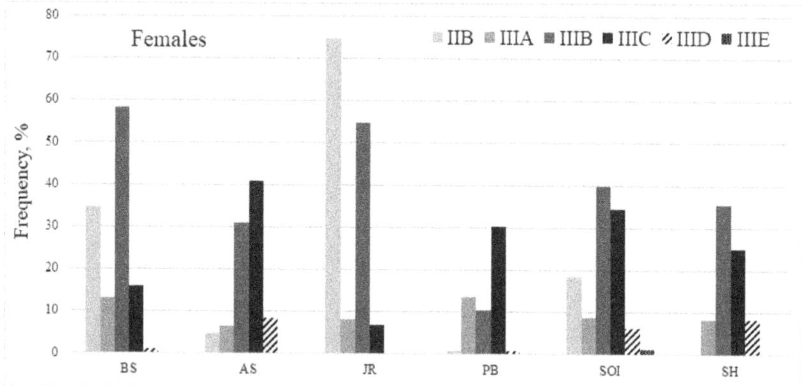

Figure 12. Proportion of Antarctic krill females at different stages of maturity (%) by area (designations according to Table 1).

Females at stage IIIA were present in all areas (6.0–13.5%). Females at stage IIID were present in all areas but in small numbers (1.0–8.3%), except for the area near James Ross Island, where females at this stage of maturity were absent.

Generally, the largest number of females was represented by pre-spawning individuals at stages IIIB (39.1%) and IIIC (29.4%) in the study areas. Females at the last stage of maturity IIID as well as females that had spawned were almost absent in the catches and accounted for 3.8% and 0.3%, respectively. Females at early stages of maturity (IIB and IIIA) were present in small numbers (17.2% and 10.2%, respectively). The majority of males in the Bransfield Strait (99.2%), near James Ross Island (96.4%), in the Powell Basin (68.8%), and near the South Orkney Islands (73.8%) comprised individuals at stage IIA (Figure 13).

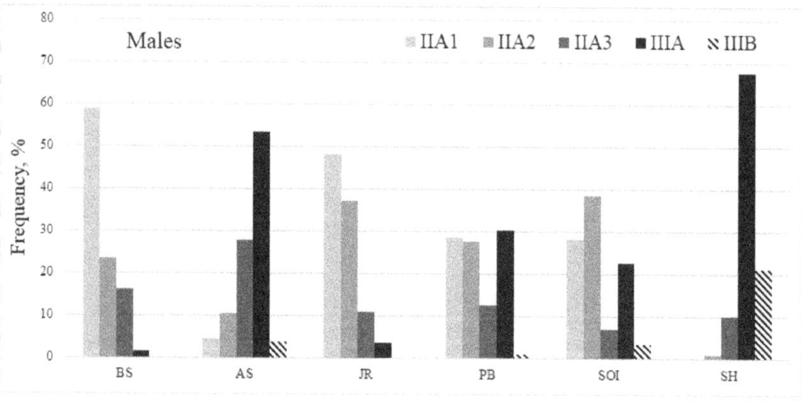

Figure 13. Proportion of Antarctic krill males at different stages of maturity (%) by area (designations according to Table 1).

Mature males prevailed in the Antarctic Sound and near Shishkov Island (IIIA accounted for 53.4% and 67.7%, respectively, and IIIB—3.8% and 21.2%, respectively). In the Powell Basin, the proportion of mature males was 30.4% (IIIA) and 0.9% (IIIB), while in the area of the South Orkney Islands, they accounted for 22.7% (IIIA) and 3.5% (IIIB).

The maturation ogives of male and female Antarctic krill from late January to mid-February are shown in Figure 14. Mass maturation (body length at 50% sexual maturity) $L_{50\%}$ occurs in 45 mm Antarctic krill for both males and females.

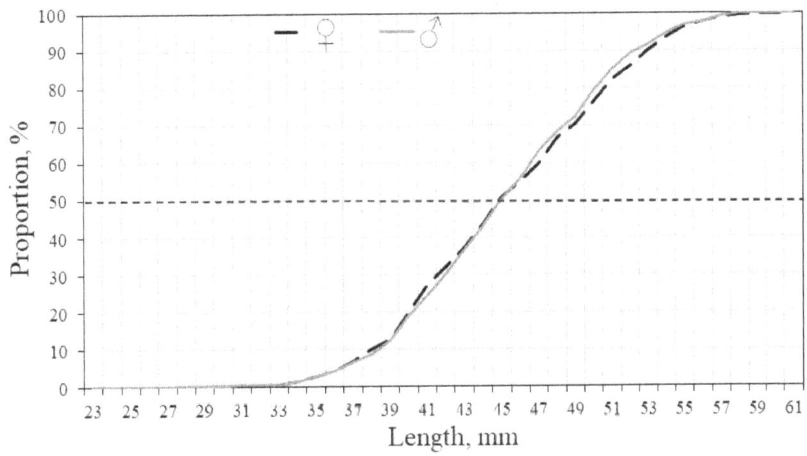

Figure 14. Maturation ogives of males (♂) and females (♀) of Antarctic krill in ASO.

Krill feeding. During the study period, krill actively fed in all areas, as evidenced by the data on the average value of stomach/intestines fullness of mature and juvenile individuals (Table 2). The stomach/intestines fullness of the juveniles was, on average, lower than that of mature males and females. The krill actively fed in the Bransfield Strait (3.7/1.9 units), in the Antarctic Sound (3.9/2.1), near the South Orkney Islands (3.3/1.4), in the area of James Ross Island (3.6/2.0) and Shishkov Island (3.7/1.3 units). The lowest feeding intensity, 1.7/0.7 units, was observed in the Powell Basin.

Table 2. Filling of the stomach/intestines (average score) of Antarctic krill in various research areas (designation of areas according to Table 1).

	BS	AS	SOI	PB	JR	SH	Average Score
Juvenile	3.6/1.8	3.4/2.2	2.5/0.5	0.5/0.2	3.4/1.7	1.0/0.0	2.7/1.3
Females	3.8/1.7	4.0/2.3	3.6/1.6	2.8/0.9	3.4/2.4	3.4/0.6	3.5/1.5
Males	3.9/2.3	3.9/2.1	3.0/1.2	1.9/0.9	3.8/2.3	3.8/1.5	3.3/1.6

Due to active feeding, almost all krill had light-green or dark-green livers (Figure 15). Krill with transparent liver was present in all study areas but accounted for only 0.3–5.0%. Krill with a yellowish liver was caught in the Powell Basin—11.7%, off the South Orkney Islands—7.3%, and also near Shishkov Island—3.4%.

Salps. The relative abundance of salps varied from 0 to 202 ind./1000 m^3. The largest concentration of salps, 201.5 ind./1000 m^3, was found in the central part of the Powell Basin in the Weddell Sea. The highest concentrations of salps were all across the section through the Powell Basin, to the east of the South Orkney Islands, in the northern and western parts of the Bransfield Strait. Krill dominated to the north of the South Orkney Islands, in the shelf zone of the southwestern part of the Powell Basin, and the south of the Bransfield Strait in the Antarctic shelf zone. The krill only areas were found in the Antarctic Sound and the southeast area of James Ross Island. Salps were also present at a ratio of 4:1 near Shishkov Island.

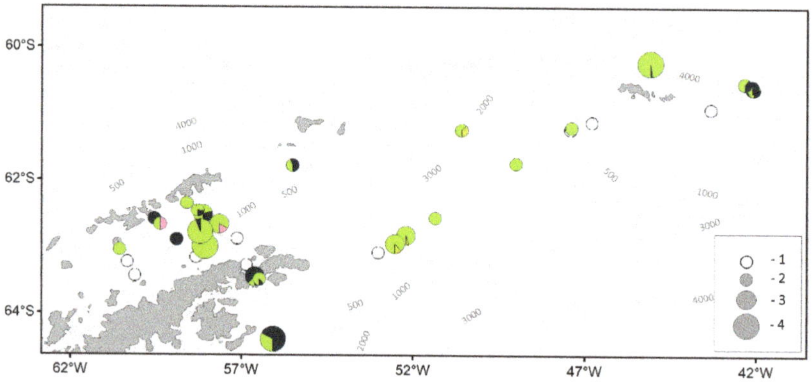

Figure 15. Distribution of Antarctic krill in study area by liver color: •—light-green, •—dark-green, •—yellow, •—milky white. Relative abundance (ind./1000 m^3): 1–0; 2–< 10; 3–11-100; 4–101-400.

The relative salp biomass varied from 0 to 72 g/1000 m^3 with its maximum value recorded in the Bransfield Strait (Figure 16). The distribution of the relative abundance and biomass of krill and salps had similar patterns.

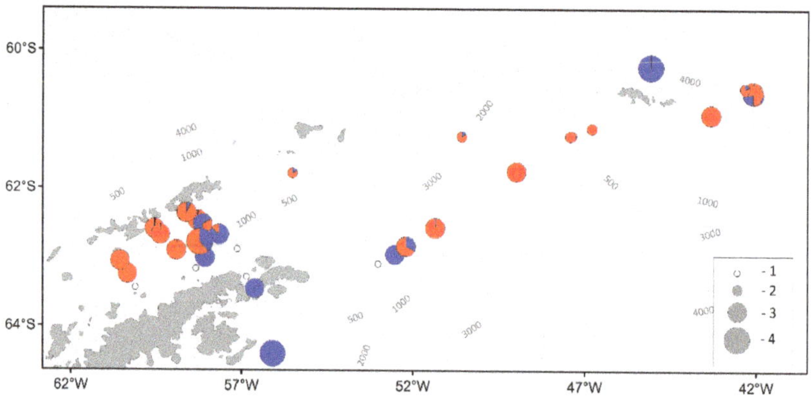

Figure 16. Spatial distribution of relative biomass (g/1000 m^3) and proportion (%) of Antarctic krill (•) and salps (•). Relative biomass (g/1000 m^3): 1–0; 2–< 10; 3–11-100; 4–101-350.

The only difference was the presence of large aggregations of small-sized krill and medium-sized salps in the shelf zone of the Antarctic Peninsula in the south of the Bransfield Strait (Figure 17), which resulted in a lower number of salps but their larger biomass.

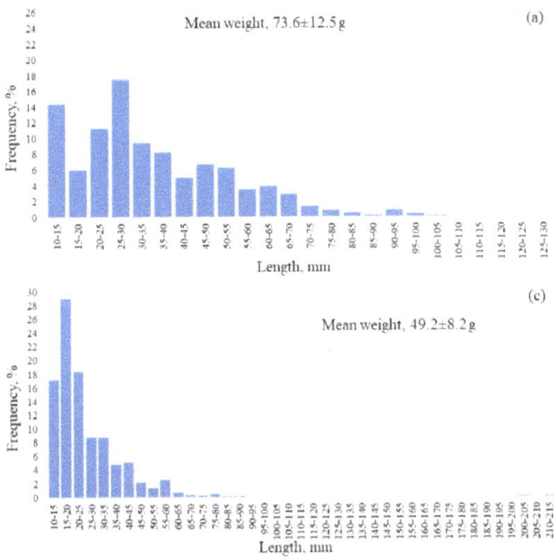

Figure 17. Size composition of salt Salpa thompsoni in different areas: (**a**)—Bransfield Strait, (**b**)—Powell Basin of the Weddell Sea, (**c**)—South Orkney Islands Area, (**d**)—Shishkov Island area.

4. Discussion

The results of our studies showed that the thermohaline structure and dynamics of the waters were in good agreement with established ideas about the hydrophysical processes in this zone [50–52] and corresponded to the data of recent years [53–55]; however, significant local differences were also observed. Thus, in the Bransfield Strait, a greater southward spread of TWW along the entire strait was recorded compared to the data of 2013 [56]; the waters of the Antarctic Sound were significantly warmer compared to the same period in 2020 [57]. There was no constant inflow of water from the Weddell Sea to the Bransfield Strait through the Antarctic Sound, the structure of the currents was variable, which was also confirmed by the recent studies in this area [58].

The spatial distribution and concentrations of chlorophyll *a* also corresponded to the results of the previous studies. In January 1995, at the ice edge in the northwestern and central parts of the Weddell Sea, chlorophyll *a* concentrations exceeded 5 mg/m^3 and reached 13 mg/m^3 [59]. In February 2015, low concentrations of chlorophyll *a*, 0.32 ± 0.02 mg/m^3, were recorded in the southeast of the South Orkney Islands and average concentrations of 1.87 ± 0.22 mg/m^3 to the north and northwest of the South Orkney Islands [60]. Obtained data most resembled the results of February–March 2008 and 2009 [61] which were as follows: concentrations of less than 0.5 mg/m^3 occurred in the central part of the Weddell Sea (AASW waters) and the zone of influence of the Weddell Sea in the Bransfield Strait (TWW waters); 0.5–1.5 mg/m^3—in the Bransfield Strait closer to the shelf of the South Shetland Islands (TBW waters); more than 2 mg/m^3 with peaks up to 4–7 mg/m^3 to the east of James Ross Island in the northwestern area of the Weddell Sea near active ice melt. Satellite estimates of the chlorophyll *a* concentration in the Antarctic Peninsula area were lower than the contact ones, which is associated with the regional optical characteristics [62,63]. Therefore, satellite maps of the chlorophyll *a* concentration should be analyzed for structural features of the pigment distribution, not the values per se. The general patterns of the spatial distribution according to the data of satellite and contact measurements of the chlorophyll *a* concentration were similar. Satellite data provide more information on the areas adjacent to the expedition route, and the ship measurements provide more accurate data for certain points of the route. Thus, in the central region of the Weddell Sea, according to our contact measurements, no high concentrations of chlorophyll *a* were recorded.

However, the satellite map showed that there were zones with increased concentrations of chlorophyll *a* nearby, most probably due to the abundance of phytoplankton near the ice edge, which corresponds well with the results of the previous studies in the northwestern Weddell Sea [59].

Hydrological data and satellite observations for our study period indicated the transfer of Antarctic krill from the Weddell Sea to the Bransfield Strait with a cold ACoC current carrying colder and saltier TWW waters from the Weddell Sea. The maximum abundance of Antarctic krill in the Bransfield Strait (the shelf of the Antarctic Peninsula), as well as in the Antarctic Sound, was associated with a combination of favourable environmental factors (temperature, salinity, and abundance of phytoplankton). In the Bransfield Strait, such conditions were formed under the influence of the TWW with the temperatures of −0.8 to 1.7 °C, which were optimal for the growth and development of both juveniles and mature krill individuals [64–66]. In the Antarctic Sound, the main biomass of Antarctic krill was also formed by krill brought from the Weddell Sea. The obtained data are consistent with the view that immature individuals tend to colder waters, unlike mature individuals [13,67]. Salinity variability seemed not to significantly affect the distribution of krill at different stages of development, for which the range of 34.5–34.6 is optimal. The maximum abundance of krill was noted around station 7336 (Figure 1), which was characterized by the minimum salinity (33.55) in the surface layer, positive values of potential temperature (0.1 °C), and high chlorophyll a concentration (>20 mg/m^3), whereas, at a depth of 200 m, the potential temperature was close to the freezing point of seawater (−1.8 °C), salinity was higher, and the concentration of chlorophyll a did not exceed 0.5 mg/m^3.

In the ASO, the average abundance and biomass of krill correlated well with the data obtained in 2019–2020. However, our results for the density of Antarctic krill aggregations in the Antarctic Sound were almost ten times higher than before [68,69]. Large accumulations of salps were registered in the Bransfield Strait, which is associated both with the influence of the warm BC current [70] and with the peculiarities of feeding and reproductive cycle of salps [28]. The findings on the presence of large aggregations of Antarctic krill in the Antarctic shelf of the Bransfield Strait are consistent with the previous studies [43,71].

The size composition of Antarctic krill varies greatly from year to year in the same study area. According to our results, small-sized krill (82%) prevailed in the Bransfield Strait, whereas, in the same ASO area, only 22% of small-sized krill were found by the 2020–21 expedition [72] and only about 40% by the 2019–20 expedition [68]. A similar situation was observed in other regions. The reasons for such discrepancies could be natural fluctuations in the size structure or differences in research methods, which therefore require standardization. In particular, a repeated grid of stations is required for correct comparisons of the obtained results and the use of large fishing gear similar in design and size since krill can slip away out of small samplers [73,74]. Moreover, the entire layer (0~200 m) of the Antarctic krill habitat needs to be sampled [75,76]. However, despite all of the above, the feeding characteristics of krill remain the main factors determiningits size compositions [6,7]. In some study areas, the catches included large individuals of krill (up to 45 mm) without clearly expressed sexually dimorphic features (insufficiently developed petasma and almost undeveloped internal genitals). The explanation of this phenomenon could be found, with high probability, in the feeding conditions of krill [77–79], which is also evidenced by the fatty acid analysis of its individuals [80]. It can be assumed that under these conditions the growth of krill individuals is ahead of their development.

The sex composition of krill differed significantly from one of the previous years. So, in the area of the Bransfield Strait and Antarctic Sound and eastward of the Antarctic Peninsula (the Weddell Sea) the number of juveniles of different sizes was significantly higher than in other researched areas, which was not observed in the recent studies [68,69,71,72]. At the same time, in January–February 2020, a significant number of spent females were found in almost all the studied areas [72], while in our study, spent females were found near the South Orkney Islands only. In ASO, the number of juveniles caught did not exceed 33%, while in the recent study this value did not exceed 10% [72]. The maturation ogives of male and female Antarctic krill indicate the same development of both genders up to the length of sexual maturity $L_{50\%}$, after which females grow more slowly than males. This pattern was also noted previously [77]. The condition of sexual maturity of krill impacts greatly on this relationship.

Antarctic krill is an active filter-feeder and feeds mainly on phytoplankton [79,81]. As was shown earlier, the decrease in the active feeding of krill occurs mainly in the autumn and winter periods, while the feeding activity increases towards the end of spring [48]. Our data also testify to the active feeding of krill in the austral summer period.

We recorded an almost complete absence of krill at the stations in the Powell Basin. This area was studied by various expeditions on different sections [68,69,71,72,82–85]. Overall, the compilation of data for research regions shows much similarity. Thus, in the Powell Basin, as well as in the waters near the South Orkney Islands, water temperature regime (with the warming up to 2 °C) is favorable for the development of salps. Earlier publications reported active salp development after winters with relatively low sea ice development [6,7]. Unlike krill, the salp propagation does not depend on the ice cover and the associated algae, which are the main prey for krill larvae and juveniles [86]. At the same time, the life cycle of salps is much shorter than that of krill, and the fluctuations in their abundance reflect the annual variability of conditions contributing to massive population growth [87–89].

5. Conclusions

The analysis of the obtained data demonstrated significant shifts in the thermohaline structure and currents of the upper 200-m water layer in the Antarctic Peninsula area. A stronger southward spread of TWW was noted along the entire Bransfield Strait. The upper layer of the Antarctic Sound waters became significantly warmer, and an extremely small amount of ice was observed. The maximum values of chlorophyll *a* concentration of more than 5 mg/m^3 were observed only on the shelf in the northwestern Weddell Sea, where the transport of icebergs and broken ice was registered. Waters with the lowest concentrations of chlorophyll a (less than 0.3 mg/m^3) were found in the central part of the Powell Basin, in the area to the east of the South Orkney Islands, and in the Bransfield Strait near the Antarctic Peninsula. Judging by the chlorophyll a concentration values, both oligotrophic and mesotrophic waters are present in the study area. Mesotrophic waters, with chlorophyll a concentrations of more than 2 mg/m^3, were observed mainly in the shelf areas in the northwestern part of the Weddell Sea and to the west of the South Orkney Islands, as well as in separate zones at the edge of ice fields in the Weddell Sea.

This results also demonstrate a close relationship between Antarctic krill and salps and a combination of such environmental factors as surface water temperature and availability of forage. The circulation of water masses within the region play into the dynamics of this interaction. Only in the Bransfield Strait and to the south of the Antarctic Sound at the boundary of packed ice, in the relatively warm and desalinated surface layer and in combination with the cold and salty subsurface water layer, were the optimal conditions for the maximum concentrations of Antarctic krill formed. The reasons for the relatively low krill reproduction in the study area may be found in the late spawning period and, consequently, the low survival rate of krill larvae. The data on Antarctic krill abundance for the latest decade have shown significant fluctuations in the ASO [8,10,15,90,91].

Thus, our results indicate the need to continue long-term monitoring of this area, where the euphausiid biomass has traditionally been one of the largest in the entire Antarctic [12]. Climate changes gradually lead to significant seasonal and interannual fluctuations in the abundance of these crustaceans against the background of an ever-increasing abundance of salps [8,14,15]. Differences in the abundance of salps and krill can be explained by differences in the ice cover in the ASO in other years, the extent of which determines the availability of food for these planktonic animals: at high ice coverage, salps are limited in food, while krill are able to "gnaw it out of the ice" [15,92]. The result indicates the replacement of the Antarctic krill populations by gelatinous zooplankton a steady long-term tendency towards a decrease in the population of Antarctic krill (from 38% to 75% per decade) and its successive replacement by salps in this water area [8,14,15,90]. Such changes may affect the structure and productivity of the Antarctic ecosystem as a whole [8,10,91,93].

Author Contributions: Conceptualization by E.Z.S. and A.M.O.; methodology and validation by D.G.B. and N.I.M.; formal analysis and original draft preparation by V.V.M. and A.M.O.; field investigations in the 87 Antarctic cruise of RV "Akademik Mstislav Keldysh" by E.S.C., O.A.Z., P.A.S., S.I.U., and D.G.B. Satellite data analysis by P.A.S. and A.N.S. All authors have read and agreed to the published version of the manuscript.

Funding: This study was conducted within the framework of the Russian state task No. FMWE-2022-0001 (IO RAS) "Assessment of the current state of natural complexes of the Atlantic sector of the Southern Ocean and their multi-period variability (ecosystems, bioproductivity, hydrophysics, hydro- and geochemistry" (A.M.O., O.A.Z.) and CTD data analysis were supported by the Russian Science Foundation grant 22-77-10004 (O.A.Z.), State Assignment KarRC FMEN-2022-0006 (D.G.B.), No. 121090800137-6 "Comprehensive studies of the current state of the ecosystem of the Atlantic sector of Antarctica" (E.Z.S., E.S.C.), Comprehensive environmental studies of the Southern Ocean, No. 0211-2019-0007 (P.A.S), No. 121041400077-1 "Functional, metabolic and toxicological aspects of the existence of hydrobionts and their populations in biotopes with different physico-chemical regime" (N.I.M., V.V.M.) and No. 121122300074-7 "Fundamental research of processes in the climate system that determine the spatial and temporal variability of the natural environment on a global and regional scale" (A.N.S.).

Informed Consent Statement: Not applicable.

Data Availability Statement: Not applicable.

Acknowledgments: The authors are sincerely grateful to the crew of the RV "Akademik Mstislav Keldysh" for their help in the fieldwork. The authors thank the head of the expedition, Morozov E.G., and the deputy head of the expedition, Molodtsova T.N. (IO RAS), for supporting and organizing trawling operations. Special thanks to A.V. Mishin, K.V. Minin, and V.L. Syomin (SIO) for working with trawls and nets that made it possible to obtain krill and salp samples. The authors are also grateful to two anonymous reviewers for valuable comments and suggestions that have significantly improved the quality of the paper.

Conflicts of Interest: The authors declare no conflict of interest.

Ethics Approval: The present research complies with international ethical norms and standards for such scientific research.

References

1. Rintoul, S.R.; Sparrow, M.; Meredith, M.P.; Wadley, V.; Speer, K.; Hofmann, E.; Summerhayes, C.; Urban, E.; Bellerby, R. (Eds.) *The Southern Ocean Observing System: Initial Science and Implementation Strategy*; SCAR: Cambridge, UK, 2012; pp. 1–74.
2. Trathan, P.N.; Hill, S.L. The Importance of Krill Predation in the Southern Ocean. In *Biology and Ecology of Antarctic Krill*; Springer: Cham, Switzerland, 2016; pp. 321–350.
3. Sologub, D.O. Modern Features of Distribution, Biology and Horizontal Migrations of Antarctic Krill (*Euphausia superba*) in the Atlantic Sector of Antarctica. Ph.D. Thesis, VNIRO Publ. House, Moscow, Russia, 2016; pp. 1–247.
4. Nicol, S.; Foster, J. The Fishery for Antarctic Krill: Its Current Status and Management Regime. In *Biology and Ecology of Antarctic Krill*; Springer: Cham, Switzerland, 2016; pp. 387–421.
5. Atkinson, A.; Siegel, V.; Pakhomov, E.A.; Rothery, P.; Loeb, V.; Ross, R.M.; Quetin, L.B.; Schmidt, K.; Fretwell, P.; Murphy, E.J.; et al. Oceanic Circumpolar Habitats of Antarctic Krill. *Mar. Ecol. Prog. Ser.* **2008**, *362*, 1–23. [CrossRef]
6. Siegel, V. Introducing Antarctic Krill Euphausia superba Dana, 1850. In *Biology and Ecology of Antarctic Krill*; Springer: Cham, Switzerland, 2016; pp. 1–20. [CrossRef]
7. Siegel, V.; Watkins, J.L. Distribution, Biomass and Demography of Antarctic Krill, *Euphausia superba*. In *Biology and Ecology of Antarctic Krill*; Springer: Cham, Switzerland, 2016; pp. 21–100.
8. Atkinson, A.; Siegel, V.; Pakhomov, E.; Rothery, P. Long-term Decline in Krill Stock and Increase in Salps within the Southern Ocean. *Nature* **2004**, *432*, 100–103. [CrossRef] [PubMed]
9. Laws, R.M. Ecology of the Southern Ocean. *Amer. Sci.* **1985**, *73*, 26–40.
10. Johnston, N.M.; Murphy, E.J.; Atkinson, A.; Andrew, J.; Constable, A.J.; Cotté, C.; Cox, M.; Daly, K.L.; Driscoll, R.; Flores, H. at al. Status, Change and Futures of Zooplankton in the Southern Ocean. *Front. Ecol. Evol.* **2022**, *9*, 624692. [CrossRef]

11. Samyshev, E.Z. Conclusion on the State of Krill Population and Pelagic Ecosystem in the Western Region of the Atlantic Part of Antarctica in the Pre-winter Period of 1998. *Bull. Ukr. Anarct. Center* **2000**, *3*, 231–236.
12. Samyshev, E.Z. *Antarctic Krill and the Structure of Planktonic Community in its Distribution Area*; USSR Nauka (Acad. of Sci. of the USSR. All-Union Hydrobiol. Soc.): Moscow, Russia; ECOSEA: Sevastopol, Ukraine, 2002; pp. 1–268.
13. Perry, F.A.; Atkinson, A.; Sailley, S.F.; Tarling, G.A.; Hill, S.G.; Lucas, C.H. Habitat Partitioning in Antarctic Krill: Spawning Hotspots and Nursery Areas. *PLoS ONE* **2019**, *14*, eo219325. [CrossRef] [PubMed]
14. Atkinson, A.; Hill, S.L.; Pakhomov, E.; Siegel, V.; Anadon, R.; Chiba, S.; Daly, K.L.; Downie, R.; Fielding, S.; Fretwell, P.; et al. KRILLBASE: A Circumpolar Database of Antarctic Krill and Salp Numerical Densities, 1926–2016. *Earth Syst. Sci. Data.* **2017**, *9*, 193–2107. [CrossRef]
15. Pakhomov, E.A.; Froneman, P.W.; Perissinoto, R. Salp/Krill Interactions in the Southern Ocean: Spatial Segregation and Implications for the Carbon Flux. *Deep Sea Res. II* **2002**, *2*, 1881–1907. [CrossRef]
16. Pakhomov, E.A.; Dubischar, C.; Strass, V.; Brichta, M.; Bathmann, U. The Tunicate *Salpa Thompsoni* Ecology in the Southern Ocean—I. Distribution, Biomass, Demography and Feeding Ecophysiology. *Mar. Biol.* **2006**, *149*, 609–623. [CrossRef]
17. Pakhomov, E.A.; Dubishar, C.; Hunt, B.P.V.; Strass, V.; Cisewski, B.; Siegel, V.; von Harbou, L.; Gurney, L.; Kitchener, J.; Bathmann, U. Biology and Life Cycles of Pelagic Tunicates in the Lazarev Sea, Southern Ocean. *Deep Sea Res. II* **2011**, *58*, 1677–1689. [CrossRef]
18. Lomakin, P.D.; Samyshev, E.Z. Oceanographic Conditions in the Area of the South Shetland Islands in March-April 1997, 1998 and Their Influence on the Distribution of Krill and Salp. *Oceanology* **2004**, *44*, 882–891.
19. Flores, H.; Atkinson, A.; Kawaguchi, S.; Krafft, B.A.; Milinevsky, G.; Nicol, S.; Reiss, C.; Tarling, G.A.; Werner, R.; Rebolledo, E.L.B.; et al. Impact of Climate Change on Antarctic Krill. *Mar. Ecol. Prog. Ser.* **2012**, *458*, 1–19. Available online: https://hal.archives-ouvertes.fr/hal-01250922 (accessed on 21 November 2022). [CrossRef]
20. Pakhomov, E.A.; Hunt, P.V. Trans-Atlantic Variability in Ecology of the Pelagic Tunicate *Salpa thompsoni* near the Antarctic Polar Front. *Deep–Sea Res. II* **2017**, *138*, 126–140. [CrossRef]
21. Groeneveld, Y., Berger, U.; Henschke, N.; Pakhomov, E.; Reiss, C.; Meyer, B. Blooms of a Key Grazer in the Southern Ocean—An Individual-Based Model of *Salpa thompsoni*. *Prog. Oceanogr.* **2020**, *185*, 102339. [CrossRef]
22. Luo, J.Y.; Stock, C.A.; Henschke, N.; Dunne, J.P.; O'Brien, T.D. Global Ecological and Biogeochemical Impacts of Pelagic Tunicates. *BioRxiv* **2022**, *205*, 102822. [CrossRef]
23. Bombosch, A. Euphausia superba or Salpa thompsoni—Who is Going to Win? 2008. Available online: https://www.coolantarctica.com (accessed on 11 October 2020).
24. Bone, Q.; Carre, C.; Rian, K.P. The Endostyle and Feeding Filter in Salps (Tunicata). *J. Mar. Biol. Ass. UK* **2000**, *80*, 523–534. [CrossRef]
25. Heron, A.C.; Benham, E.E. Individual Growth Rates of Salps in Three Populations. *J. Plankt. Res.* **1984**, *6*, 811–828. [CrossRef]
26. Lüskow, F.; Pakhomov, E.A.; Stukel, M.R.; Décima, M. Biology of *Salpa thompsoni* at the Chatham Rise, New Zeland: Demography, Growth, and Diel Vertical Migration. *Mar. Biol.* **2020**, *167*, 175. [CrossRef]
27. Henschke, N.; Blain, S.; Cherel, Y.; Cotte, C.; Espinasse, B.; Brian, P.V.; Hunt, B.P.V.; Pakhomov, E.A. Population Demographics and Growth Rate of *Salpa Thompsoni* on the Kerguelen Plateau. *J. Mar. Syst.* **2021**, *214*, 103489. [CrossRef]
28. Everett, J.; Baird, M.; Suthers, I. Three-Dimensional Structure of a Swarm of the Salp *Thalia democratica* within a Cold-Core Eddy off Southeast Australia. *J. Geophys. Res.* **2011**, *116*, C12046. [CrossRef]
29. Alcaraz, M.; Almeda, R.; Duarte, C.M.; Horstkotte, B.; Lasternas, S.; Agustí, S. Changes in the C, N, and P Cycles by the Predicted Salps-Krill Shift in the Southern Ocean. *Front. Mar. Scien.* **2014**, *1*, 1–13. [CrossRef]
30. Morozov, E.G.; Flint, M.V.; Orlov, A.M.; Frey, D.I.; Molodtsova, T.N.; Krechik, V.A.; Latushkin, A.A.; Salyuk, P.A.; Murzina, S.A.; Minin, K.V.; et al. Oceanographic and Ecosystem Studies in the Atlantic Sector of Antarctica (Cruise 87 of the Research Vessel Akademik Mstislav Keldysh). *Oceanology* **2022**, *62*, 721–723. [CrossRef]
31. Cross-Calibrated Multi-Platform Ocean Surface Wind. PO. DAAC. Available online: https://www.remss.com (accessed on 20 August 2022).
32. Copernicus Marine Environment Monitoring Service. CMEMS. Available online: https://www.copernicus.eu (accessed on 20 July 2022).
33. Ocean Color WEB. Available online: https://oceancolor.gsfc.nasa.gov/l3/ (accessed on 20 August 2022).
34. Hu, C.; Lee, Z.; Franz, B. Chlorophyll a Algorithms for Oligotrophic Oceans: A Novel Approach Based on Three-Band Reflectance Difference. *J. Geophys. Res.* **2012**, *117*, C01011. [CrossRef]
35. O'Reilly, J.E.; Maritorena, S.; Mitchell, B.G.; Siegel, D.A.; Carder, K.L.; Garver, S.A.; Kahru, M.; McClain, C. Ocean Color Chlorophyll Algorithms for SeaWiFS. *J. Geophys. Res.* **1998**, *103*, 24937–24953. [CrossRef]
36. Visbeck, M. Deep Velocity Profiling Using Lowered Acoustic Doppler Current Profiler: Bottom Track and Inverse Solution. *J. Atmos. Ocean. Tech.* **2002**, *19*, 794–807. [CrossRef]
37. Egbert, G.D.; Erofeeva, S.Y. Efficient Inverse Modeling of Barotropic Ocean Tides. *J. Atmos. Ocean. Tech.* **2002**, *19*, 183–204. [CrossRef]
38. Koblents-Mishke, O.I. Extractive and Non-Extractive Methods for the Determination of Photosynthetic Pigments in Samples. In *Modern Methods for Quantifying the Distribution of Marine Plankton*; Vinogradov, M.E., Ed.; Nauka: Moscow, Russia, 1983; pp. 114–125.

39. Nagornyi, I.G.; Salyuk, P.A.; Maior, A.Y.; Doroshenkov, I.M. A Mobile Complex for On-Line Studying Water Areas and Surface Atmosphere. *Instrum. Exp. Tech.* **2014**, *57*, 68–71. [CrossRef]
40. Bouchard, C.; Mollard, S.; Suzuki, K.; Robert, D.; Fortier, L. Contrasting the Early Life Histories of Sympatric Arctic Gadids Boreogadus Saida and Arctogadus Glacialis in the Canadian Beaufort Sea. *Polar Biol.* **2016**, *39*, 1005–1022. [CrossRef]
41. Kobyliansky, S.G.; Orlov, A.M.; Gordeeva, N.V. Composition of Deepsea Pelagic Ichthyocenes of the Southern Atlantic, from Waters of the Range of the Mid-Atlantic and Walvis Ridges. *J. Ichthyol.* **2010**, *50*, 932–50949. [CrossRef]
42. Bongo Plankton Net. Available online: https://www.nhbs.com/bongo-plankton-net (accessed on 20 August 2022).
43. Siegel, V.; Kawaguchi, S.; Ward, P.; Litvinov, F.; Sushin, V.; Loeb, V.; Watkins, J. Krill Demography and Large-Scale Distribution in the Southwest Atlantic During January/February 2000. *Deep Sea Res. II* **2004**, *51*, 1253–1273. [CrossRef]
44. Morris, D.J.; Watkins, J.L.; Ricketts, C.; Buchholz, F.; Priddle, J. An Assessment of the Merits of Length and Weight Measurements of Antarctic Krill *Euphausia superba*. *Brit. Ant. Surv. Bull.* **1988**, *79*, 27–50.
45. Anonymous. *Scientific Observers Manual*; CCAMLR: Hobart, Australia, 2011; pp. 1–232.
46. Petrov, A.F.; Shust, K.V.; Piyanova, S.; Uryupova, E.; Gordeev, I.I.; Sitov, A.M.; Demina, S.N. *Guidelines for the Collection and Processing of Fishing and Biological Data on Aquatic Bioresources of Antarctica for the Russian Scientific Observers in the CCAMLR Convention Area*; VNIRO Publ. House: Moscow, Russia, 2014; pp. 45–53.
47. Makarov, R.R.; Denys, C.J. Stages of Sexual Maturity of *Euphausia superba* Dana. *BIOMASS Handbook* **1980**, *11*, 1–11.
48. Meyer, B.; Auerswald, L.; Siegel, V.; Spahic, S.; Pape, C.; Fach, B.A.; Teschke, M.; Lopata, A.L.; Fuentes, V. Seasonal Variation in Body Composition, Metabolic Activity, Feeding and Growth of Adult Krill *Euphausia superba* in the Lazarev Sea. *Mar. Ecol. Prog. Ser.* **2010**, *398*, 1–18. [CrossRef]
49. Watkins, J.L.; Morris, D.J.; Ricketts, C. Nocturnal Changes in the Mean Length of a Euphausiid Population: Vertical Migration, Net Avoidance or Experimental Error? *Mar. Biol.* **1985**, *86*, 123–127. [CrossRef]
50. Tokarczyk, R. Classification of Water Masses in the Bransfield Strait and Southern Part of the Drake Passage Using a Method of Statistical Multidimensional Analysis. *Polish Polar Res.* **1987**, *8*, 333–336.
51. Orsi, A.H.; Nowlin, W.D.; Whitworth III, T. On the Circulation and Stratification of the Weddell Gyre. *Deep-Sea Res. I* **1993**, *40*, 169–203. [CrossRef]
52. Heywood, K.J.; Garabato, A.C.N.; Stevens, D.P.; Muench, R.D. On the Fate of the Antarctic Slope Front and the Origin of the Weddell Front. *J. Geophys. Res.* **2004**, *109*, C06021. [CrossRef]
53. Dorschel, B.; Gutt, J.; Huhn, O.; Bracher, A.; Huntemann, M.; Huneke, W.; Gebhardt, C.; Schroder, M.; Herr, H. Environmental Information for a Marine Ecosystem Research Approach for the Northern Antarctic Peninsula (RV Polarstern expedition PS81, ANT-XXIX/3). *Polar Biol.* **2016**, *39*, 765–787. [CrossRef]
54. Krechik, V.A.; Frey, D.I.; Morozov, E.G. Peculiarities of Water Circulation in the Central Part of the Bransfield Strait in January 2020. *Dokl. Earth Sci.* **2021**, *496*, 92–95. [CrossRef]
55. Morozov, E.G.; Flint, M.V.; Spiridonov, V.A. *Antarctic Peninsula Region of the Southern Ocean*; Springer: Cham, Switzerland, 2021; pp. 1–455. [CrossRef]
56. Huneke, W.G.; Huhn, O.; Schroeder, M. Water Masses in the Bransfield Strait and Adjacent Seas, Austral Summer 2013. *Polar Biol.* **2016**, *39*, 789–798. [CrossRef]
57. Krek, A.V.; Krek, E.V.; Krechik, V.A. The Circulation and Mixing Zone in the Antarctic Sound in February 2020. In *Antarctic Peninsula Region of the Southern Ocean*; Springer: Cham, Switzerland, 2021; pp. 83–99. [CrossRef]
58. van Caspel, M.; Hellmer, H.H.; Mata, M.M. On the Ventilation of Bransfield Strait Deep Basins. *Deep-Sea Res. II* **2017**, *149*, 25–30. [CrossRef]
59. Kang, S.H.; Kang, J.S.; Lee, S.; Chung, K.H.; Kim, D.; Park, M.G. Antarctic Phytoplankton Assemblages in the Marginal Ice Zone of the Northwestern Weddell Sea. *J. Plankt. Res.* **2001**, *23*, 333–352. [CrossRef]
60. Nunes, S.; Latasa, M.; Delgado, M.; Emelianov, M.; Simó, R.; Estrada, M. Phytoplankton Community Structure in Contrasting Ecosystems of the Southern Ocean: South Georgia, South Orkneys and Western Antarctic Peninsula. *Deep-Sea Res. I* **2019**, *151*, 103059. [CrossRef]
61. Mendes, C.R.B.; de Souza, M.S.; Garcia, V.M.T.; Leal, M.C.; Brotas, V.; Garcia, C.A.E. Dynamics of Phytoplankton Communities During Late Summer Around the Tip of the Antarctic Peninsula. *Deep-Sea Res. I.* **2012**, *65*, 1–14. [CrossRef]
62. Ferreira, A.; Brito, A.C.; Mendes, C.R.; Brotas, V.; Costa, R.R.; Guerreiro, C.V.; Sá, C.; Jackson, T. OC4-SO: A New Chlorophyll-a Algorithm for the Western Antarctic Peninsula Using Multi-Sensor Satellite Data. *Remote Sens.* **2022**, *14*, 1052. [CrossRef]
63. Pereira, E.S.; Garcia, C.A.E. Evaluation of Satellite-Derived MODIS Chlorophyll Algorithms in the Northern Antarctic Peninsula. *Deep-Sea Res. II* **2018**, *149*, 124–137. [CrossRef]
64. Meredith, M.P.; Renfrew, I.A.; Clarke, A.; King, J.C.; Brandon, M.A. Impact of the 1997/98 ENSO on the Upper Waters of Marguerite Bay, Western Antarctic Peninsula *J. Geophys. Res.* **2004**, *109*, C09013. [CrossRef]
65. Atkinson, A.; Shreeve, R.; Hirst, A.; Rothery, P.; Tarling, G.; Pond, D.; Korb, R.; Murphy, E.; Watkins, J.L. Natural Growth Rates in Antarctic Krill (*Euphausia Superba*): II. Predictive Models Based on Food, Temperature, Body Length, Sex, and Maturity Stage. *Limnol. Oceanogr.* **2006**, *51*, 973–987. [CrossRef]
66. Tarling, G.A.; Cuzin-Roudy, J.; Thorpe, S.E.; Shreeve, R.S.; Ward, P.; Murphy, E.J. Recruitment of Antarctic Krill *Euphausia Superba* in the South Georgia Region: Adult Fecundity and the Fate of Larvae. *Mar. Ecol. Prog. Ser.* **2007**, *331*, 161–179. [CrossRef]

67. Wiedenmann, J.; Cresswell, K.; Mangel, M. Temperature-Dependent Growth of Antarctic Krill: Predictions for a Changing Climate from a Cohort Model. *Mar. Ecol. Prog. Ser.* **2008**, *358*, 191–202. [CrossRef]
68. Spiridonov, V.A.; Zalota, A.K.; Yakovenko, V.A.; Gorbatenko, K.M. Population Composition and Transport of Antarctic Krill Juveniles in the Powell Basin (Northwestern Part of the Weddell Sea) in January 2020. *Tr. VNIRO* **2020**, *181*, 33–51. [CrossRef]
69. Yakovenko, V.A.; Spiridonov, V.A.; Gorbatenko, K.M.; Shadrin, N.V., Samyshev, E.Z.; Minkina, N.I. Macro- and Mesozooplankton in the Powell Basin (Antarctica): Species Composition and Distribution of Abundance and Biomass in February 2020. In *Antarctic Peninsula Region of the Southern Ocean*; Springer: Cham, Switzerland, 2021; pp. 131–141. [CrossRef]
70. Hereu, C.M.; Suárez-Morales, E.; Lavaniegos, B.E. Record of the Rare Oceanic Salp *Helicosalpa Komaii* (Tunicata: Thaliacea: Salpida) in the Northeast Pacific. *Rev. Mex. Biodiver.* **2014**, *85*, 624–629. [CrossRef]
71. Kasatkina, S.M.; Abramov, A.M.; Sokolov, M.Y. Biomass and Distribution of Antarctic Krill in the Antarctic Atlantic Area in January-February 2020. *Tr. AtlantNIRO* **2021**, *5*, 49–61.
72. Sytov, A.M.; Kozlov, D.A. Dimensional Composition and Biological Characteristics of Antarctic Krill Euphausia Superba in the Antarctic Part of the Atlantic in January-March 2020. *Tr. AtlantNIRO* **2021**, *5*, 101–115.
73. Krag, L.A.; Herrmann, B.; Iversen, S.A.; Engas, A.; Nordrum, S.; Krafft, B.A. Size Selection of Antarctic Krill (*Euphausia superba*) in Trawls. *PLoS ONE* **2014**, *9*, e102168. [CrossRef]
74. Wang, Z.; Tang, H.; Herrmann, B.; Xu, L. Catch Pattern for Antarctic Krill (*Euphausia superba*) of Different Commercial Trawls in Similar Times and Overlapping Fishing Grounds. Grounds. *Front. Mar. Sci.* **2021**, *8*, 670663. [CrossRef]
75. Cox, M.J.; Borchers, D.L.; Demer, D.A.; Cutter, G.R.; Brierley, A.S. Estimating the Density of Antarctic Krill (*Euphausia Superba*) from Multibeam Echo-Sounder Observations Using Distance Sampling Methods. *J. R. Stat. Soc. Ser. C App. Statist.* **2011**, *60*, 301–316. [CrossRef]
76. Kasatkina, S.M. Methodical Aspects of Acoustic Survey for Antarctic Krill in the CCAMLR Convention are. *Tr. AtlantNIRO* **2021**, *5*, 39–48, ISSN: 2541-9692.
77. Bargmann, H.E. The Development and Life History of Adolescent and Adult Krill *Euphausia superba*. *Discovery Rep.* **1945**, *XXIII*, 103–178.
78. Fraser, F.G. On the Development and Distribution of the Young Stages of Krill (*Euphausia superba*). *Discovery Rep.* **1936**, *XIV*, 1–192. [CrossRef]
79. Marr, J.W.S. The Natural History and Geography of the Antarctic Krill (*Euphausia superba* Dana). *Discovery Rep.* **1962**, *XXXII*, 33–464.
80. Shvetsov, V.V.; Makarov, R.R. On the Biology of Antarctic Krill. *Tr. VNIRO.* **1969**, *LXVI*, 177–206.
81. Barkley, E. Nahrung und Filterapparat des Walkrebschens *Euphausia superba* Dana. *Z. Fisch. Deren Hilfswiss.* **1940**, *1*, 65–156.
82. Brotz, L.; Cheung, W.W.L.; Kleisner, K.; Pakhomov, E.; Pauly, D. Increasing Jellyfish Populations: Trends in Large Marine Ecosystems. *Hydrobiol.* **2012**, *690*, 3–20. [CrossRef]
83. Constable, A.J.; Melbourne-Thomas, J.; Corney, S.P.; Arrigo, K.R.; Barbraud, C.; Barnes, D.K.A.; Bindoff, N.L.; Boyd, P.W.; Brandt, A.; Costa, D.P. at al. Climate Change and Southern Ocean Ecosystems I: How Changes in Physical Habitats Directly Affect Marine Biota. *Glob. Change Biol.* **2014**, *20*, 3004–3025. [CrossRef] [PubMed]
84. Henschke, N.; Everett, J.D.; Richardson, A.J.; Suthers, I.M. Rethinking the Role of Salps in the Ocean. *Trends Ecol. Evol.* **2016**, *31*, 720–733. [CrossRef]
85. Shnar, V.N.; Kasatkina, S.M. Long-Term Variability of Environmental Conditions and Distribution of Antarctic Krill *Euphausia Superba* in the Sub-Region of the Antarctic Peninsula in 1970-2020. *Tr. AtlantNIRO* **2021**, *5*, 101–110.
86. Atkinson, A.; Hill, S.L.; Pakhomov, E.A.; Siegel, V.; Reiss, C.S.; Loeb, V.J.; Steinberg, D.K.; Schmidt, K.; Tarling, G.A.; Gerrish, L.; et al. Krill (*Euphausia superba*) Distribution Contracts Southward During Rapid Regional Warming. *Nat. Clim. Chang.* **2019**, *9*, 142–147. [CrossRef]
87. Loeb, V.; Siegel, V.; Holm-Hansen, O.; Hewitt, R.; Fraser, W.; Trivelpiece, W.; Trivelpiece, S. Effects of Sea-Ice Extent and Krill or Salp Dominance on the Antarctic Food Web. *Nature* **1997**, *387*, 897–900. [CrossRef]
88. Ross, R.; Quetin, L.; Newberger, T.; Shaw, T.; Jones, J.; Oakes, S.; Moore, K. Trends, Cycles, Interannual Variability for Three Pelagic Species West of the Antarctic Peninsula 1993–2008. *Mar. Ecol. Prog. Ser.* **2014**, *515*, 11–32. [CrossRef]
89. Minkina, N.I.; Samyshev, E.Z.; Pakhomov, E.A.; Melnikov, V.V. Temporal and Satial Variability of Energy Exchange in Antarctic Salps. *Res. Square* **2022**. [CrossRef]
90. Atkinson, A.; Siegel, V.; Pakhomov, E.A.; Jessopp, M.J.; Loeb, V. A Re-Appraisal of the Total Biomass and Annual Production of Antarctic Krill. *Deep-Sea Res. I* **2009**, *56*, 727–740. [CrossRef]
91. Murphy, E.J.; Watkins, J.L.; Trathan, P.; Reid, K.; Meredith, M.P.; Thorpe, S.E.; Fleming, A.H. Spatial and Temporal Operation of the Scotia Sea Ecosystem: A Review of Large-Scale Links in a Krill Centred Food Web. *Philosophical Trans. R. Soc. B. Biol. Sci.* **2007**, *362*, 113–148. [CrossRef] [PubMed]
92. Loeb, V.J.; Santora, J.A. Population Dynamics of *Salpa Thompsoni* Near the Antarctic Peninsula: Growth Rates and Interannual Variations in Reproductive Activity (1993–2009). *Prog. Oceanogr.* **2012**, *96*, 93–107. [CrossRef]
93. Kasyan, V.V.; Bitiutskii, D.G.; Voronin, V.P.; Zuev, O.A.; Kalinina, O.Y.; Kolbasova, G.D.; Mishin, A.V.; Murzina, S.A.; Kolbasova, G.D.; Voronin, V.P.; et al. Composition and Distribution of Plankton Communities in the Atlantic Sector of the Southern Ocean. *Diversity* **2022**, *14*, 923. [CrossRef]

water

MDPI

Article

Influence of Hydrological Factors on the Distribution of Methane Fields in the Water Column of the Bransfield Strait: Cruise 87 of the R/V "Academik Mstislav Keldysh", 7 December 2021–5 April 2022

Andrei Kholmogorov, Nadezhda Syrbu * and Renat Shakirov

V.I. Il'ichev Pacific Oceanological Institute FEB RAS, 690041 Vladivostok, Russia
* Correspondence: syrbu@poi.dvo.ru

Abstract: Within the framework of the expedition research "Complex studies of the Antarctic marine ecosystem in the areas of the transport and interaction of water masses in the Atlantic sector of Antarctica, the Scotia Sea and the Drake Strait" (cruise 87 of the R/V "Academik Mstislav Keldysh", 7 December 2021–5 April 2022), the distribution of gas-geochemical fields of methane in the Bransfield Strait was studied in detail for the first time. The connection of the methane distribution in water with the complex hydrological regime of the strait has been revealed. Elevated values of methane concentrations brought to the Bransfield Strait in the warm current flow from the Bellingshausen Sea have been established. Low concentrations of methane also mark the cold waters of the Weddell Sea, which carry out the transit of water masses into the Atlantic Ocean. The research was carried out within the framework of the theme FWMM-2022-033 "Integrated environmental studies of the Southern Ocean" AAAA17-117030110035-4 and international obligations of the Russian Federation as a party to the Antarctic Treaty and the Convention on the Conservation of Antarctic Marine Living Resources.

Keywords: methane; temperature; salinity; current system; the Bransfield Strait; South Ocean

Citation: Kholmogorov, A.; Syrbu, N.; Shakirov, R. Influence of Hydrological Factors on the Distribution of Methane Fields in the Water Column of the Bransfield Strait: Cruise 87 of the R/V "Academik Mstislav Keldysh", 7 December 2021–5 April 2022. *Water* **2022**, *14*, 3311. https://doi.org/10.3390/w14203311

Academic Editor: Domenico Cicchella

Received: 15 September 2022
Accepted: 15 October 2022
Published: 20 October 2022

1. Introduction

Currently, the discussion about the role of methane as the most sensitive indicator of the evolutionary climate change processes remains one of the most discussed topics in ocean sciences [1–4]. Information about the amount of methane flow into the atmosphere is extremely contradictory. Models from insignificant emissions to explosive emissions leading to a global catastrophe are considered. Data concerning dynamics of the methane flow are limited, and little is known about the factors controlling it [5]. The contribution of the oceans to global methane emissions into the atmosphere remains very uncertain. Various studies have recognized that open oceans are a source of methane into the atmosphere contributing from 0.005% to 3% of global methane production [6], although the main mechanism of methane production remains unclear. According to some estimates, the flow of methane from the World Ocean into the atmosphere is 6 million tons of CH_4 per year on the ocean area (3.6×1014 m^2) with an average flow of methane into the atmosphere of 2.5 mol/(km^2 $\times$ day) [7].

As climate and chemical composition of the atmosphere transform, the World Ocean also meets global changes. An increase in the concentration of carbon dioxide in the atmosphere directly leads to ocean acidification [8,9]. Another global process is ocean deoxygenation [10]. Recent biogeochemical studies have shown that all of the above processes directly affect the carbonate system and the hydrochemical characteristics of the waters of the Southern Ocean [11–14]. In this regard, the Southern Ocean has its own characteristics. The most important hydrological feature is the Antarctic circumpolar current and the large depth of the upper mixed water layer exceeding 400 m [15,16]. Due to intensive vertical circulation, methane is involved in the deep layers of the ocean [13].

Bottom waters in the deep-sea channels of the Atlantic spread from their sources on the shelf of Antarctica, flowing from one basin to another. The process of such overflow often concentrates channels, faults and passages connecting the abyssal of individual basins. At the same time, due to intensive interaction with the bottom in the channels, a significant transformation of the thermohaline properties of bottom waters occurs. These circumstances make expedition research in the Southern Ocean especially important from the point of view of studying the large-scale meridional circulation of the World Ocean.

In addition, the analysis of available geological and geophysical materials revealed that in the research area there are active permeability zones (fault zones, crushing zones, rifts, etc.) for gases and fluids of both a deep nature associated with rifting, and prospects for detecting near-surface active seeps of reduced gases, in particular methane. There are certain prerequisites for gas hydrates detection, so it is possible to predict the presence of gas "torches"—jet outflows of gas bubbles from the seabed into the water column. In other areas of the World Ocean, it has been established that such sites are often hotbeds of active development of biological communities involved in the food chain of underwater biocenoses. There is practically no specific information about gas or gas-fluid manifestations in the research area. Almost all Antarctic coastal waters are large sedimentary basins of a significant hydrocarbon deposits potential. The largest of them are the basins of the Weddell Seas (1.5 million square kilometers), Ross (1.26 million square kilometers), Mawson (710 thousand square kilometers), Commonwealth (580 thousand square kilometers) and Riser-Larsen (400 thousand square kilometers). According to modern data based on complex geophysical studies, the projected hydrocarbon resources in all marine sedimentary basins of Antarctica could reach (taking into account deep-water areas) some tens of billions of tons.

In general, in the research region there are geological objects that have a strong influence on the hydrodynamic and hydrological regime of the water area (ridges, depressions, rifts, etc.). Detailed geophysical and geological studies of these structures have not yet been completed, so the bottom relief and the structure of orographic barriers are relevant issues for the study of the hydrodynamic regime of the research area in the future. Our paper presents the results of detailed gas-geochemical studies of methane distribution in the water column of the Bransfield Strait, Antarctica Strait and the Weddell Sea in cruise 87 of the R/V "Academik Mstislav Keldysh", 7 December 2021–5 April 2022. Data obtained allowed us to assess the influence of currents on the distribution of methane fields in the water column and outline prospects for further research.

2. Materials and Methods

Gas-geochemical research makes it possible to assess the variability of the methane content and a number of its homologues (ethane, propane, butane) in seawater along the horizons, to identify the sources of gas flow, as well as to obtain information for further assessment of the relationship of background and anomalous gas-geochemical fields with morphostructural and tectonic elements of the geological structure.

Sampling and analysis in the studied areas of the Southern Ocean (the Bransfield Strait, Antarctica Strait and the Weddell Sea) in cruise 87 was carried out in the period 19 January–3 February 2022.

Water sampling from the board of the R/V "Academik Mstislav Keldysh" was carried out by a 24-position Rosset system equipped with an INDRONAUT CTD probe (Italy). Water sampling was carried out by a cassette of NISKIN system bathometers (23 bathometers) on various horizons, according to the vertical distribution of temperature, salinity and other hydrological parameters identified during CTD-profiling.

The method of equilibrium concentrations "HeadSpace" was used to analyze the methane content in all water samples [17]. The water from the bathometers was taken by the "triple overflow" method into pre-sterilized medical glass bottles with a volume of 68 mL, which were hermetically sealed with sterile rubber stoppers without access to the bottles of atmospheric air. Needles from a medical syringe were used to remove excess water.

When creating the gas phase, 12 mL of helium was injected into the bottles. Pure helium of the brand "6.0" was used as the gas phase. Then, the samples were intensively mixed for 120 min at a speed of 200 rpm using a LOIP LS-110 shaker (Russia). Before the analysis, the equilibrium gas phase was extracted with a disposable medical syringe (12 mL) to inject the sample into the gas chromatograph. Methane concentrations were analyzed in the shipboard laboratory.

A "CrystalLux4000M" (Russia) gas chromatograph, equipped with a flame ionization detector (FID) and two detectors for thermal conductivity (TCD, sensitivity of 10^{-5}%), was used to analyze the gas composition. Certified calibration gas mixtures were used to calibrate the instruments. The error of this method is less than 5%. The concentrations of methane dissolved in seawater were calculated according to the method [18] in modification [19] using the calculated solubility constants of methane. Microsoft Excel spreadsheets were used to calculate and systematize the data obtained during the cruise. For operational visualization and interpretation, gas-geochemical data obtained after analysis on a chromatograph were entered into geoinformation applications (Arc GIS 10.4.1, Ocean Data View 5.2.1).

Sampling, preparation of samples, and analytical studies were carried out according to the certified method adopted in the Gas Geochemistry Laboratory of V.I. Il'ichev Pacific Oceanological Institute (State Standard Certificate No. 58 for the Laboratory Passport PS 1.051-21).

3. Results

The Bransfield Strait is located in the southern Atlantic Ocean (Southern Ocean) between the archipelago of the South Shetland Islands and the Antarctic Peninsula. The southwestern part of the strait is bounded by the Gerlash Strait. The extended deep-water section, reaching 1500 m depth, abruptly passes into the shallow shelf zone of the South Shetland Islands and the Antarctic Peninsula with a depth of 300 to 700 m. According to modern studies, the strait enters a rift zone marked by increased seismic activity, while the central part of the strait is characterized by a sharp decrease in relief, probably of tectonic or volcanic-tectonic nature [20]. The South Shetland Islands are of volcanic origin, while there are areas of recent activity in this area, for example, the Deception Island caldera [21]. Generalization of the results of complex geological and geophysical studies in the Southern Ocean, in particular in the area of its junction with the Atlantic Ocean, shows that this region is the area of existence of large fragments of the destroyed intercontinental bridge between South America and West Antarctica, which has experienced cracking and local rift spreading. The dynamics of its development was determined by the heating of the lithosphere under the influence of extensive uplift of mantle matter (plume) in conditions of moderate separation between the continental massifs of South America and West Antarctica. This was accompanied by regional magmatism of the type of areal oceanic platobasalts, local magmatism (such as SOX basalts) in young local rifts, crushing and collapse of crustal blocks in the Drake Strait, as well as sequential immersion of such blocks in the basin of the Scotia Sea.

Brief hydrological characteristics of the research area Bransfield Strait meets a collision of warm waters of the Atlantic Circumpolar Current (ACC) penetrating through the deep straits between the islands of Brabant, Lowe and Smith, and cold surface waters of the Weddell Sea entering the Bransfield Strait through the Antarctic Sound and the shelf zone of Joinville Island [22–24]. The interaction of the waters of the ACC and the waters of the Weddell Sea results in the formation of a coastal boundary current in the Bransfield Strait, directed along the southern coast of the South Shetland Islands to the northeast. The boundary current of the Bransfield Strait has a significant impact on the physicochemical properties of water in this area and participates in the transfer of biogenic elements [25] necessary for the development of marine biota at all trophic levels. This area has an important bioresource value, since it is the area where the growth and development of Antarctic krill to adults takes place. Krill is an important ecosystem link in the Southern Ocean, representing a food source for many species of fish, penguins and marine mammals [26]. During the Australian summer (December–February), the water area of the Bransfield Strait is free of ice cover, there are practically no icebergs and large fragments of ice that interfere with navigation, which makes the Strait a convenient testing ground for a number of geological, oceanological and hydrobiological studies of this part of the Southern Ocean.

3.1. Bransfield Strait

Located between the South Shetland Islands and the Antarctic Peninsula, the Bransfield Strait has an asymmetric cross-section and consists of (from north to south): a narrow (less than 10 km) shelf zone, a deep-water central basin with a width of 15–20 to 80 km (Figure 1) and the wide shelf of the Antarctic Peninsula (the Bransfield platform).

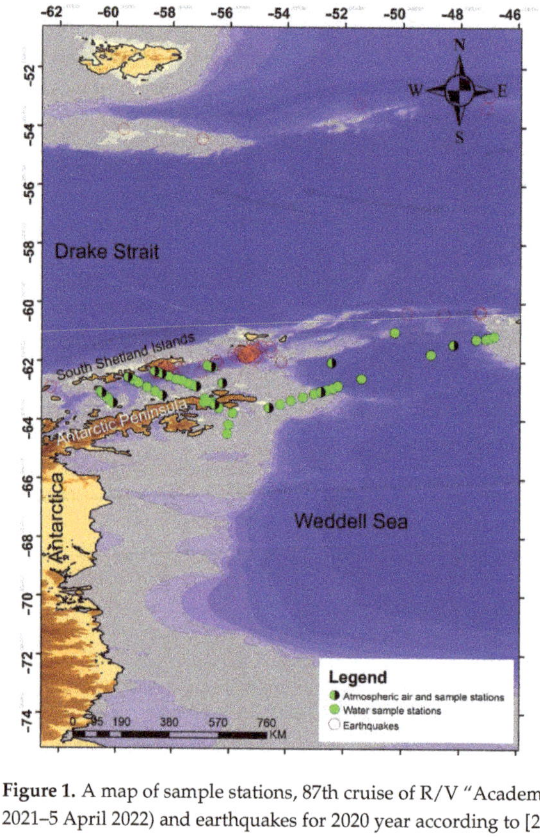

Figure 1. A map of sample stations, 87th cruise of R/V "Academik Mstislav Keldysh" (7 December 2021–5 April 2022) and earthquakes for 2020 year according to [27].

The Bransfield Strait meets a particular interest as a rift zone with increased seismic and modern volcanic activity, as well as an area with recently discovered areas of hydrothermal activity. Earlier, the study of methane distribution profiles in the water column of the Bransfield Strait was carried out in a limited volume in its western part, and the main attention of researchers was focused on analyzing the patterns of methane distribution in the Weddell Sea.

The western part of the Bransfield Strait, with a strong coastal influence and extensive shelf zones, are considered as highly productive areas [28]. Wefer et al. (1988) [29] associated these regions with episodic high production events and documented very high fluxes of settling material to the sediment in the Bransfield Strait.

The Bransfield Strait is of a particular interest as a rift zone with increased seismic and modern volcanic activity, as well as an area with recently discovered areas of hydrothermal activity. Earlier, the study of methane distribution profiles in the water column of the Bransfield Strait was carried out in a limited volume in its western part, and the main attention of researchers was focused on analyzing the patterns of methane distribution in the Weddell Sea.

During cruise 87 of the R/V "Akademik Mstislav Keldysh", 7 December 2021–5 April 2022 (AMK87), 3 sections were made in the Bransfield Strait, including 24 gas-geochemical stations (200 samples). Analysis of seawater samples taken from bathometers in the depth range 2–1767 m revealed methane concentrations of 2.47–10.95 nM/L (Figures 2–4).

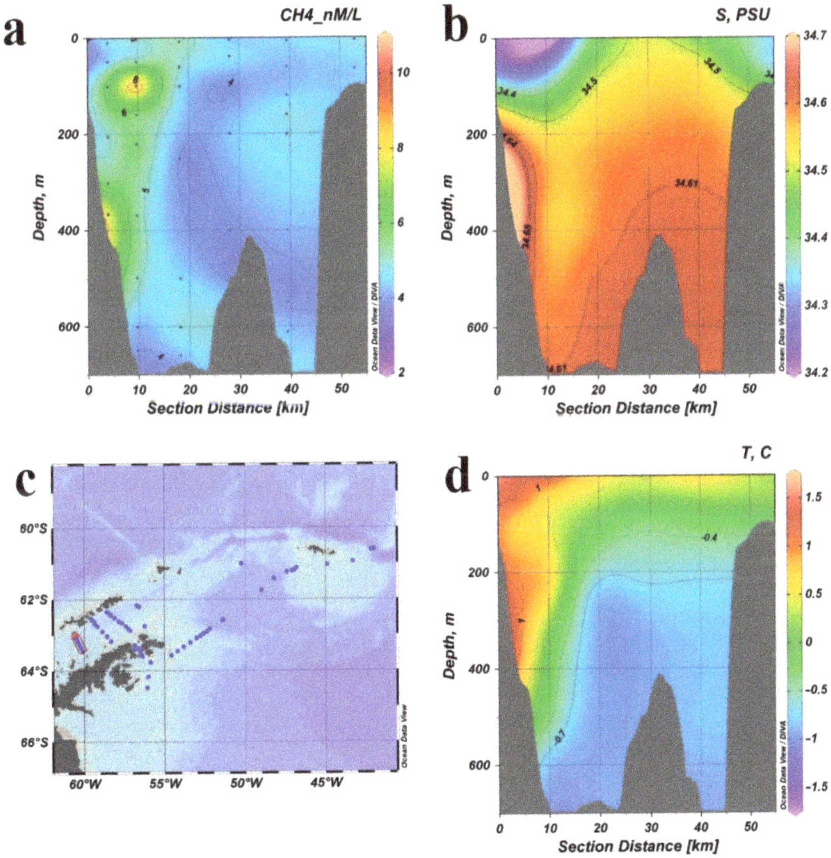

Figure 2. Distribution of methane concentration (**a**), temperature (**d**) and salinity (**b**) in the western section (**c**) in the Bransfield Strait.

The maximum methane concentrations (10.95 nM/L) are observed on the slope of the trough of the Strait near the South Shetland Islands in the eastern (Figure 4a,c) and central (Figures 4c and 5a) sections at the horizons of 0–500m. Figures 2a, 3a and 4a clearly show a core of elevated methane concentrations centered at a depth of 400 m and having a horizontal scale of about 10 km. This may be due to the transfer of methane-saturated waters by the Bransfield Strait current, directed to the East. The presence of this jet carrying elevated concentrations of methane is confirmed by temperature (Figures 2d, 3d and 4d) and salinity (Figures 2b, 3b and 4b) data.

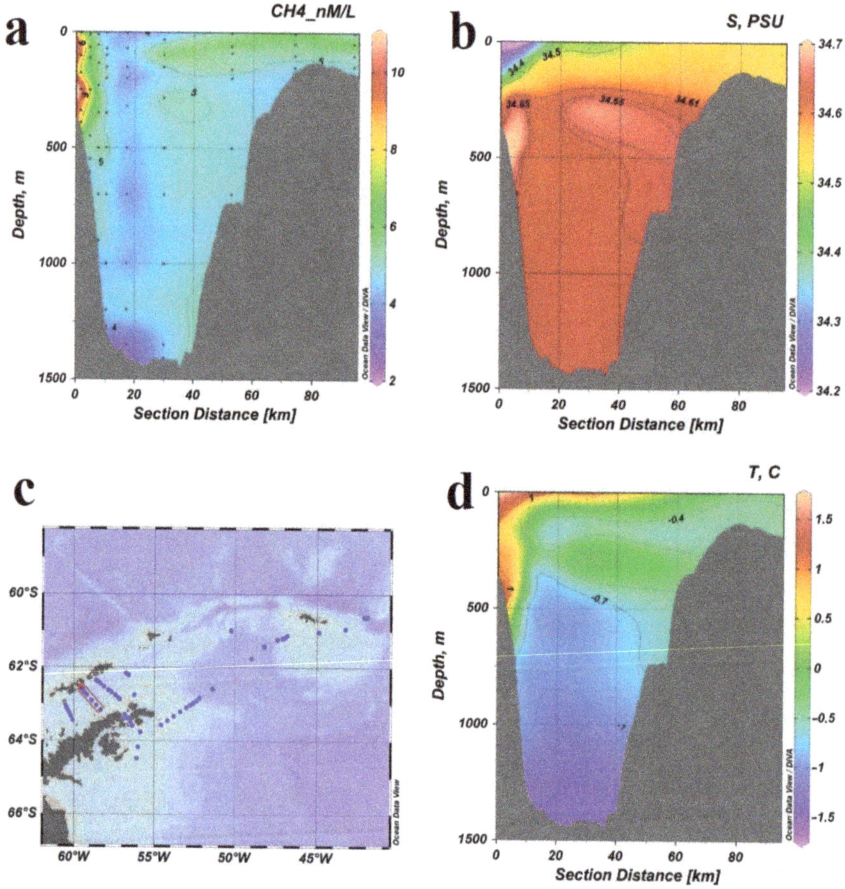

Figure 3. Distribution of methane concentration (**a**), temperature (**d**) and salinity (**b**) in the central section (**c**) in the Bransfield Strait.

In the surface and subsurface (depth of 0–200 m) water layers in the central and eastern sections (Figures 3a and 4a), an area of increased concentrations (7.63 nm/L) is allocated, which may also be associated with the hydrological regime—the existence of vortices on the central axis of the strait (Figure 5).

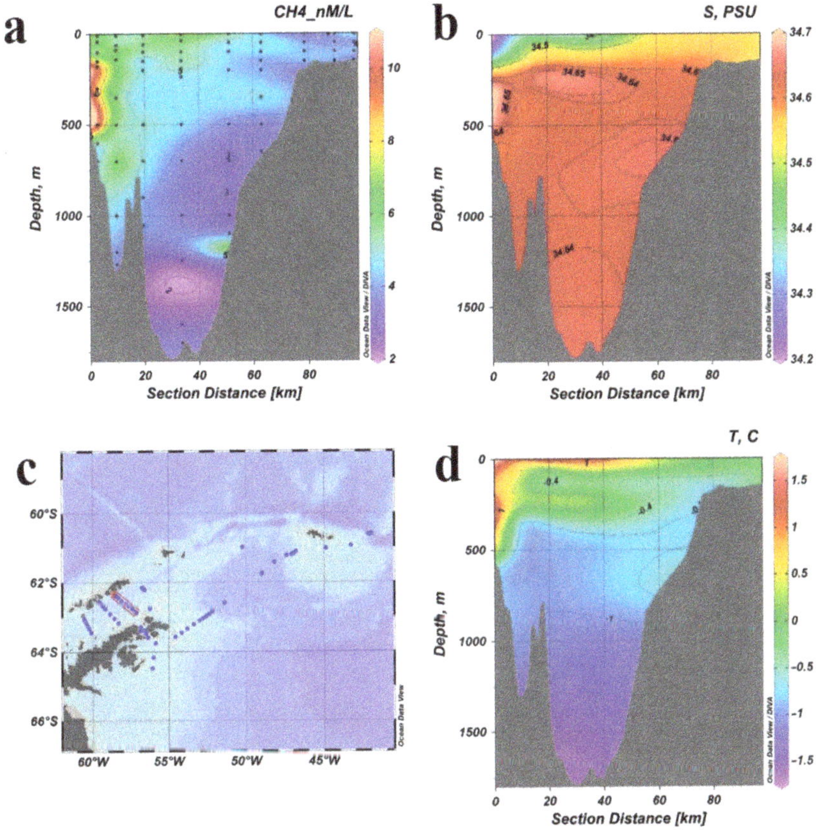

Figure 4. Distribution of methane concentration (**a**), temperature (**d**) and salinity (**b**) in the eastern section (**c**) in the Bransfield Strait.

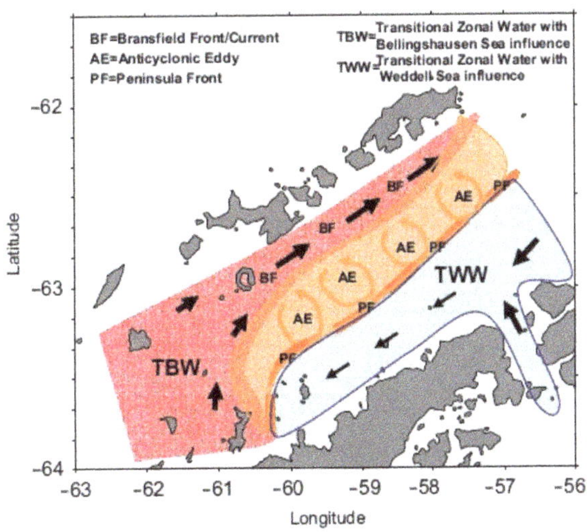

Figure 5. Diagram of the main components of the Bransfield Strait surface current system [30].

The minimum methane concentrations (2.47 nm/L) were found at the bottom of the strait—in a water layer with a depth of more than 1000 m (Figures 2–4), which may be caused by the transit of Deep Water of the Bransfield Strait (GWB), the upper limit of its distribution is considered to be the values of potential temperature −1.0 °C and salinity 34.5 eps [31].

The methane concentration distribution in the Bransfield Strait is largely controlled by the hydrological regime, the scheme of methane transit at different horizons through the strait is confirmed by modern hydrological studies [32], as well as studies of Russian and foreign colleagues in previous periods [22,30–32], the flow diagram of the strait in the section is shown in Figure 6.

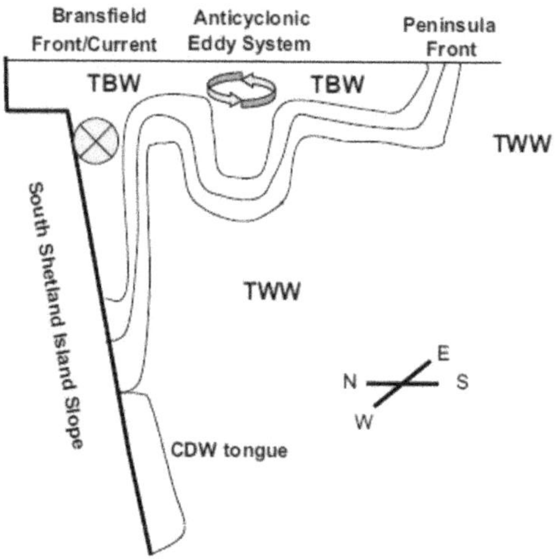

Figure 6. Cross-sectional diagram of the main components of the Bransfield Strait current system [30].

The central section of the Strait (Figure 3c) geographically repeated the section made during the expedition 79 of the R/V "Akademik Mstislav Keldysh", 2020. Gas-geochemical studies of 2020 revealed a zone of increased concentrations of methane at the bottom of the strait, a local gas discharge site of hydrothermal nature associated with a fault system at coordinates 62°40′3″ S. 59°21′40″ W. [33]. Our paper describes that a methane-containing hydrothermal fluid causes an increase of the methane content in sea water from 10 to 43 nmol/L in a wide range of depths of the water column. The bottom water layer with a high methane content extends over considerable distances from the hydrothermal outlet. Additionally, the largest number of thermophilic methanotrophic bacteria found in the bottom water layer was found at this site. The detection of thermophilic microorganisms in the bottom layer confirms the endogenous nature of the methane source [34].

However, the cruise 87 of the R/V "Academik Mstislav Keldysh" (2021–2022) did not detect this anomaly or any result of its activity. The instability of the endogenous sources of the strait is probably controlled by the seismic situation of the region and needs further study.

3.2. *Antarctic Sound*

Analysis of seawater samples taken from bathometers in the Antarctic Sound The analyses showed a different distribution of methane (Figures 7a and 8a). In the Antarctic Sound 7 hydrological stations (52 samples) were carried out at horizons of 4–949m, forming 2 sections:) across (Figure 7c) and along (Figure 8c) the sound. The cross section between the Antarctic Peninsula and Joinville Island is characterized by elevated methane concentrations (Figure 7a) at the extreme eastern (7.28–11.4 nm/L) and western (5.78–7.93 nm/L). The main water mass of the central part of the sound contains average methane concentrations (4.11–6.55 nm/L) unevenly distributed in the water column (Figure 7a).

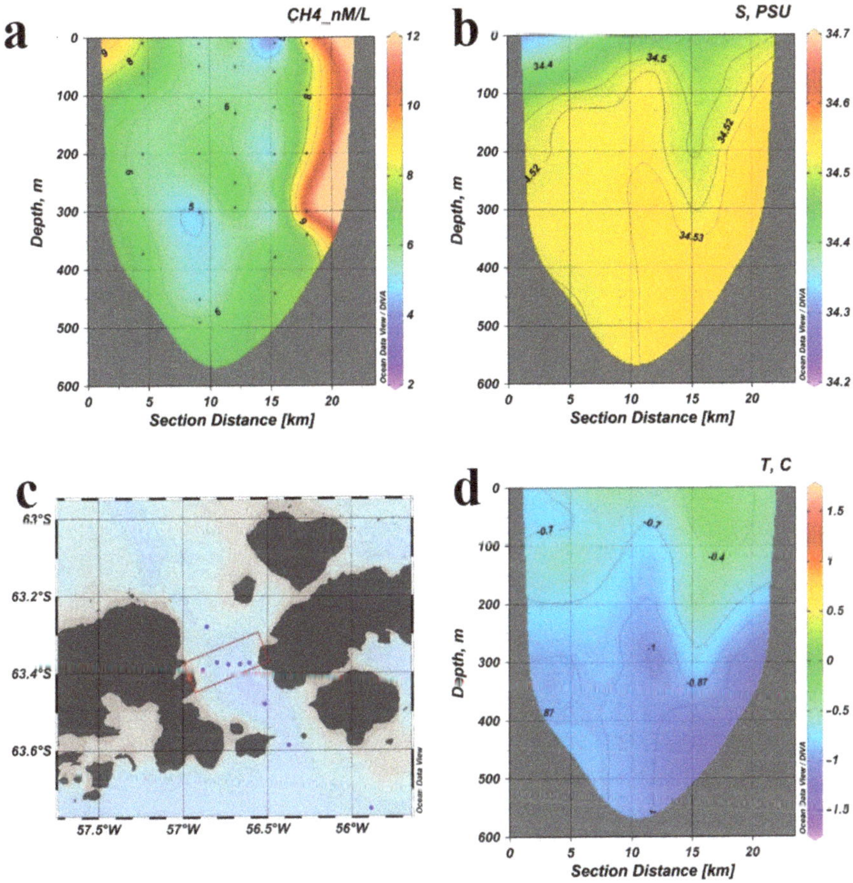

Figure 7. Distribution of methane concentration (**a**), temperature (**d**) and salinity (**b**) on a cross section (**c**) in the Antarctic Sound.

A local bottom field of elevated methane concentrations (up to 7.33 nM/L) (Figure 8a), which presents here probably due to endogenous processes was revealed on the longitudinal section (Figure 8c). This area of elevated concentrations was also determined during the 79th cruise of R/V "Academik Mstislav Keldysh", 2020 [33].

The most important result of the research in the Antarctic Sound is the discovery of a constant stable zone with an increased concentration of methane in the bottom water layer in the deep part of the Sound. It is likely that the increase of the methane concentration in the bottom water layer is associated with the movement of water masses from the deep-water part of the Antarctic Sound, where there are not yet mapped areas of gas discharge spots.

The Antarctic Sound transports water masses with low concentrations of methane from the Weddell Sea to the Bransfield Strait [33], where these waters spread westward along the Antarctic Peninsula (see Figure 5) and form methane transit fields. Figures 7b,d and 8b,d show the transit of cold (−0.7–−0.4 °C) waters with "low" (34.5 psu) salinity from the Weddell Sea to the Bransfield Strait (Figures 2b,d, 3b,d and 4b,d).

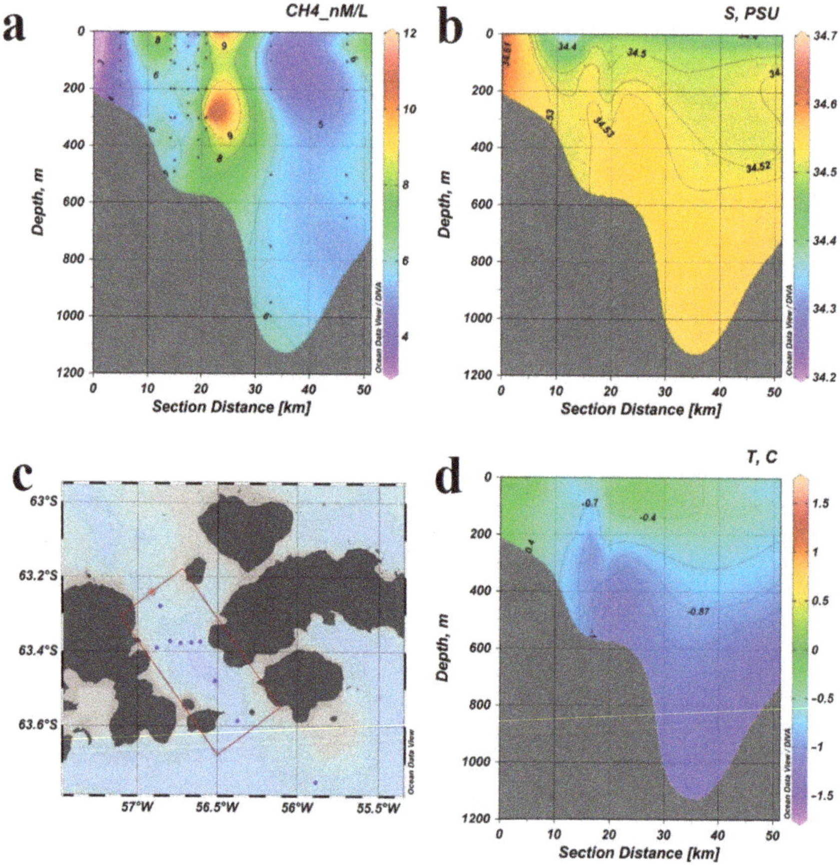

Figure 8. Distribution of methane concentration (**a**), temperature (**d**) and salinity (**b**) on the longitudinal section (**c**) in the Antarctic Sound.

3.3. Weddell Sea

A section (about 80 km) of 3 hydrological CTD stations was made in the Weddell Sea in the direction south of the Antarctic Sound (Figure 9c), the depth range was 5–371 m, 19 samples were taken.

The section distinguishes 3 layers: two—with increased methane concentrations (0–200 m and 250–350 m) and a bottom layer of low methane concentrations (Figure 9a) in cold subzero waters (Figure 9b) of "low" salinity about 34.5psu (Figure 9b).

A 500 km section (Figure 10c) was made through the Weddell Sea (63.6° S, 54.6° E–61.1° S, 46.8° E), the depth range was 11–3270 m, including 16 hydrological stations (135 samples). As gas-geochemical studies have shown (Figure 10a) that the main body of water is characterized by low concentrations of methane (1.56–3.45 nm/L) observed in the deep-water part of the basin, and its western part is deeper than the isobate of 2000 m has slightly higher values (up to 3.95 nm/L) compared with the eastern one. Average concentrations are observed in the surface water layer. At the extreme eastern station of the section, elevated concentrations of methane were recorded (8.95 nm/L in the bottom layer of 516 m and 6.5 nm/L in the layer of 10–200 m); however, at the horizon of 396 m, the concentration of methane was 2.8 nm/L, which may be caused by the influence of an undercurrent, confirmed by temperature drop (Figure 10d), existing under the subsurface layer of low salinity (Figure 10b).

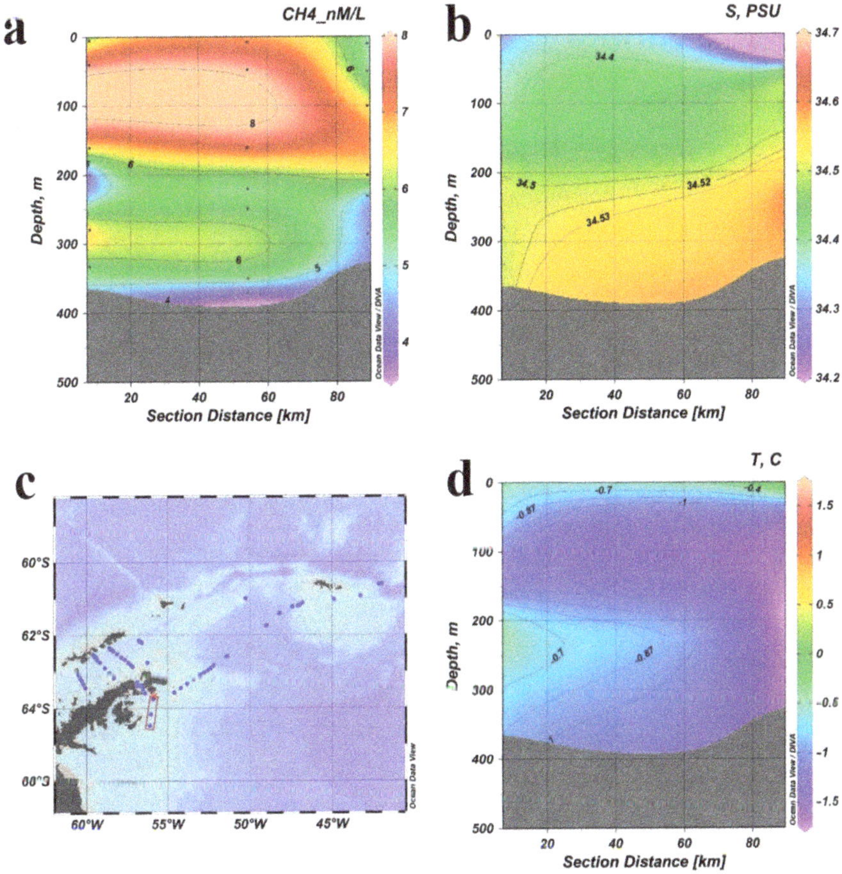

Figure 9. Distribution of methane concentration (**a**), temperature (**d**) and salinity (**b**) on the section (**c**) in the Weddell Sea south of the Antarctic Sound.

3.4. Methane in the Atmosphere

The Southern Ocean is one of the most biologically productive oceanic regions in the world and characterized by high biomasses of zooplankton, Antarctic krill, and salps [35], all of which having the potential to produce CH_4. Although some researchers have measured CH_4 concentrations in the Southern Ocean [36–40], accurate assessment of the sea–air flux of CH_4 has been limited by the sparsity of available data.

We have taken 14 samples of atmospheric air on the study area (Figure 11). The range of methane concentration was 1.96–2.24 ppm. It is a rather high value, as a global atmospheric CH_4 level is about 1.9 ppm according to [41,42].

Despite low methane mole fraction in the atmosphere (1.803 ppm in 2011), CH_4 is the second largest contributor (17%, after carbon dioxide) to the total radiative forcing caused by the well-mixed greenhouse gases [43]. The continuous increase of atmospheric CH_4 in the past decades implies an imbalance between CH_4 sources and sinks and prompts urgent questions about the causes [44].

Atmospheric methane concentrations are probably controlled by active volcanism and wind regime of the region and are a sphere of the further research. Gille et al. (2014) [45]

reported that low temperatures of the surface waters in the Kerguelen region correlate with high wind speeds, and that wind-mixing of the upper ocean there resulted in entrainment of cold water into the mixed layer and euphotic zone [46].

In the Southern Ocean, strong vertical mixing is induced by near-surface westerly winds that causes convergent flow at intermediate depths; such mixing partially offsets upwelling [47]. The flow of methane into the atmosphere from the sea surface is controlled not only by physical factors, but also by the availability of organic matter and oxygen [48].

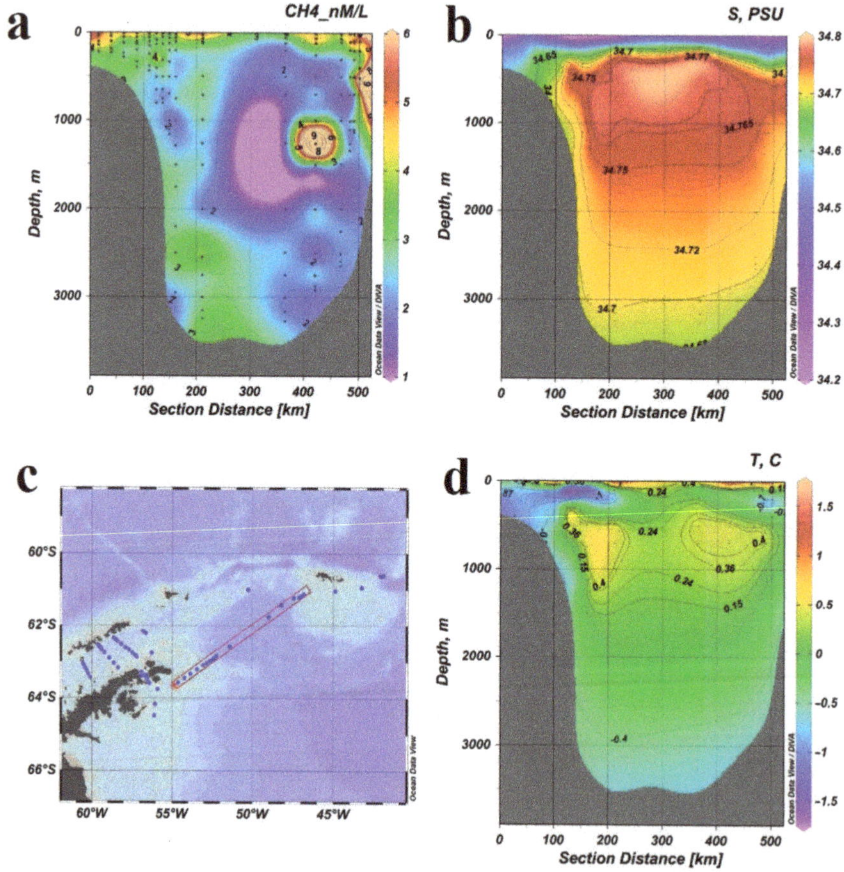

Figure 10. Distribution of methane concentration (**a**), temperature (**d**) and salinity (**b**) on the 500 km section (**c**) through the Weddell Sea.

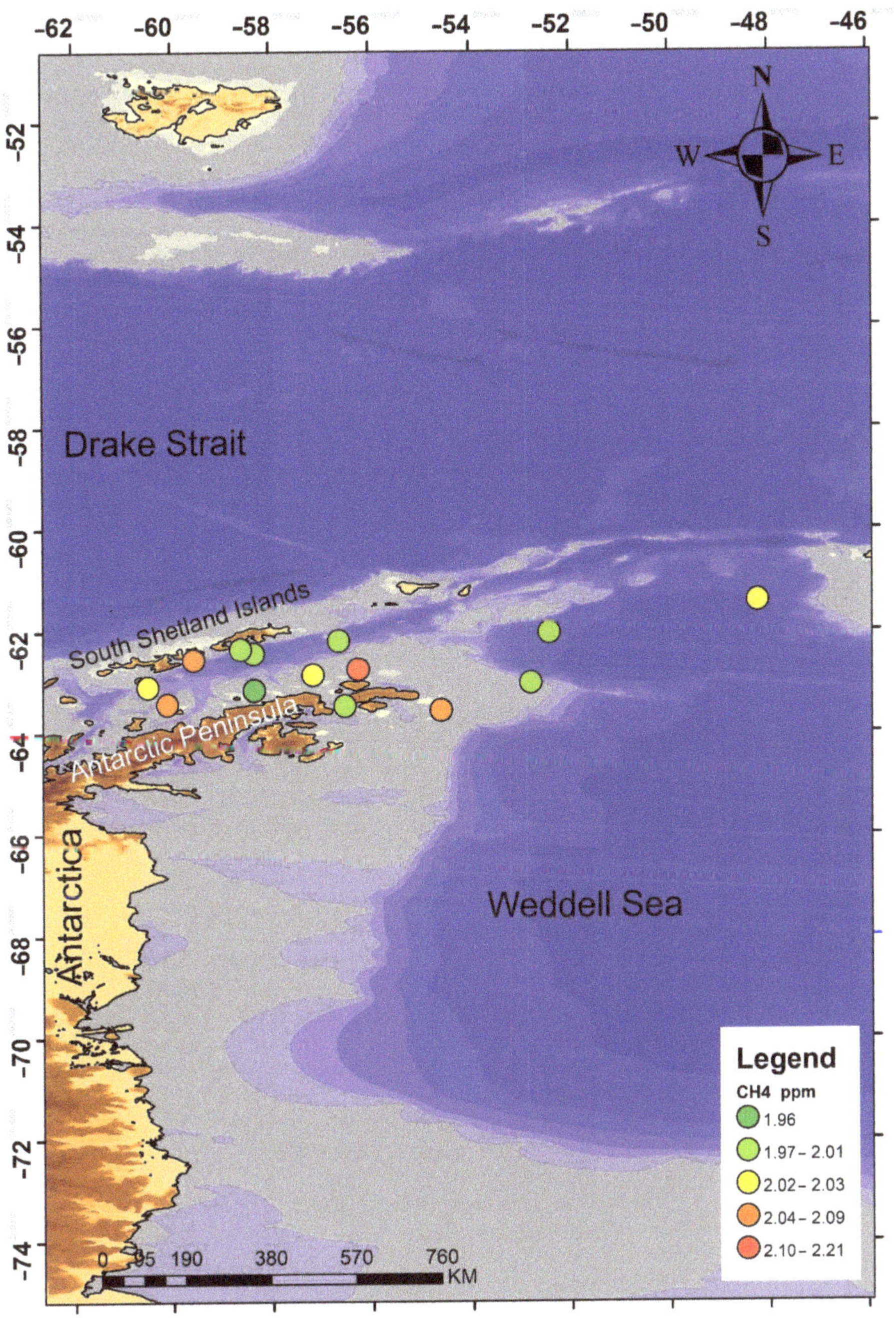

Figure 11. Methane concentration in the atmosphere.

4. Discussion

During the research in the Southern Ocean as part of cruise 87 of the R/V "Academik Mstislav Keldysh", 7 December 2021–5 April 2022, local anomalies of methane concentrations and their sources were not detected. The vast majority of the water mass is characterized by low concentrations of methane, which indicates the absence of intensive and stable endogenous methane sources. In general, the area is characterized by low rates of methane synthesis and oxidation. At all stations, bacterial utilization of methane prevailed over its synthesis, which indicates the presence of a permanent source of methane [34].

The stable methane emission sources presence assumption was substantiated earlier basing on the fact, that Bransfield Strait is located between the archipelago of the South Shetland Islands and the Antarctic Peninsula. It is of particular interest as a rift zone with increased seismic and modern volcanic activity, which are expressed in a special gas-geochemical regime. The most important result of the Antarctic Strait research is the discovery of a stable zone with an increased methane concentration on two sections in the sea bottom layer in the deep part of the Strait, which is associated with the movement of water masses from the deep part of the Antarctic Strait, where there are not yet mapped areas of bottom gas discharge. A single methane sip was detected at a depth of 1397 m, characterized by a high level of methane content (43 nmol/L) in the bottom water horizon. It was also found that the methane-enriched (10.29–14.46 nmol/L) bottom layer of seawater extends to a distance of 30–35 km from the central part of the methane anomaly [49]

Most currently active Antarctic volcanoes are controlled by the West Antarctic Rift System (WARS). Three of them are located on Victoria Land (Mount Erebus, Mount Melbourne and Mount Rittman), and the fourth is Deception Island in the South Shetland Islands group, located in the Bransfield Strait.

It is likely that the observed by us in 2022 increased methane concentration is related to the water masses transfer from the deep-water part of the Antarctic Sound, where there are areas of gas emission on the bottom could occurs.

Thus, a comparison of gas-geochemical, hydrological and microbiological data indicates the absence of a permanent active focused source of methane at the bottom of the Bransfield Strait.

Apparently, the new source of elevated methane concentrations at the bottom of the Bransfield Strait, discovered in the cruise 79 of R/V "Academik Mstislav Keldysh", is local and unstable one and is associated with a gas discharge site, most likely representing a hydrothermal outlet. The methane anomaly found [33] is associated with endogenous processes in the riftogenic seismically active zone, and the localization of the source is due to the peculiarities of the tectonic structure of the bottom of the Bransfield Strait and requires more detailed further study. According to the data obtained, it can be concluded that in the Atlantic sector of the Southern Ocean there may be areas of rapid and long-range transport of anomalous methane fields in the water column, possibly up to the surface.

Studies of gas-geochemical fields of the bottom water layer provide the most accurate detection of jet methane outputs from bottom sediments. In connection with the above results, one can assume that in the Atlantic sector of the Southern Ocean there may be areas of long-range transit of anomalous methane fields in the water column laterally. Methane can achieve the surface water layer just with a powerful bubble jet. Their height sometimes reaches 2 km in different areas of the World Ocean [50]. Therefore, further gas-geochemical studies in this region remain highly relevant and it can be concluded that systematic complex expeditionary studies are required. Low regional activity of lithosphere gas discharge made it possible to study the distribution pattern of methane dissolved in water and its transit through a complex system of currents of the studied region. As a result, the location of dissolved methane particles is controlled by a complex hydrological regime of the region: the transit of water masses from the Weddell Sea through the Antarctic Strait to the Bransfield Strait, circulation in the Bransfield Strait (vortices of the central axis of the Strait, a stream along the South Shetland Islands).

5. Conclusions

Climate changes and the chemical composition of the atmosphere result in global changes process of the World Ocean. Oceans play a key role in climate regulation, especially in buffering (neutralizing) the effects of increasing levels of greenhouse gases in the atmosphere and rising global temperatures. The regulatory processes carried out by the oceans alter as a response to climate change, and positive feedbacks to the ocean may exacerbate climate change.

The author's research has shown that the processes causing these changes is directly affected by many characteristics of the waters of the Southern Ocean, especially the Atlantic sector. This, in turn, has a strong impact on the state of the ecosystem of this important fishing area. Complex and active hydrodynamics caused by the Antarctic circumpolar current, intense vertical circulation, orographic barriers, the development of contrasting fronts, areas of demolition of terrigenous material with signs of chemical anomalies, etc. This has led to the fact that conducting complex oceanological studies to assess the current state of the Southern Ocean environment at the junction of disciplines has allowed us to obtain a number of important and unique scientific results.

Detailed gas-geochemical studies have been carried out and methane concentrations in the water column of the Bransfield Strait and Antarctica have been studied. A study of the Bransfield Strait area, where a local bottom-level methane anomaly was detected in 2020, showed the absence of elevated concentrations, which may indicate a high intensity of the "biological filter", as well as instability of the lithosphere discharge. During the expedition, no active sources of methane output were found, which made it possible to assess the influence of currents on the distribution of methane fields in the water column and to outline prospects for further research. The increased values of methane concentration in the atmosphere of the Southern Ocean indicate a significant contribution of the region to the global debit of greenhouse gases, which implies further careful study of the sources of gas discharge not only in the water area, but also in the land areas of Antarctica.

It is advisable to carry out a complex of geological and geophysical studies of the sea bottom, including lithological testing, geochemical, hydroacoustic and bathymetric studies of the bottom ecosystem in the subsequent stages of research of the ecosystem of the Atlantic sector of the Southern Ocean.

Author Contributions: Conceptualization, R.S. and A.K.; methodology, N.S.; software, A.K.; validation, R.S. and N.S.; formal analysis, A.K.; investigation, A.K.; resources, A.K.; data curation, N.S.; writing—original draft preparation, A.K.; writing—review and editing, N.S.; visualization, A.K.; supervision, R.S.; project administration, N.S.; funding acquisition, R.S. All authors have read and agreed to the published version of the manuscript.

Funding: The work was carried out within the framework of the theme FWMM-2022-033 "Integrated environmental studies of the Southern Ocean" AAAA17-117030110035-4. Research for this paper was carried out according to the goals of the United Nations Decade of Ocean Science for Sustainable Development GEOMIR project (ID 164, the head is R.B. Shakirov).

Data Availability Statement: CTD profiles used in the publication are available in open access through Mendeley Data service (http://dx.doi.org/10.17632/69v8599btr.1, accessed on 14 September 2022). The GEBCO2021 data are available at the official service https://www.gebco.net/data_and_products/gridded_bathymetry_data/gebco_2021/, accessed on 14 September 2022.

Conflicts of Interest: The authors declare no conflict of interest.

References

1. Ruppel, C.D.; Kessler, J.D. The interaction of climate change and methane hydrates. *Rev. Geophys.* **2017**, *55*, 126–168. [CrossRef]
2. Bange, H.W.; Bartell, U.H.; Rapsomanikis, S.; Andreae, M.O. Methane in the Baltic and North Seas and a reassessment of marineemissions of methane. *Glob. Biogeochem. Cycles* **1994**, *8*, 465–480. [CrossRef]
3. Saunois, M.; Bousquet, P.; Poulter, B.; Peregon, A.; Ciais, P.; Cana-dell, J.G.; Dlugokencky, E.J.; Etiope, G.; Bastviken, D.; Houweling, S.; et al. The global methane budget 2000–2012. *Earth Syst. Sci. Data* **2016**, *8*, 697–751. [CrossRef]

4. Weber, T.; Wiseman, N.A.; Kock, A. Global ocean methane emissions dominated by shallow coastal waters. *Nat. Commun.* **2019**, *10*, 4584. [CrossRef]
5. Shakirov, R.B.; Mau, S.; Mishukova, G.I.; Obzhirov, A.I.; Shakirova, M.V.; Mishukova, O.V. The features of methane fluxes in the western and eastern Arctic: A review. Part I. *Geosyst. Transit. Zones* **2020**, *4*, 4–25. [CrossRef]
6. Conrad, R.; Seiler, W. Methane and hydrogen in seawater (Atlantic Ocean). *Deep-Sea Res. Part A* **1988**, *35*, 1903–1917. [CrossRef]
7. Watanabe, S.; Higashitani, N.; Tsurushima, N.; Tsunogai, S. Methane in the Western North Pacific. *J. Oceanogr.* **1995**, *51*, 39–60. [CrossRef]
8. Doney, S.C.; Fabry, V.J.; Feely, R.A.; Kleypas, J.A. Ocean Acidification: The Other CO_2. *Probl. Annu. Rev. Mar. Sci.* **2009**, *1*, 169–192. [CrossRef]
9. Cai, W.; Hu, X.; Huang, W.; Murrell, M.C.; Lehrter, J.C.; Lohrenz, S.E.; Gong, G.C. Acidification of subsurface coastal waters enhanced by eutrophication. *Nat. Geosci.* **2011**, *4*, 766–770. [CrossRef]
10. Falkowski, P.G. The biological and geological contingencies for the rise of oxygen on Earth. *Photosynth. Res.* **2011**, *107*, 7–10. [CrossRef]
11. Majkut, J.D.; Carter, B.R.; Frölicher, T.L.; Dufour, C.O.; Rodgers, K.B.; Sarmiento, J.L. An observing system simulation for Southern Ocean carbon dioxide uptake. *Phil. Trans. R. Soc. A* **2014**, *372*, 1–17. [CrossRef] [PubMed]
12. Williams, N.L.; Feely, R.A.; Sabine, C.L.; Dickson, A.G.; Swift, J.H.; Talley, L.D.; Russell, J.L. Quantifying anthropogenic carbon inventory changes in the Pacific sector of the Southern Ocean. *Mar. Chem.* **2015**, *174*, 147–160. [CrossRef]
13. Carter, B.R.; Feely, R.A.; Mecking, S.; Cross, J.N.; Macdonald, A.M.; Siedlecki, S.A.; Talley, L.D.; Sabine, C.L.; Millero, F.J.; Swift, J.H.; et al. Two decades of Pacific anthropogenic carbon storage and ocean acidification along Global Ocean Ship-based Hydrographic Investigations Program sections P16 and P02. *Glob. Biogeochem. Cycles* **2017**, *31*, 306–327. [CrossRef]
14. Hancock, A.M.; Davidson, A.T.; McKinlay, J.; McMinn, A.; Schulz, K.G.; van den Enden, R.L. Ocean acidification changes the structure of an Antarctic coastal protistan community. *Biogeosciences* **2018**, *15*, 2393–2410. [CrossRef]
15. Dong, S.; Sprintall, J.; Gille, S.T.; Talley, L. Southern Ocean mixed-layer depth from Argo float profiles. *J. Geophys. Res.* **2008**, *113*, C06013. [CrossRef]
16. Holte, J.; Talley, L. A new algorithm for finding mixed layer depths with applications to Argo data and subantarctic mode water formation. *J. Atmos. Ocean. Technol.* **2009**, *26*, 1920–1939. [CrossRef]
17. Vereshchagina, O.F.; Korovitskaya, E.V.; Mishukova, G.I. Methane in water columns and sediments of the north western Sea of Japan. *Deep-Sea Res. II* **2013**, *86–87*, 25–33. [CrossRef]
18. Yamamoto, S.; Alcauskas, J.B.; Crozier, T.E. Solubility of methane in distilled water and sea water. *J. Chem. Eng. Data* **1976**, *21*, 78–80. [CrossRef]
19. Wiesenburg, D.A.; Guinasso, N.L. Equilibrium solubility of methane, carbon monooxide, and hydrogen in water and sea water. *J. Chem. Eng. Data* **1979**, *24*, 356–360. [CrossRef]
20. Teterin, D.E. Geodynamic evolution of the Drake Strait in the Post-Miocene time, western Antarctica, South Ocean]. *Earth Physics.* **2011**, *8*, 3–19. (In Russian)
21. Antoniades, D.; Giralt, S.; Geyer, A.; Álvarez-Valero, A.M.; Pla-Rabès, S.; Granados, I.; Liu, E.J.; Toro, M.; Smellie, J.L.; Oliva, M. The timing and widespread effects of the largest Holocene volcanic eruption in Antarctica. *Sci. Rep.* **2018**, *8*, 1–11.
22. Morozov, E.G. Currents in the Bransfield Strait. *Dokl. Earth Sci.* **2007**, *415*, 823–825. (In Russian) [CrossRef]
23. Zhou, M.; Niiler, P.P.; Hu, J.-H. Surface currents in the Bransfield and Gerlache Straits, Anarctica. *Deep-Sea Res. I.* **2002**, *49*, 267–280. [CrossRef]
24. Zhou, M.; Dorland, R. Aggregation and vertical migration of Euphausia superba. *Deep Sea Res. Part II Top. Stud. Oceanogr.* **2004**, *51*, 2119–2137. [CrossRef]
25. Gordon, A.L.; Mensch, M.; Zhaoqian, D.; Smethie, W.M., Jr.; de Bettencourt, J. Deep and bottom water of the Bransfield Strait eastern and central basins. *J. Geophys. Res. Athmospheres* **2000**, *105*, 11337–11346. [CrossRef]
26. Samyshev, E.Z. *Antarctic Krill and Plankton Community Structure in Its Range*; Nauka Publication: Moscow, Russia, 1991; p. 168. (In Russian)
27. USGS. Available online: https://earthquake.usgs.gov/ (accessed on 11 July 2022).
28. Huntley, M.; Karl, D.M.; Niiler, P.; Holm-Hansen, O. Research on Antarctic Coastal Ecosystem Rates (RACER): An interdisciplinary field experiment. *Deep-Sea Res.* **1991**, *38*, 911–941. [CrossRef]
29. Wefer, G.; Fischer, G.; Utterer, D.F.; Gersonde, R. Seasonal particle flux in the Bransfield Strait, Antarctica. *Deep-Sea Res.* **1988**, *35*, 891–898. [CrossRef]
30. Sangrà, P.; Gordo, C.; Hernández-Arencibia, M.; Marrero-Díaz, A.; Rodríguez-Santana, A.; Stegner, A.; Martinez-Marrero, A.; Pelegrí, J.; Pichon, T. The Bransfield Current System. *Deep Sea Res. Part I Oceanogr. Res. Pap.* **2011**, *58*, 390–402. [CrossRef]
31. Garcıa, M.; Castro, C.; Rıos, A.; Doval, M.D.; Rosón, G.; Gomis, D.; López, O. Water Masses and Distribution of Physico-chemical Properties in the Western Bransfield Strait and Gerlache Strait during Austral Summer 1995/96. *Deep Sea Res. Part II Top. Stud. Oceanogr.* **2002**, *49*, 585–602. [CrossRef]
32. Krechika, V.A.; Freya, D.I.; Morozova, E.G. Features of circulation in the central part of the central part of the Bransflield Strait in January 2020. *Dokl. Earth Sci.* **2021**, *496*, 101–105. (In Russian)
33. Polonik, N.S.; Ponomareva, A.L.; Shakirov, R.B.; Obzhirov, A.I. Methane Distribution in Antarctic Sound (Southern Ocean). *Oceanology* **2021**, *61*, 469–471. [CrossRef]

34. Ponomareva, A.L.; Polonik, N.S.; Kim, A.V.; Shakirov, R.B. Detection of Thermophilic Methanotrophic Microbial Communities in the Water Column of the Bransfield Strait (Antarctica). In *Antarctic Peninsula Region of the Southern Ocean. Advances in Polar Ecology*; Morozov, E.G., Flint, M.V., Spiridonov, V.A., Eds.; Springer: Cham, Switzerland, 2021; Volume 6, pp. 209–215. [CrossRef]
35. Knox, G.A. *Biology of the Southern Ocean*; Taylor & Francis Group, LLC, CRC Press: Boca Raton, FL, USA, 2007; pp. 147–158.
36. Lamontagne, R.A.; Swinnert, J.W.; Linnenbo, V.J.; Smith, W.D. Methane concentrations in various marine environments. *J. Geophys. Res.* **1973**, *78*, 5317–5324. [CrossRef]
37. Tilbrook, B.D.; Karl, D.M. Methane sources, distributions and sinks from California coastal waters to the oligotrophic North Pacific gyre. *Mar. Chem.* **1995**, *49*, 51–64. [CrossRef]
38. Bates, T.S.; Kelly, K.C.; Johnson, J.E.; Gammon, R.H. A reevaluation of the open ocean source of methane to the atmosphere. *J. Geophys. Res. Atmos.* **1996**, *101*, 6953–6961. [CrossRef]
39. Heeschen, K.U.; Keir, R.S.; Rehder, G.; Klatt, O.; Suess, E. Methane dynamics in the Weddell Sea determined via stable isotope ratios and CFC-11. *Glob. Biogeochem.Cycles* **2004**, *18*, GB2012. [CrossRef]
40. Yoshida, O.; Inoue, H.Y.; Watanabe, S.; Suzuki, K.; Noriki, S. Dissolved methane distribution in the South Pacificand the Southern Ocean in austral summer. *J. Geophys. Res. Ocean.* **2011**, *116*, C07008. [CrossRef]
41. Global CH4 Levels. Available online: http://www.methanelevels.org/ (accessed on 11 July 2022).
42. Yokota, T.; Yoshida, Y.; Eguchi, N.; Ota, Y.; Tanaka, T.; Watanabe, H.; Maksyutov, S. Global Concentrations of CO_2 and CH_4 Retrieved from GOSAT: First Preliminary Results. *Sola* **2009**, *5*, 160–163. [CrossRef]
43. IPCC. *Climate Change 2013: The Physical Science Basis in Contribution of Working Group I to the Fifth Assessment Report of the Intergovernmental Panel on Climate Change*; Cambridge University Press: Cambridge, UK, 2013.
44. Kirschke, S.; Bousquet, P.; Ciais, P.; Saunois, M.; Canadell, J.G.; Dlugokencky, E.J.; Bergamaschi, P.; Bergmann, D.; Blake, D.R.; Bruhwiler, L.; et al. Three decades of global methane sources and sinks. *Nat. Geosci.* **2013**, *6*, 813–823. [CrossRef]
45. Gille, S.T.; Carranza, M.M.; Cambra, R.; Morrow, R. Wind-induced upwelling in the Kerguelen Plateau region. *Biogeosciences* **2014**, *11*, 6389–6400. [CrossRef]
46. Bui, O.T.N.; Kameyama, S.; Yoshikawa-Inoue, H.; Ishii, M.; Sasano, D.; Uchida, H.; Tsunogai, U. Estimates of methane emissions from the Southern Ocean from quasi-continuous underway measurements of the partial pressure of methane in surface seawater during the 2012/13 austral summer. *Tellus B Chem. Phys. Meteorol.* **2018**, *70*, 1–15. [CrossRef]
47. Marshall, J.; Speer, K. Closure of the meridionaloverturning circulation through Southern Ocean upwelling. *Nature Geosci.* **2012**, *5*, 171–180. [CrossRef]
48. Codispoti, L.A.; Brandes, J.A.; Christensen, J.P.; Devol, A.H.; Naqvi, S.W.A.; Paerl, H.W.; Yoshinari, T. The oceanic fixed nitrogen and nitrous oxide budgets: Moving targets as we enter the anthropocene? *Sci. Mar.* **2001**, *65*, 85–105. [CrossRef]
49. Polonik, N.S.; Ponomareva, A.L.; Shakirov, R.B. Deep water methane anomaly in Bransfield Straw (Antarctic). *Bull. Irkutsk. State Univ. Ser. Earth Sci.* **2020**, *32*, 61–76. [CrossRef]
50. Shakirov, R.B.; Obzhirov, A.I.; Salomatin, A.S.; Makarov, M.M. New data on lineament control of modern centers of methane degassing in east Asian seas. *Dokl. Earth Sci.* **2017**, *477*, 1287–1290. [CrossRef]

water

MDPI

Article

Cryptophyte and Photosynthetic Picoeukaryote Abundances in the Bransfield Strait during Austral Summer

Vladimir Mukhanov [1,*], Evgeny Sakhon [1], Alexander Polukhin [2], Vladimir Artemiev [2], Eugene Morozov [2] and An-Yi Tsai [3,4]

1 A. O. Kovalevsky Institute of Biology of the Southern Seas, Russian Academy of Sciences, 299011 Sevastopol, Russia; sachon@mail.ru
2 Shirshov Institute of Oceanology, Russian Academy of Sciences, 117997 Moscow, Russia; aleanapol@gmail.com (A.P.); artemiev195@yandex.ru (V.A.); egmorozov@mail.ru (E.M.)
3 Institute of Marine Environment and Ecology, National Taiwan Ocean University, Keelung 202-24, Taiwan; anyitsai@mail.ntou.edu.tw
4 Center of Excellence for the Oceans, National Taiwan Ocean University, Keelung 202-24, Taiwan
* Correspondence: v.s.mukhanov@ibss-ras.ru; Tel.: +7-978-7239140

Abstract: A remarkable shift in the species composition and size distribution of the phytoplankton community have been observed in coastal waters along the Antarctic Peninsula over the last three decades. Smaller photoautotrophs such as cryptophytes are becoming more abundant and important for the regional ecosystems. In this study, flow cytometry was used to quantify the smallest phytoplankton in the central Bransfield Strait and explore their distribution across the strait in relation to physical and chemical properties of the two major water masses: the warmer and less saline Transitional Zonal Water with Bellingshausen Sea influence (TBW), and the cold and salty Transitional Zonal Water with Weddell Sea influence (TWW). Pico- and nano-phytoplankton clusters were distinguished and enumerated in the cytograms: photosynthetic picoeukaryotes, cryptophytes (about 9 μm in size), and smaller (3 μm) nanophytoplankton. It was shown that nanophytoplankton developed higher abundances and biomasses in the warmer and less saline TBW. This biotope was characterized by a more diverse community with a pronounced dominance of Cryptophyta in terms of biomass. The results support the hypothesis that increasing melt-water input can potentially support spatial and temporal extent of cryptophytes. The replacement of large diatoms with small cryptophytes leads to a significant shift in trophic processes in favor of the consumers such as salps, which able to graze on smaller prey.

Keywords: cryptophytes; nanophytoplankton; picophytoplankton; photosynthetic picoeukaryotes; phytoplankton bloom; Bransfield Strait; Antarctic Peninsula; flow cytometry

Citation: Mukhanov, V.; Sakhon, E.; Polukhin, A.; Artemiev, V.; Morozov, E.; Tsai, A.-Y. Cryptophyte and Photosynthetic Picoeukaryote Abundances in the Bransfield Strait during Austral Summer. *Water* **2022**, *14*, 185. https://doi.org/10.3390/w14020185

Academic Editors: Michele Mistri, Maria Moustaka-Gouni and Jun Yang

Received: 16 November 2021
Accepted: 29 December 2021
Published: 10 January 2022

1. Introduction

Diatoms, haptophytes (predominantly *Phaeocystis antarctica*), and cryptophytes represent the major taxonomic groups of phytoplankton in the Southern Ocean [1–5]. Summer phytoplankton blooms in coastal waters of the Antarctic Peninsula are usually associated with a shallow upper mixed layer (UML) that keeps phytoplankton under favorable light conditions and better supply of dissolved iron (e.g., [6]). As a rule, diatoms and/or *P. antarctica* make the largest contribution to the blooms, but the role of cryptophytes has been reported to substantially increase, especially in water areas of melting glaciers [7–9]. The mechanisms behind Cryptophyta bloom formation are still poorly understood. Lower water temperatures in the surface layer, narrowing of the UML and a decrease in water salinity appear to favor the development of the phenomenon [9,10].

Publicly available data provide evidence of the transformation of the Antarctic phytoplankton community as a probable consequence of global warming [10]. The transformation covers not only taxonomic composition but also the size structure of the community. The replacement of large diatoms with small cryptophytes leads to a significant shift in trophic processes in favor of the consumers able to graze on smaller prey. Since Antarctic krill is not one of them [11], such a trend gives rise to a well-founded concern. A decrease in the food supply of krill and a substitution of the latter by other zooplankton (such as salps) imply a reduction in the carbon pool available for higher trophic levels by almost 70% [12].

It was shown in a number of experimental studies that Antarctic krill feed on small prey (about 10 μm in size) with low efficiency [13,14], while the optimal size of its food objects is about 50 μm due to the features of the filtration apparatus of euphausiids [13]. This well explains why krill exhibit positive selectivity for diatoms and avoid cryptophytes when feeding on complex prey mixtures [11].

On the contrary, salps, a food competitor of Antarctic krill, are satisfied with a wide range of taxonomic and size composition of phytoplankton prey thus demonstrating much lower feeding selectivity [11]. Against the background of their ongoing expansion to higher latitudes [12,15], their potential ability to utilize the biomass of cryptophytes during phytoplankton bloom periods poses additional threats to Antarctic krill populations, and hence, also for higher consumers including penguins, seals, and whales which feed on krill [16]. Thus, an increasing role of Cryptophyta in the Antarctic waters implies restrictions in food supply for krill and strengthening of its competitor's position.

The phenomenon of increasing cryptophyte abundances in the Antarctic waters have been initially explained by advection [7], sedimentation of large diatoms [17], grazing [2], and tolerance of cryptophytes to lower salinity waters [8]. Recently, Mendes et al. [18] studied the prevalence of cryptophytes in the coastal waters of the western Antarctic Peninsula and hypothesized that the phenomenon can be due to their pigment protection capability and, as a result, an ability to successfully grow under high irradiance exposure in strongly stratified shallow mixing surface layers. In this context, our major objective was to characterize the nanophytoplankton assemblage (2 to 20 μm in size) by flow cytometry (FCM), identify the Cryptophyta component, and analyze their distribution in the water masses of the Bransfield Strait, differing significantly from each other and the rest of the western Antarctic Peninsula shelf by their physical and chemical properties. Special attention was paid to nanophytoplankton of the transitional water mass with Bellingshausen Sea influence, which is warmer and fresher due to melting ice. We tested the hypothesis that an increase in melt-water input can potentially promote the spatial and temporal extent of cryptophytes.

Additionally, we used FCM to enumerate the smallest known component of the phytoplankton community, picophytoplankton (0.2 to 2.0 μm size fraction), which are recognized as an important player in the Southern Ocean ecosystem. In this part of the work, we focused on revealing spatial patterns of picophytoplankton abundance in relation to the water masses and frontal zones in the Bransfield Strait as the frontal features are among the factors other than temperature that control the abundance and distribution of phytoplankton.

2. Materials and Methods

2.1. Sampling Sites and Methods

The oceanographic and microbiological data were collected on 21 January 2020 over a transect across the central Bransfield Strait (BS) (7 stations, Figure 1) during 79th cruise of the R/V "Akademik Mstislav Keldysh". The length of the transect from Greenwich Island (South Shetland Islands) to the shelf of the Antarctic Peninsula was 93 km. Water samples were taken at 5 to 7 depths from the surface to 190 m, depending on the hydrological structure and chlorophyll *a* (Chl-*a*) fluorescence distribution.

The water area sufficiently differs from the rest of the western Antarctic Peninsula shelf by the inflow of cold and saline Weddell Sea shelf water [18,19]. According to Tokarczyk's classification [20], the water masses in the BS are the Transitional Zonal Water with Bellingshausen Sea influence (TBW; water temperature $T > 1$ °C, salinity $S < 34.1$ psu), and the Transitional Zonal Water with Weddell Sea influence (TWW; $T < 1$ °C, $S > 34.1$ psu).

Figure 2 represents water temperature profiles at the stations (6587–6595) along the transect in the BS and the front between TBW and TWW. These waters are separated from each other by a shallow hydrographic front which has a clear surface thermal signature. The 1 °C isotherm serves as a good boundary between TBW and TWW in summer. TBW occupies only a 50-m thick upper layer along the South Shetland Islands whereas the rest of the Bransfield basin is occupied by TWW [21].

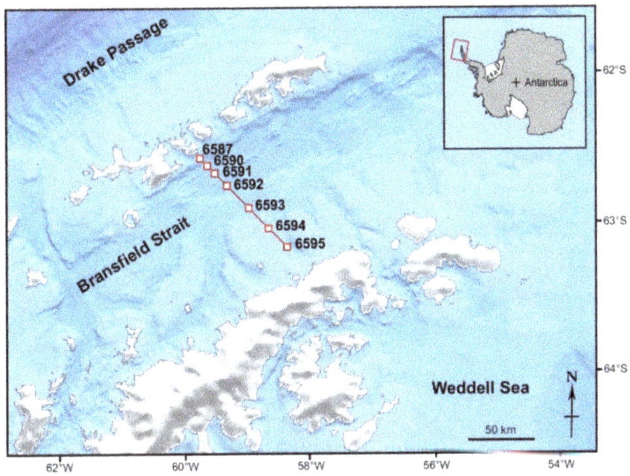

Figure 1. Location of oceanographic stations in the Bransfield Strait (79th cruise of R/V "Akademik Mstislav Keldysh", January 2020) where water samples were collected at 5 to 7 depths in austral mid-summer.

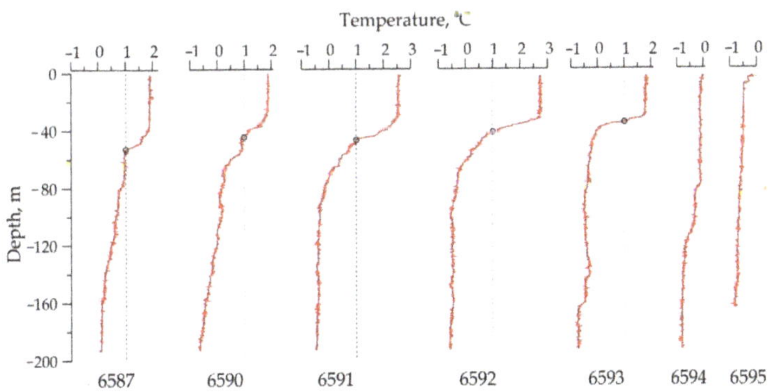

Figure 2. Water temperature profiles at the stations (6587–6595) along the transect in the Bransfield Strait. The depths of 1 °C isotherm (marked with empty circles) corresponded to the front between TBW and TWW.

2.2. Physical and Chemical Oceanographic Data

Temperature, conductivity, and pressure data were collected using a SBE911 CTD attached to a Carrousel system with 24 5-L Niskin bottles for water sampling. Current velocity profiles were measured by a Workhorse Sentinel ADCP. Profile measurements of Chl-*a* concentration and photosynthetically active radiation (PAR) were conducted using a PUM-200 transparency meter equipped with a Minitracka-II fluorimeter (Chelsey Instruments Ltd., Surrey, UK), and a LI-COR radiometer equipped with two LI-192 sensors.

Dissolved oxygen and nutrients were immediately measured onboard in all of the samples using a modified Winkler method [22] and standard colorimetric methods [23–26].

2.3. Flow Cytometry

Water samples were fixed (2% formaldehyde fin. conc.) and frozen on board in liquid nitrogen for their storage and transportation to a stationary laboratory before flow-cytometric analysis. No fractionation was employed with the exception of screening the sample through the 70-µm nylon mesh to avoid clogging the fluidic system of the instrument by larger objects (detritus, microalgae, zooplankton, etc.). Taxonomic composition and abundance of the micro-phytoplankton were not studied.

Cytometric gating and sizing of pico- and nanophytoplankton were performed using a Cytomics^{TM} FC 500 flow cytometry system (Beckman Coulter Inc., Brea, CA, USA) and Flowing Software v. 2.5.0 (Perttu Terho, Turku Centre for Biotechnology, University of Turku, Finland, https://bioscience.fi/services/cell-imaging/flowing-software/, released 4 November 2013). The standard forward light scatter (FSC), orange fluorescence (FL2, 575 nm), and red fluorescence (FL4, 675 nm) channels were used for the acquisition and analysis.

Two nanophytoplankton clusters were distinguished and enumerated, namely cryptophytes (CP) and other nanophytoplankton (NP). Cryptophytes contain a distinctive set of pigments, the orange fluorescing phycobilins (although there are also colorless members, [27]. These cryptomonads were identified by their bright orange fluorescence while NP did not produce it (Figure 3). The light microscopy observation of cryptomonad cells confirmed the flow cytometry data. The FL4 and FL2 values measured in individual cells served a measure of their intracellular Chl-*a* and phycoerythrin (PE) contents, respectively [28]. Picophytoplankton were dominated by photosynthetic picoeukaryotes (PPE) while abundances of picocyanobacteria were extremely low or undetectable. For this reason, we identified and enumerated PPE only, according to Marie et al. [29] (Figure 3).

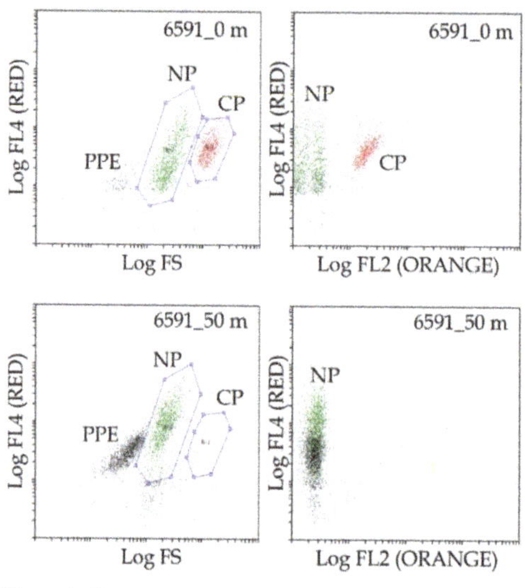

Figure 3. Gating cryptophytes (CP), other nanophytoplankton (NP) and photosynthetic picoeukaryotes (PPE) in the space of forward light scattering (FS is cell size), red autofluorescence (FL4 is Chl-*a*) and orange autofluorescence (FL2 is phycoerythrin) in two samples (station 6591 at 0 and 50 m).

Cell size measurements were calibrated in terms of equivalent spherical diameter (ESD, µm), using a set of standard polystyrene beads (Polysciences, Inc., Warrington, PA, USA) in the size range between 0.5 and 10 µm; calibration curves were obtained for each flow cytometry protocol by plotting the ESD as a function of the FSC signal (coefficient of determination $r^2 > 0.8$ in all cases). Mean ESD and cell volume were calculated for every cell population from the calibration curves [30]. To allow for a better comparison with earlier published data, nanophytoplankton abundances and biovolumes were converted to cellular carbon (C). Carbon conversion factors were shown to vary depending on taxonomy, cell size, and growth-regulating factors such as light and nutrient availability [31]. In this study, carbon content of pico- and nanophytoplankton was estimated from the mean ESD using the conversion factors of 237 fg C μm^{-3} [32] and 196.5 fg C μm^{-3} [33], respectively.

2.4. Statistics

Multivariate statistical analysis (cluster analysis and non-metric multi-dimensional scaling, nMDS) of microbiological and hydrological-hydrochemical data collected in the upper 190-m layer was carried out using the PAST3 software [34]. Mapping the spatial distribution of the variables and graph plotting were performed using Surfer v. 10 and Grapher v. 8 software (Golden Software, Inc., Golden, CO, USA), respectively.

In total, 24 biological, hydrological and hydrochemical variables were measured in the study site and investigated by the methods of multivariate statistical analysis (Table 1). The 12 microbiological variables included in this list characterized both the structure (*CPN, CPN%, CPB%, NPN, NPB, CP-ESD, NP-ESD, PPEN, PPEB, PPE-ESD*) and some functional characteristics of the communities namely, intracellular content of photosynthetic pigments in the photoautotrophic cells (*CP-CHL, CP-PE, NP-CHL, PPE-CHL*).

Table 1. Symbols of variables used in this study with their definitions, units, and descriptive statistics (n = 39). R. u. is relative units, CP is cryptophyte, NP is nanoplankton, PPE is photosynthetic picoeukaryotes, ESD is equivalent spherical diameter, SD is standard deviation.

Symbol	Definition, Units	Min	Max	Mean ± SD
T	Water temperature, °C	−0.79	2.77	0.59 ± 1.19
S	Water salinity, psu	34.1	34.4	34.3 ± 0.1
PO4	Phosphates, μM	1.56	4.37	3.09 ± 0.58
Si	Silicates, μM	67.9	112.3	81.3 ± 10.1
NO3	Nitrates, μM	22.6	35.4	29.4 ± 3.7
NO2	Nitrites, μM	0.11	0.23	0.17 ± 0.04
NH4	Ammonium, μM	1.32	3.81	2.13 ± 0.58
DO	Dissolved oxygen, mL L^{-1}	6.44	7.98	7.44 ± 0.36
FLUOR	Chl-*a* fluorescence, r. u.	0.10	0.49	0.24 ± 0.12
CPN	Cryptophyte abundance, 10^6 cells L^{-1}	0.00	1.55	0.35 ± 0.46
CP-ESD	Mean CP ESD, μm	6.70	11.59	9.42 ± 1.77
CPB	CP biomass, μg C L^{-1}	0	179	33 ± 45
CP-CHL	Intracellular Chl-*a* content in CP, r. u. cell $^{-1}$	0	163	69 ± 39
CP-PE	Intracellular phycoerythrin content in CP, r. u. cell $^{-1}$	0	206	31 ± 39
NPN	NP abundance, 10^6 cells L^{-1}	0.19	3.95	1.71 ± 1.07
NP-ESD	Mean NP ESD, μm	2.14	3.67	2.88 ± 0.38
NPB	NP biomass, μg C L^{-1}	0.4	15.7	4.3 ± 3.2
NP-CHL	Intracellular Chl-*a* content in NP, r. u. cell $^{-1}$	8	220	74 ± 52
TNPB	Total NP+CP biomass, μg C L^{-1}	0.4	194.2	37.3 ± 47.7
CPN%	Proportion of CP in the total abundance, %	0	49	14 ± 12
CPB%	Proportion of CP in the total biomass, %	0	96	74 ± 21
PPEN	PPE abundance, 10^6 cells L^{-1}	0.06	4.18	0.87 ± 1.03
PPE-ESD	Mean PPE ESD, μm	1.72	2.09	1.90 ± 0.11
PPEB	PPE biomass, μg C L^{-1}	0.04	3.50	0.72 ± 0.83

3. Results

3.1. Description of Oceanographic Features

Vertical distribution of water temperature and salinity across the BS (Figure 4) corresponded to the typical BS circulation system, with the Southern BS Current carrying waters advected from the Weddell Sea and meeting the relatively warm and less saline waters from the western Antarctic Peninsula through the Gerlache Strait Current at the western end of the BS [35,36]. Together they feed into the eastward flowing Bransfield Current, which is a baroclinic jet directed from the southwest to the northeast and transporting TBW along the southern slope of the Southern Shetland Islands [35–39]. The highest Bransfield Current velocity (0.45 m s^{-1}) was observed at station 6590 in the surface layer. The inflow of cold and saline Weddell Sea Shelf Water was recorded at station 6594 over the entire range of water depths (0 to 200 m) with the maximum velocity of about 0.15 m s^{-1}.

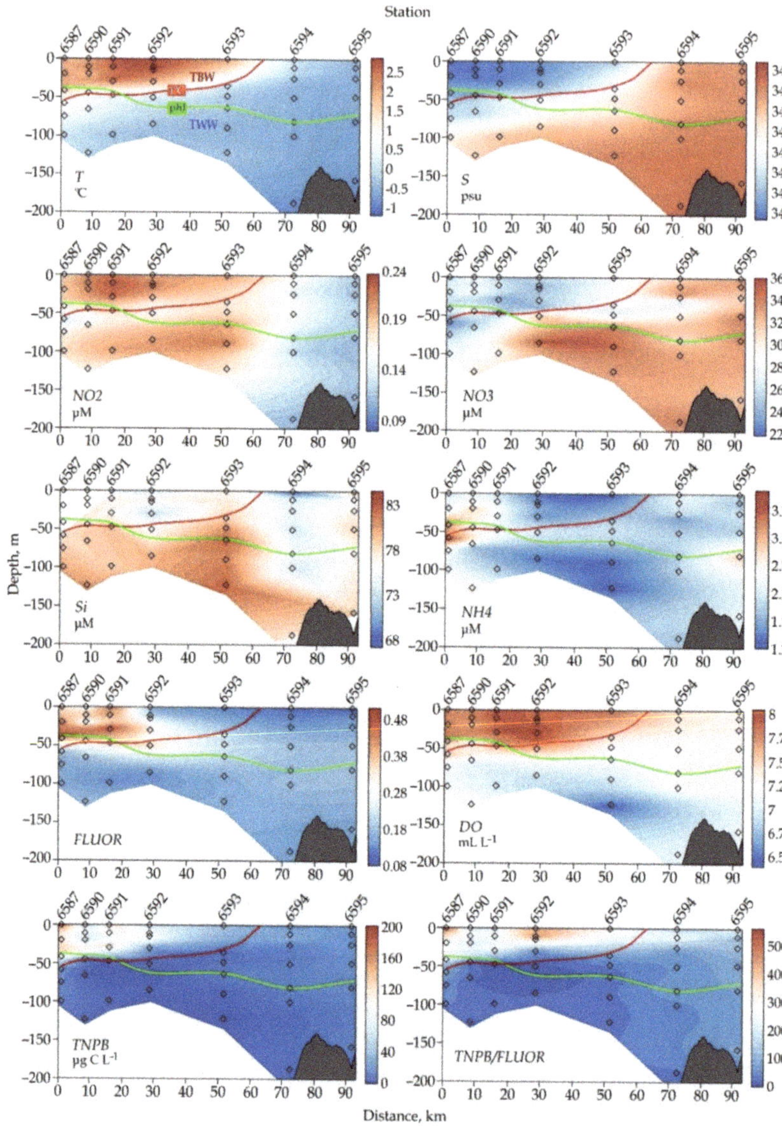

Figure 4. Vertical distribution of water temperature (*T*), salinity (*S*), nutrients (*NO2*, *NO3*, *NH4*, *Si*), dissolved oxygen (*DO*), Chl-*a* fluorescence (*FLUOR*), total nanophytoplankton biomass (*TNPB*), and *TNPB/FLUOR* ratio along the transect in the Bransfield Strait during austral mid-summer, January 2020. Green and red lines represent the bottom of the euphotic layer (phl) and 1 °C isotherm (as a boundary between TBW and TWW; explanations are in the text), respectively. Description of the variables is given in Table 1.

At the surface, warmer and less saline TBW occupied most of the BS (from station 6587 to 6593, Figure 4). Between stations 6593 and 6594, the 2 water masses were separated from each other in the surface layer by the Bransfield Front [37,38]. Being a conventional boundary between the water masses in the BS [20], the 1 °C isotherm did not deepen below

50 m (station 6587) and also corresponded to the maximum temperature gradient (marked in Figure 2 with empty circles).

The offshore waters of the South Shetland Islands are subject to local heating and meltwater inflow [40], which also affects the oxygen distribution in the northern part of the BS. During the study period, near-surface oxygen concentration reached 8.0 mL L^{-1} (Table 1, Figure 4), with oxygen saturation ranging between 100% and 104%. High ammonium nitrogen concentrations (1.0 to 3.8 μM) (Figure 4) were apparently associated with both organic matter oxidation as a result of active phytoplankton bloom and runoff of meltwater from the shores of the South Shetland Islands [19]. Concentrations of nitrite nitrogen (about 0.2 μM) also indicated an active decomposition of organic matter against the background of high primary production in TBW [19] (Figure 4). Nitrate nitrogen concentration varied between 22 and 35 μM, with the highest values observed below 50-m depth in the central part of the transect. In the upper 200-m layer, dissolved silicate concentration was the highest (up to 112 μM) in the northern part of the strait (Figure 4).

The euphotic layer was within the upper 40 m in the northern part of the transect and extended to about 80 m in the TWW water mass at station 6594 (Figure 4). The highest Chl-*a* fluorescence values were measured at the northern stations (6587, 6590, 6591) at the depths between 0 and 50 m within TBW (Figure 4). The maximum of Chl-*a* (1.28 μg L^{-1}) and high *DO* values (up to 8 mL L^{-1}) were observed at station 6591, providing evidence of phytoplankton bloom in the water area [19,41]. At the same time, meltwater conditions of TBW may also affect bio-optical properties of the surface waters in the BS [42].

3.2. Picophytoplankton

Abundance and biomass of PPE averaged $0.87 \pm 1.03 \times 10^6$ cells L^{-1} and 0.72 ± 0.83 μg C L^{-1}, respectively (Table 1). They preferred deeper, colder layer below the <1 °C isotherm, thus, demonstrating their association with TWW (Figure 5). The highest PPE abundance and biomass were observed at 50-m depth at station 6593. In the patch, these values reached 4.18×10^6 cells L^{-1} and 3.5 μg C L^{-1}, respectively. In the surface layer, PPE were relatively abundant at the front between TWW and TBW.

3.3. Nanophytoplankton

CP were significantly less abundant ($0.35 \pm 0.46 \times 10^6$ cells L^{-1}; ± SD is presented here and further) than NP ($1.71 \pm 1.07 \times 10^6$ cells L^{-1}), however, their cells were much larger (about 9.5 μm versus 3 μm in NP). Hence, their average carbon biomass (33 ± 45 μg C L^{-1}) significantly exceeded those of NP (4.3 ± 3.2 μg C L^{-1}) (Table 1). The average cryptophyte ESD measured by flow cytometry (9.4 ± 1.8 μm) slightly exceeded the estimates (8 ± 2 μm) obtained earlier for Antarctic cryptophytes, using light microscopy [43]. Contribution of Cryptophyta to the total nanophytoplankton abundance and biomass reached 49% and 96%, respectively, with the average values of 14 ± 12% and 74 ± 21% (Table 1). The Cryptophyta biomass maximum (about 180 μg C L^{-1}) was observed in TBW in the northern part of the transect (station 6587).

Patches of the highest CP abundance and biomass were observed in the euphotic layer around the jet of the Bransfield Current (station 6590, surface layer) while in TWW (i.e., below the 1 °C isotherm), cryptophytes were scarce or undetectable, especially in the deeper layers (Figure 5). The NP demonstrated similar distribution pattern with the highest abundances at 35–50 m depth at station 6591 (Figure 5). The maximum contribution of CP to the total nanophytoplankton biomass was recorded farther south in TBW (stations 6591 and 6592) at shallow depths (about 10 m) where NP were not abundant (Figure 5).

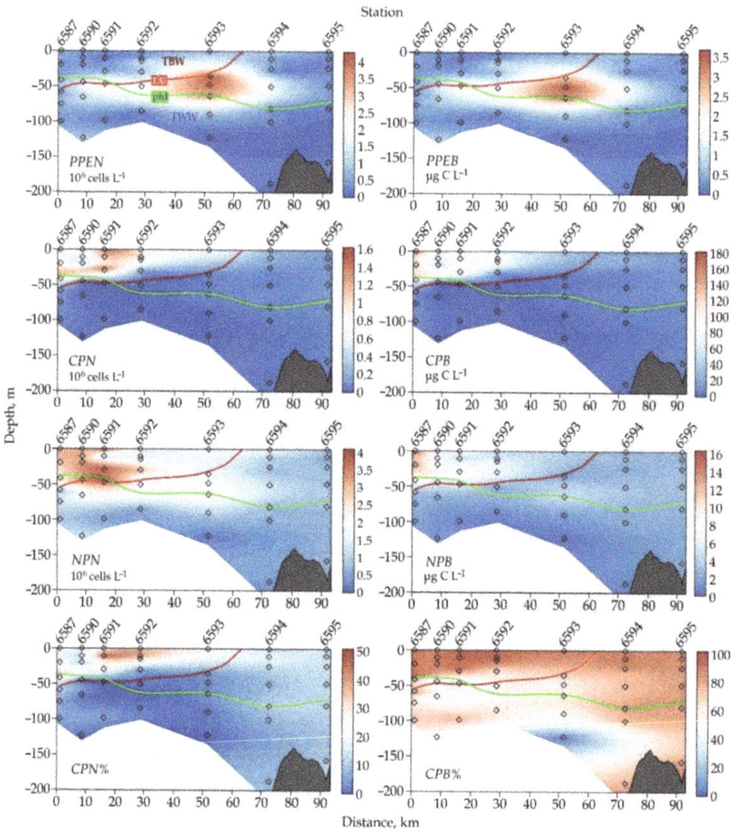

Figure 5. Vertical distribution of photosynthetic picoeukaryote (*PPEN, PPEB*), cryptophytes (*CPN, CPB, CPN%, CPB%*), and other nanophytoplankton (*NPN, NPB*) along the transect in the Bransfield Strait during austral mid-summer, January 2020. Green and red lines represent the bottom of the euphotic layer (phl) and 1 °C isotherm (as a boundary between TBW and TWW; explanations are in the text), respectively. Description of the variables is in Table 1.

4. Discussion

4.1. Picoeukaryotes

Undetectable levels of prokaryotic picophytoplankton in the study site were not surprising since these microorganisms tend to be scarce or absent in polar region. Their abundance is often inversely related to picoeukaryotes, which are favored by more physically active mixed layers (e.g., [44]). According to the data synthesis by Buitenhuis et al. [45], picoeukaryotes dominate by >75% poleward of 40°. Thus, in the Antarctic waters, picophytoplankton are mostly represented by picoeukaryotes. Moreover, they may occasionally contribute up to 100% of the autotrophic biomass in subsurface waters of the BS [46].

The picoeukaryote concentrations reported in this study were relatively low compared to the data collected earlier in the summer picophytoplankton community [45–47]. In the BS waters, PPE abundance varied from 6 to 14 × 10^6 cells L^{-1} during summer 2000 [46] that was significantly higher than our estimates. According to Lin et al. [48], the highest PPE summer abundance appeared offshore in the Southwest Atlantic Ocean and changed from 0.2 to 18.2 × 10^6 cells L^{-1}, with an average of 3.2 × 10^6 cells L^{-1}. PPE biomass estimates reported in the above study (0.03 to 38.2 μg C L^{-1}) also substantially exceeded our results, with neither water temperature nor salinity having significant correlation with the picoeukaryote biomass.

Picoeukaryotes demonstrated statistically significant ($p < 0.05$) correlation with water temperature (0,35), salinity (–0.39), phosphates (0.41), and total Chl-*a* (0.45) (Table 2) but this did not provide any explanation of the low PPE abundance and their patchy distribution in the BS waters. Contrary to expectations from the correlation analysis, the PPE 'patch' was situated in the colder and more saline TWW just below the front between the water masses. This may support a suggestion that the nanophytoplankton over-competed the PPE in TBW, forcing them to stay outside the Cryptophyta bloom. Other factors such as frontal dynamics and light penetration also may be important. In particular, despite picophytoplankton ability to utilize low irradiance [49], they may be light-limited in the Antarctic turbid coastal waters or deeply mixed oceanic waters [50,51].

Table 2. Spearman's correlation between the studied variables. Statistically significant ($p < 0.05$) indicators are in bold. The description of the variables is in Table 1.

Variable	*T*	*S*	*PO4*	*Si*	*NO3*	*NO2*	*NH4*	*DO*	*FLUOR*
CPN	**0.59**	**−0.53**	−0.22	−0.22	**−0.45**	0.29	0.20	**0.78**	**0.37**
CP-ESD	−0.24	0.02	−0.14	0.28	0.29	−0.18	0.27	−0.15	−0.11
CPB	**0.54**	**−0.52**	−0.23	−0.16	**−0.40**	0.25	0.21	**0.74**	**0.36**
CP-CHL	−0.12	−0.06	**−0.32**	0.27	0.09	**−0.41**	**0.62**	0.04	−0.04
CP-PE	0.10	−0.06	−0.01	−0.13	0.03	0.00	0.28	0.11	−0.03
NPN	**0.74**	**−0.81**	0.32	0.24	**−0.59**	**0.56**	0.32	**0.60**	**0.73**
NP-ESD	−0.02	0.01	−0.32	−0.17	−0.10	−0.26	0.08	0.31	0.06
NPB	**0.68**	**−0.75**	0.12	0.10	**−0.63**	**0.42**	0.25	**0.73**	**0.73**
NP-CHL	0.01	−0.14	0.04	**0.44**	−0.16	−0.12	**0.31**	−0.15	0.12
TNPB	**0.59**	**−0.57**	−0.19	−0.14	**−0.45**	0.27	0.22	**0.77**	**0.42**
CPN%	**0.43**	**−0.33**	−0.31	**−0.32**	−0.30	0.15	0.07	**0.65**	0.15
CPB%	**0.47**	**−0.41**	−0.30	−0.25	−0.30	0.25	0.15	**0.65**	0.19
PPEN	**0.35**	**−0.39**	**0.41**	0.13	−0.30	0.24	0.17	0.21	**0.45**
PPE-ESD	**−0.85**	**0.78**	−0.28	−0.04	**0.62**	**−0.68**	−0.04	**−0.62**	**−0.47**
PPEB	0.06	−0.11	**0.34**	0.09	−0.09	−0.03	0.17	0.02	0.26

4.2. *Cryptophyta*

The distribution patterns of Chl-*a* and the total biomass of nanophytoplankton (*TNPB*) estimated by flow cytometry were similar, with the highest values at the same stations and depths (Figure 4). This provided evidence that nanophytoplankton was a major, if not dominant, component of the phytoplankton in the study site. To verify this suggestion, we estimated a potential contribution of CP to the total Chl-*a*, using different conversion factors: 0.2 pg Chl-*a* cell^{-1} [52] and 1.5 pg Chl-*a* cell^{-1} [53]. In the high Chl-*a* spot at station 6591 (1.28 μg Chl-*a* L^{-1}), the approximation of CP biomass in terms of Chl-*a* ranged from 0.31 to as high as 2.32 μg Chl-*a* L^{-1}, depending on which conversion factor is used. Additionally, we used the *TNPB*-to-*FLUOR* ratio as a measure of the nanophytoplankton contribution to the total phytoplankton biomass. The highest ratio values were observed in TBW with the maximum at station 6592 (Figure 4). Thus, the nanophytoplankton and CP could potentially dominate phytoplankton in TBW, where the phytoplankton bloom occurred.

The Cryptophyta biomass maximum (about 180 μg C L^{-1}) reported in our study is an outstanding characteristic for Antarctic waters. The importance of cryptophytes in different basins of the Southern Ocean has been reported earlier [1,2,54–57] but their maximum concentrations were much lower (between 0.3 and 6.0 $\times$ 10^6 cells L^{-1}) than our estimates. Garibotti et al. [3] have surprisingly revealed peak concentrations in the region of Anvers Island, reaching 369 μg C L^{-1}, i.e., close to our estimates. Therefore, our results widen the area of Cryptophyta blooms to the central Bransfield Strait and support the conclusion of the above authors that cryptophytes are a distinctive and annually recurrent component of Antarctic phytoplankton. Their blooms are increasingly becoming a peculiarity of the continental shelf west of the Antarctic Peninsula.

4.3. Linkage between Nanophytoplankton and Environmental Variables

All of the quantitative characteristics of the nanophytoplankton (*CPN, NPN, CPB, NPB, TNPB*) demonstrated a significant positive correlation with water temperature (*T*), dissolved oxygen (*DO*), and Chl-*a* fluorescence (*FLUOR*), while negative correlations were found with salinity (*S*) and nitrates (*NO3*) (Table 2). These environmental variables gradually changed with depth and in the transition from TBW to TWW, that explained well the above dependencies: in general, the nanophytoplankton abundances also decreased from the surface (TBW) to deeper layers (TWW) and from the northern (TBW) to the southern (TWW) stations along the transect. Thus, both CP and NP preferred warmer and fresher TBW, that supported earlier published observations [2,41,51,57–59]. The contribution of CP to the total nanophytoplankton abundance and biomass appeared the highest in TBW (Figure 5).

Chl-*a* fluorescence was significantly correlated with the total nanophytoplankton, especially with the NP concentrations (Table 2), most likely due to the higher NP abundance (*NPN*) and intracellular Chl-*a* content (*NP-CHL*) compared with the same characteristics in CP (Table 1). Intracellular Chl-*a* values in both CP and NP had a positive relationship with the ammonium concentration (*NH4*). Significant correlations have also been revealed between *NP-CHL* and *Si* (0.44), *CP-CHL* and *PO4* (−0.32), *CP-CHL* and *NO3* (−0.41) (Table 2).

It should be noted that a strong relationship between the nanophytoplankton abundance (*CPN, NPN*) and water temperature was found over the entire set of data (see the regression lines in Figure 6a,b). However, within each of the biotopes, TBW and TWW, the nature of the relationship was different. The *CPN* kept strong dependence (r^2 = 0.65) on temperature within TBW but the other links were lost (see the regressions represented with dashed lines in Figure 6a,b). Temperature did not appear to be a key factor controlling the NP abundance.

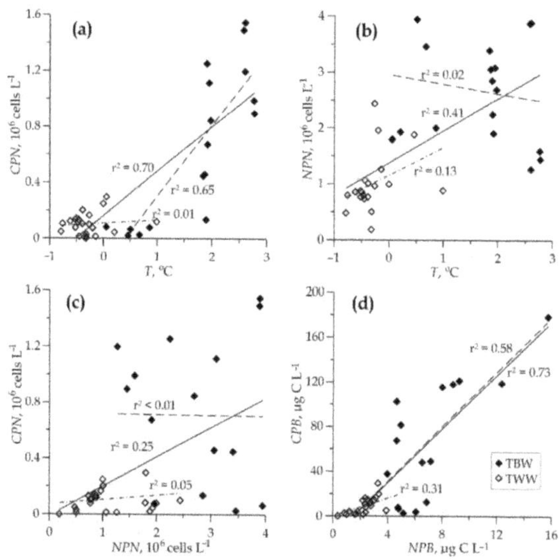

Figure 6. Relationships between: (**a**,**b**) the abundance of each group (*CPN, NPN*) and water temperature (*T*); (**c**) the cryptophyte (*CPN*) and other nanophytoplankton (*NPN*) abundances; (**d**) the cryptophyte (*CPB*) and other nanophytoplankton (*NPB*) biomasses. Samples from TBW (above 1 °C isotherm) and TWW (below 1 °C isotherm) water masses are represented with filled and empty symbols, respectively. Solid lines are regressions for the entire set of data, dashed ones are for individual biotopes. Description of the variables is in Table 1.

The CP and NP exhibited significant inter-correlation in terms of their biomass (r^2 = 0.73) but not of their abundance (r^2 = 0.25) (Figure 6c,d) that could provide evidence of their competition for the same resources. Yet again, the linkage between *CPB* and *NPB* were strong (r^2 = 0.58) in TBW, and poor (r^2 = 0.31) in TWW.

4.4. Comparing Nanophytoplankton from TBW and TWW

Cluster analysis allowed us to classify the nanophytoplankton samples, depending on the community structure (Figure 7a). A total of 4 sample groups were identified, 3 of which (the clusters 2 to 4) were within the TBW biotope (see the scheme of the cluster distribution in the water column in Figure 7b). The sample collected in the surface layer at station 6587 formed the cluster 4, owing to the outstanding microbial characteristics: the highest biomasses and average ESD of CP (178.5 µg C L^{-1}; 11.6 µm) and NP (15.7 µg C L^{-1}; 3.7 µm). The cluster 3 included 'patches' of nanophytoplankton at stations 6587 and 6591 within the TBW water mass (upper 50 m). In the cluster 2, *CPN* and *CPN%* significantly decreased. In the TWW biotope, the community was represented by the cluster 1 with low or undetectable microbial abundances (Figure 7b). Thus, the physical and chemical properties of the water masses deeply influenced the nanophytoplankton structure. The TBW biotope was characterized by much greater cell abundance and by a more diverse community with a pronounced dominance of Cryptophyta in terms of biomass.

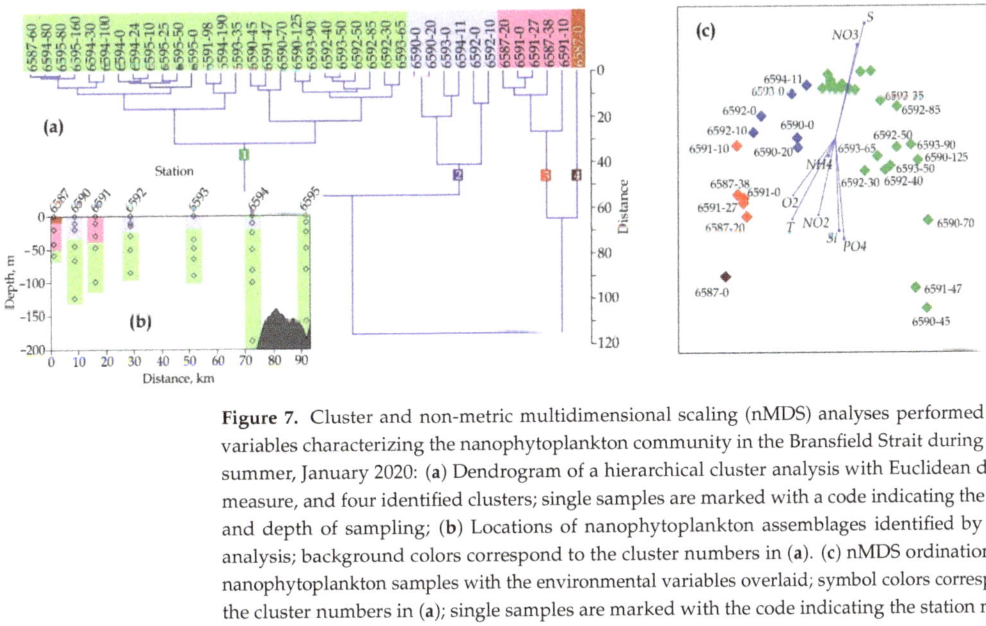

Figure 7. Cluster and non-metric multidimensional scaling (nMDS) analyses performed on the variables characterizing the nanophytoplankton community in the Bransfield Strait during austral summer, January 2020: (**a**) Dendrogram of a hierarchical cluster analysis with Euclidean distance measure, and four identified clusters; single samples are marked with a code indicating the station and depth of sampling; (**b**) Locations of nanophytoplankton assemblages identified by cluster analysis; background colors correspond to the cluster numbers in (**a**). (**c**) nMDS ordination of the nanophytoplankton samples with the environmental variables overlaid; symbol colors correspond to the cluster numbers in (**a**); single samples are marked with the code indicating the station number and the depth of sampling.

The ordination of the nanophytoplankton samples by nMDS produced similar results (Figure 7c) however, nMDS did not reveal a clear distinction between the sample clusters 1–4 revealed by the cluster analysis (shown in Figure 7c) in different colors). The clouds of dots corresponding to both the biotopes, TBW (the clusters 2–4) and TWW (the cluster 1), were distributed along the environmental vectors, which indicated a strong environmental control of nanophytoplankton. Distribution of the TBW-TWW fronts, anticyclonic eddies and their thermohaline structures, and nutrients are among the factors, which can control the community across the BS [60–62].

4.5. Cryptophyta Distribution Pattern in the Bransfield Strait

In general, the spatial distribution patterns of Chl-*a* fluorescence that we observed in the central BS were well aligned with the earlier long-term observations: the highest Chl-*a* was usually found close to the South Shetland Islands with the UML depth being negatively related to Chl-*a* [63]; at the well-stratified sites (TBW), with the euphotic layer varying from 40 to 50 m, the highest Chl-*a* was at the surface, while the weakly stratified sites (TWW) were associated with the low Chl-*a* and deeper euphotic layer [64].

However, the spatial patterns of cryptophytes in the coastal areas of the Antarctic Peninsula are more complicated. In the Eastern BS, Gonçalves-Araujo et al. [9] observed a dominance of microplanktonic diatoms and higher Chl-*a* within shallow upper mixed layers in TBW, close to the South Shetland Islands, while TWW was primarily characterized by lower Chl-*a* within well-mixed water column and dominated by nanoplanktonic flagellates (including haptophytes and cryptophytes). This contradicted our observations of coincident maxima of the Chl-*a* fluorescence and the cryptophyte concentration within the TBW surface layer with lower salinity and higher temperature (>1 °C).

According to Schofield et al. [65], peak populations of cryptophytes were observed in the coastal areas of the West Antarctic Peninsula with lower salinity (33.75 to 32.5 psu) and colder water (−1 to 1 °C), while they were absent at higher temperatures (>1 °C) and salinities (>33.75 psu). On the contrary, Mendes et al. [18] provided different data on the distribution of cryptophyte biomass in the Gerlache Strait, which was mostly dependent on the sea surface temperature, rather than salinity. The authors found that cryptophytes were strongly associated with high temperature, water column stability, and Chl-*a*, and negatively associated with nutrient concentrations, UMLD, and salinity, which was in good agreement with our observations and a few earlier reports [2,51,57,59,64].

An explanation of the phenomenon may consist in early retreat of sea ice, which, together with the increase in sea surface temperature, leads to the development of a shallow mixing layer and strong vertical water column stratification. The latter confines phytoplankton near the surface under high irradiance [66]. Mendes et al. [18] hypothesized that the dominance of cryptophytes in the coastal waters of the Antarctic Peninsula is just associated with their ability to grow under high irradiance exposure due to their pigment protection capability.

Thus, the increase in melt-water input may potentially enlarge the spatial and temporal extent of cryptophytes [8]. Our study supports this suggestion, providing evidence of extremely high cryptophyte biomass within low-salinity, high-temperature TBW. Historical data also show an emerging dominance of cryptophytes and other phytoflagellates in low-saline coastal waters [43,56,57,67–70]. As a consequence of these processes, zooplankton community structure undergoes cascade changes [16,71] leading to an increase in salp abundance and concurrent variations in penguin migration patterns [72]. If this trend persists for a long time, the food web shifts will eventually impact biogeochemical cycling in the Antarctic coastal waters.

5. Conclusions

In this study, we found higher abundances and biomasses of nanophytoplankton in the warmer and fresher transitional water mass with Bellingshausen Sea influence (TBW). This biotope was characterized by a more diverse community with a pronounced dominance of cryptophytes in terms of biomass (up to 180 μg C L^{-1}). Patches of the highest cryptophyte abundance were observed in the euphotic layer around the jet of the Bransfield Current. In the transitional water mass with Weddell Sea influence (TWW), they were scarce or undetectable, especially in the deeper layers. Our results widen the area of Cryptophyta blooms to the central Bransfield Strait and support the conclusion that these small eukaryotes are a distinctive and annually recurrent component of Antarctic phytoplankton.

In the study site, picophytoplankton were represented by picoeukaryotes and characterized by low abundances (about 0.87×10^6 cells L^{-1}) and patchy distribution. They preferred deeper, colder layer, thus, demonstrating their association with TWW and the frontal zone between the water masses.

Our results support the hypothesis that increasing melt-water input can potentially increase spatial and temporal extent of cryptophytes in the Antarctic region.

Author Contributions: Conceptualization, V.M.; methodology, V.M., E.S., A.P., V.A. and E.M.; formal analysis, V.M. and A.-Y.T.; data curation, V.M., A.P. and E.M.; writing—original draft preparation, V.M.; writing—review and editing, V.M. and A.-Y.T. All authors have read and agreed to the published version of the manuscript.

Funding: This research was conducted in the frame of the Russian state assignments No. AAAA-A19-119100290162-0 (V.M. and E.S.) and 0128-2021-0017 (E.M., A.P. and V.A.), and supported by RFBR, grant number 21-55-52001 (V.M. and E.S.), and the Ministry of Science and Technology, ROC (Taiwan), grant number NSC 109-2611-M-019-013 (A.-Y.T.).

Conflicts of Interest: The authors declare no conflict of interest.

References

1. Rodriguez, F.; Varela, M.; Zapata, M. Phytoplankton assemblages in the Gerlache and Bransfield Straits (Antarctic Peninsula) determined by light microscopy and CHEMTAX analysis of HPLC pigment data. *Deep Sea Res. Part II Top. Stud. Oceanogr.* **2002**, *49*, 723–747. [CrossRef]
2. Garibotti, I.A.; Vernet, M.; Ferrario, M.E.; Smith, R.C.; Ross, R.M.; Quetin, L.B. Phytoplankton spatial distribution patterns along the western Antarctic Peninsula (Southern Ocean). *Mar. Ecol. Prog. Ser.* **2003**, *261*, 21–39. [CrossRef]
3. Garibotti, I.A.; Vernet, M.; Ferrario, M.E. Annually recurrent phytoplanktonic assemblages during summer in the seasonal ice zone west of the Antarctic Peninsula (Southern Ocean). *Deep Sea Res. Part I Oceanogr. Res. Pap.* **2005**, *52*, 1823–1841. [CrossRef]
4. Kozlowski, W.A.; Deutschman, D.; Garibotti, I.; Trees, C.; Vernet, M. An evaluation of the application of CHEMTAX to Antarctic coastal pigment data. *Deep Sea Res. Part I Oceanogr. Res. Pap.* **2011**, *58*, 350–364. [CrossRef]
5. Mendes, C.R.B.; de Souza, M.S.; Garcia, V.M.T.; Leal, M.C.; Brotas, V.; Garcia, C.A.E. Dynamics of phytoplankton communities during late summer around the tip of the Antarctic Peninsula. *Deep Sea Res. Part I Oceanogr. Res. Pap.* **2012**, *65*, 1–14. [CrossRef]
6. Prézelin, B.B.; Hofmann, E.E.; Mengelt, C.; Klinck, J.M. The linkage between Upper Circumpolar Deep Water (UCDW) and phytoplankton assemblages on the west Antarctic Peninsula continental shelf. *J. Mar. Res.* **2000**, *58*, 165–202. [CrossRef]
7. Moline, M.A.; Prezelin, B.B. Long-term monitoring and analyses of physical factors regulating variability in coastal Antarctic phytoplankton biomass, *in situ* productivity and taxonomic composition over subseasonal, seasonal and interannual time scales. *Mar. Ecol. Prog. Ser.* **1996**, *145*, 143–160. [CrossRef]
8. Moline, M.A.; Claustre, H.; Frazer, T.K.; Schofield, O.; Vernet, M. Alteration of the food web along the Antarctic Peninsula in response to a regional warming trend. *Glob. Chang. Biol.* **2004**, *10*, 1973–1980. [CrossRef]
9. Gonçalves-Araujo, R.; de Souza, M.S.; Tavano, V.M.; Garcia, C.A.E. Influence of oceanographic features on spatial and interannual variability of phytoplankton in the Bransfield Strait, Antarctica. *J. Mar. Syst.* **2015**, *142*, 1–15. [CrossRef]
10. Mendes, C.R.B.; Tavano, V.M.; Leal, M.C.; de Souza, M.S.; Brotas, V.; Garcia, C.A.E. Shifts in the dominance between diatoms and cryptophytes during three late summers in the Bransfield Strait (Antarctic Peninsula). *Polar Biol.* **2013**, *36*, 537–547. [CrossRef]
11. Haberman, K.L.; Ross, R.M.; Quetin, L.B. Diet of the Antarctic krill (*Euphausia superba* Dana): II Selective grazing in mixed phytoplankton assemblages. *J. Exper. Mar. Biol. Ecol.* **2003**, *283*, 97–113. [CrossRef]
12. Moline, M.A.; Claustre, H.; Frazer, T.K.; Grzymski, J.; Vernet, M. Changes in phytoplankton assemblages along the Antarctic Peninsula and potential implications for the Antarctic food web. In *Antarctic Ecosystems: Models for Wider Ecological Understanding*; Davidson, W., Howard-Williams, C., Broady, P., Eds.; University of Canterbury: Christchurch, New Zealand, 2001.
13. McClatchie, S.; Boyd, C.M. Morphological study of sieve efficiencies and mandibular surfaces in the Antarctic krill, Euphausia superba. *Can. J. Fish. Aquat. Sci.* **1983**, *40*, 955–967. [CrossRef]
14. Boyd, C.M.; Heyraud, M.; Boyd, C.N. Feeding of the Antarctic krill Euphausia superba. *J. Crustacean Biol.* **1984**, *4*, 123–141. [CrossRef]
15. Montes-Hugo, M.; Doney, S.C.; Ducklow, H.W.; Fraser, W.; Martinson, D.; Stammerjohn, S.E.; Schofield, O. Recent changes in phytoplankton communities associated with rapid regional climate change along the western Antarctic Peninsula. *Science* **2009**, *323*, 1470–1473. [CrossRef] [PubMed]
16. Loeb, V.; Siegel, V.; Holm-Hansen, O.; Hewitt, R.; Fraser, W.; Trivelpiece, W.; Trivelpiece, S. Effects of sea-ice extent and krill or salp dominance on the Antarctic food web. *Nature* **1997**, *387*, 897–900. [CrossRef]
17. Castro, C.G.; Ríos, A.F.; Doval, M.D.; Pérez, F.F. Nutrient utilisation and chlorophyll distribution in the Atlantic sector of the Southern Ocean during Austral summer 1995–96. *Deep Sea Res.* **2002**, *49*, 623–641. [CrossRef]
18. Mendes, C.R.B.; Tavano, V.M.; Dotto, T.S.; Kerr, R.; De Souza, M.S.; Garcia, C.A.E.; Secchi, E.R. New insights on the dominance of cryptophytes in Antarctic coastal waters: A case study in Gerlache Strait. *Deep Sea Res. Part II Top. Stud. Oceanogr.* **2018**, *149*, 161–170. [CrossRef]

19. Polukhin, A.A.; Morozov, E.G.; Tishchenko, P.P.; Frey, D.I.; Artemiev, V.A.; Borisenko, G.V.; Vidnichuk, A.V.; Marina, E.N.; Medvedev, E.V.; Popov, O.S.; et al. Water structure in the Bransfield Strait (Antarctica) in January 2020: Hydrophysical, optical and hydrochemical features. *Oceanology* **2021**, *61*, 632–644. [CrossRef]
20. Tokarczyk, R. Classification of water masses in the Bransfield Strait and southern part of the Drake Passage using a method of statistical multidimensional analysis. *Pol. Res.* **1987**, *8*, 333–366.
21. Garcıa, M.; Castro, C.; Rıos, A. Water masses and distribution of physico-chemical properties in the Western Bransfield Strait and Gerlache Strait during Austral summer 1995/96. *Deep Sea Res. Part II Top. Stud. Oceanogr.* **2002**, *49*, 585–602. [CrossRef]
22. Bordovsky, O.K.; Chernyakova, A.M. *Modern Methods of Hydrochemical Studies of the Ocean*; IO RAS: Moscow, Russia, 1992; pp. 1–200. (In Russian)
23. Solorzano, L. Determination of ammonia in natural waters by phenol-hypochlorite method. *Limnol. Oceanogr.* **1969**, *14*, 799–801.
24. Strickland, J.D.H.; Parsons, T.R. *A Practical Handbook of Seawater Analysis*, 2nd ed.; Bulletin of Fisheries Research Board of Canada: Ottawa, ON, Canada, 1972; pp. 1–328.
25. Grasshoff, K.; Kremling, K.; Ehrhardt, M. *Methods of Seawater Analysis*; John Wiley & Sons: New York, NY, USA, 2009; pp. 1–599.
26. Parsons, T.R. *A Manual of Chemical & Biological Methods for Seawater Analysis*; Elsevier: Amsterdam, The Netherlands, 2013; pp. 1–188.
27. Hoef-Emden, K.; Marin, B.; Melkonian, M. Nuclear and nucleomorph SSU rDNA phylogeny in the Cryptophyta and the evolution of cryptophyte diversity. *J. Mol. Evol.* **2002**, *55*, 161–179. [CrossRef] [PubMed]
28. Six, C.; Thomas, J.C.; Brahamsha, B.; Lemoine, Y.; Partensky, F. Photophysiology of the marine cyanobacteria *Synechococcus* sp. WH8102, a new model organism. *Aquat. Microb. Ecol.* **2004**, *35*, 17–29. [CrossRef]
29. Marie, D.; Simon, N.; Vaulot, D. Phytoplankton cell counting by flow cytometry. In *Algal Culturing Techniques*; Andersen, R.A., Ed.; Elsevier: Amsterdam, The Netherlands, 2005; pp. 253–267.
30. Olson, R.J.; Zettler, E.R.; DuRand, M.D. Phytoplankton analysis using flow cytometry. In *Handbook of Methods in Aquatic Microbial Ecology*; CRC Press: Boca Raton, FL, USA, 1993; pp. 175–186.
31. Alvarez, E.; Nogueira, E.; López-Urrutia, Á. In vivo single-cell fluorescence and size scaling of phytoplankton chlorophyll content. *Appl. Environ. Microbiol.* **2017**, *83*, 1–16. [CrossRef] [PubMed]
32. Worden, A.; Nolan, J.; Palenik, B. Assessing the dynamics and ecology of marine picophytoplankton: The importance of the eukaryotic component. *Limnol. Oceanogr.* **2004**, *49*, 168–179. [CrossRef]
33. Garrison, D.; Gowing, M.; Hughes, M. Microbial food web structure in the Arabian Sea: A US JGOFS study. *Deep Sea Res. Part II* **2000**, *47*, 1387–1422. [CrossRef]
34. Hammer, O.; Harper, D.A.T.; Ryan, P.D. PAST: Paleontological statistics software package for education and data analysis. *Paleontol. Electron.* **2001**, *4*, 1–9.
35. Zhou, M.; Niiler, P.P.; Zhu, Y.; Dorland, R.D. The western boundary current in the Bransfield Strait, Antarctica. *Deep Sea Res. I* **2006**, *53*, 1244–1252. [CrossRef]
36. Niiler, P.P.; Amos, A.; Hu, J.-H. Water masses and 200 m relative geostrophic circulation in the western Bransfield Strait region. *Deep Sea Res.* **1991**, *38*, 943–959. [CrossRef]
37. Garcia, M.A.; López, O.; Sospedra, J.; Espino, M.; Gracia, V.; Morrison., G.; Arcilla, A.S. Mesoscale variability in the Bransfield Strait region (Antarctica) during Austral summer. *Ann. Geophys.* **1994**, *12*, 856–867. [CrossRef]
38. Sangrà, P.; Gordo, C.; Hernández-Arencibia, M.; Marrero-Díaz, A.; Rodríguez-Santana, A.; Stegner, A.; Pichon, T. The Bransfield current system. *Deep Sea Res. I* **2011**, *58*, 390–402. [CrossRef]
39. Jiang, M.; Charette, M.A.; Measures, C.I.; Zhu, Y.; Zhou, M. Seasonal cycle of circulation in the Antarctic Peninsula and the off-shelf transport of shelf waters into southern Drake Passage and Scotia Sea. *Deep Sea Res. Part II Top. Stud. Oceanogr.* **2013**, *90*, 15–30. [CrossRef]
40. Morozov, E.G. Currents in Bransfield Strait. *Dokl. Earth Sci.* **2007**, *415*, 984–986. [CrossRef]
41. Holm-Hansen, O.; Mitchell, B.G. Spatial and temporal distribution of phytoplankton and primary production in the western Bransfield Strait region. *Deep Sea Res. II* **1991**, *39*, 961–980. [CrossRef]
42. Dierssen, H.M.; Smith, R.C. Case 2 Antarctic coastal waters: The bio-optical properties of surface meltwater. In Proceedings of the Ocean Optics XV, Kailua-Kona, HI, USA, 16–20 October 2000.
43. McMinn, A.; Hodgson, D. Summer phytoplankton succession in Ellis Fjord, eastern Antarctica. *J. Plankton Res.* **1993**, *15*, 925–938. [CrossRef]
44. Bouman, H.A.; Ulloa, O.; Barlow, R.; Li, W.K.; Platt, T.; Zwirglmaier, K.; Sathyendranath, S. Water-column stratification governs the community structure of subtropical marine picophytoplankton. *Environ. Microbiol. Rep.* **2011**, *3*, 473–482. [CrossRef]
45. Buitenhuis, E.T.; Li, W.K.; Vaulot, D.; Lomas, M.W.; Landry, M.R.; Partensky, F.; McManus, G.B. Picophytoplankton biomass distribution in the global ocean. *Earth Syst. Sci. Data* **2012**, *4*, 37–46. [CrossRef]
46. Agawin, N.S.; Agustí, S.; Duarte, C.M. Abundance of Antarctic picophytoplankton and their response to light and nutrient manipulation. *Aquat. Microb. Ecol.* **2002**, *29*, 161–172. [CrossRef]
47. Weber, L.H.; El-Sayed, S.Z. Contribution of the net-, nano and picoplankton standing crop and primary productivity in the Southern Ocean. *J. Plankton Res.* **1987**, *9*, 973–994. [CrossRef]
48. Lin, L.; He, J.; Zhao, Y.; Zhang, F.; Cai, M. Flow cytometry investigation of picoplankton across latitudes and along the circum Antarctic Ocean. *Acta Oceanol. Sin.* **2012**, *31*, 134–142. [CrossRef]

49. Morris, I.; Glover, H. Physiology of photosynthesis by marine coccoid cyanobacteria—Some ecological implications. *Limnol. Oceanogr.* **1981**, *26*, 957–961. [CrossRef]
50. Sakshaug, E.; Holm-Hansen, O. Photoadaptation in Antarctic phytopfankton: Variations in growth rate, chemical composition and P versus I curves. *J. Plankton Res.* **1986**, *8*, 459–473. [CrossRef]
51. Mitchell, B.G.; Holm-Hansen, O. Observations of modeling of the Antarctic phytoplankton crop in relation to mixing depth. *Deep Sea Res. Part A Oceanogr. Res. Pap.* **1991**, *38*, 981–1007. [CrossRef]
52. Vaillancourt, R.D.; Brown, C.W.; Guillard, R.R.L.; Balch, W.M. Light scattering properties of marine phytoplankton: Relationships to cell size, chemical composition and taxonomy. *J. Plankton Res.* **2004**, *26*, 191–212. [CrossRef]
53. Eriksen, N.T.; Poulsen, B.R.; Iversen, J.L. Dual sparging laboratory-scale photobioreactor for continuous production of microalgae. *J. Appl. Phycol.* **1998**, *10*, 377–382. [CrossRef]
54. Jacques, G.; Panouse, M. Biomass and composition of size fractionated phytoplankton in the Weddell-Scotia Confluence area. *Polar Biol.* **1991**, *5*, 315–328. [CrossRef]
55. Kopczynska, E.E. Distribution of microflagellates and diatoms in the sea-ice zone between Elephant Island and the South Orkney Islands (December 1988–January 1989). *Pol. Polar Res.* **1991**, *12*, 515–528.
56. Buma, A.G.J.; Gieskes, W.W.C.; Thomsen, H.A. Abundance of Cryptophyceae and chlorophyll b-containing organisms in the Weddell-Scotia Confluence area in the spring of 1988. In *Weddell Sea Ecology*; Hempel, G., Ed.; Springer: Berlin/Heidelberg, Germany, 1992; pp. 43–52.
57. Kang, S.-H.; Lee, S. Antarctic phytoplankton assemblage in the western Bransfield Strait region, February 1993: Composition, biomass, and mesoscale distributions. *Mar. Ecol. Prog. Ser.* **1995**, *129*, 253–267. [CrossRef]
58. Holm-Hansen, O.; Mitchell, B.G.; Hewes, C.D.; Karl, D.M. Phytoplankton blooms in the vicinity of Palmer Station, Antarctica. *Polar Biol.* **1989**, *10*, 49–57. [CrossRef]
59. Smith, W.O., Jr.; Sakshaug, E. Polar Phytoplankton. In *Polar Oceanography, Part B*; Smith, W.O., Jr., Ed.; Academic Press: London, UK, 1990; pp. 477–525.
60. García-Muñoz, C.; Lubián, L.M.; García, C.M.; Marrero-Díaz, Á.; Sangrà, P.; Vernet, M. A mesoscale study of phytoplankton assemblages around the South Shetland Islands (Antarctica). *Polar Biol.* **2013**, *36*, 1107–1123. [CrossRef]
61. Mura, M.P.; Satta, M.P.; Agustí, S. Water-mass influences on summer Antarctic phytoplankton biomass and community structure. *Polar Biol.* **1995**, *15*, 15–20. [CrossRef]
62. Sangrà, P.; García-Muñoz, C.; García, C.M.; Marrero-Díaz, Á.; Sobrino, C.; Mouriño-Carballido, B.; Hernández-Arencibia, M. Coupling between upper ocean layer variability and size-fractionated phytoplankton in a non-nutrient-limited environment. *Mar. Ecol. Prog. Ser.* **2014**, *499*, 35–46. [CrossRef]
63. Hewes, C.D.; Reiss, C.S.; Holm-Hansen, O. A quantitative analysis of sources for summertime phytoplankton variability over 18 years in the South Shetland Islands (Antarctica) region. *Deep Sea Res. Part I Oceanogr. Res. Pap.* **2009**, *56*, 1230–1241. [CrossRef]
64. Holm-Hansen, O.; Hewes, C.D.; Villafane, V.E.; Helbling, E.W.; Silva, N.; Amos, T. Distribution of phytoplankton and nutrients in relation to different water masses in the area around Elephant Island, Antarctica. *Polar Biol.* **1997**, *18*, 145–153. [CrossRef]
65. Schofield, O.; Saba, G.; Coleman, K.; Carvalho, F.; Couto, N.; Ducklow, H.; Montes-Hugo, M. Decadal variability in coastal phytoplankton community composition in a changing West Antarctic Peninsula. *Deep Sea Res. Part I Oceanogr. Res. Pap.* **2017**, *124*, 42–54. [CrossRef]
66. Moreau, S.; Ferreyra, G.A.; Mercier, B.; Lemarchand, K.; Lionard, M.; Roy, S.; Mostajir, B.; Van Hardenberg, B.; Demers, S. Variability of the microbial community in the western Antarctic Peninsula from late fall to spring during a low ice cover year. *Polar Biol.* **2010**, *33*, 1599–1614. [CrossRef]
67. Ferreyra, G.A.; Tomo, A.P. Variacion estational de las diatomeas planctonicas en Puerto Paraiso-I. *Contrib. Del Inst. Antart. Argentio* **1979**, *264*, 149–184.
68. Whitaker, T.M. Primary production of phytoplankton off Signy Island, South Orkneys, the Antarctic. *Proc. R. Soc. Lond. Ser. B Biol. Sci.* **1982**, *214*, 169–189.
69. Krebs, W.N. Ecology of neritic marine diatoms, Arthur Harbor, Antarctica. *Micropaleontology* **1983**, *29*, 267–297. [CrossRef]
70. Kopczynska, E.E. Dominance of microflagellates over diatoms in the Antarctic areas of deep vertical mixing and krill concentrations. *J. Plankton Res.* **1992**, *14*, 1031–1054. [CrossRef]
71. Walsh, J.J.; Dieterle, D.A.; Lenes, J. A numerical analysis of carbon dynamics of the Southern Ocean phytoplankton community: The roles of light and grazing in effecting both sequestration of atmospheric CO_2 and food availability to larval krill. *Deep Sea Res.* **2001**, *48*, 1–48. [CrossRef]
72. Ainley, D.G.; Ballard, G.; Emslie, S.D. Adélie penguins and environmental change. *Science* **2003**, *300*, 429–430. [CrossRef] [PubMed]

MDPI
St. Alban-Anlage 66
4052 Basel
Switzerland
Tel. +41 61 683 77 34
Fax +41 61 302 89 18
www.mdpi.com

Water Editorial Office
E-mail: water@mdpi.com
www.mdpi.com/journal/water

www.ingramcontent.com/pod-product-compliance
Lightning Source LLC
Chambersburg PA
CBHW040750250926
51850CB00007BB/1463
* 9 7 8 3 0 3 6 5 8 2 6 9 6 *